Essentials of Statistical Inference

Essentials of Statistical Inference is a modern and accessible treatment of the procedures used to draw formal inferences from data. Aimed at advanced undergraduate and graduate students in mathematics and related disciplines, as well as those in other fields seeking a concise treatment of the key ideas of statistical inference, it presents the concepts and results underlying the Bayesian, frequentist and Fisherian approaches, with particular emphasis on the contrasts between them. Contemporary computational ideas are explained, as well as basic mathematical theory.

Written in a lucid and informal style, this concise text provides both basic material on the main approaches to inference, as well as more advanced material on modern developments in statistical theory, including: contemporary material on Bayesian computation, such as MCMC, higher-order likelihood theory, predictive inference, bootstrap methods and conditional inference. It contains numerous extended examples of the application of formal inference techniques to real data, as well as historical commentary on the development of the subject. Throughout, the text concentrates on concepts, rather than mathematical detail, while maintaining appropriate levels of formality. Each chapter ends with a set of accessible problems.

Based to a large extent on lectures given at the University of Cambridge over a number of years, the material has been polished by student feedback. Some prior knowledge of probability is assumed, while some previous knowledge of the objectives and main approaches to statistical inference would be helpful but is not essential.

G. A. YOUNG is Professor of Statistics at Imperial College London.

R. L. SMITH is Mark L. Reed Distinguished Professor of Statistics at University of North Carolina, Chapel Hill.

Essentials of Statistical Inference

G. A. Young
Department of Mathematics, Imperial College London

R. L. Smith
*Department of Statistics and Operations Research,
University of North Carolina, Chapel Hill*

CAMBRIDGE
UNIVERSITY PRESS

CAMBRIDGE
UNIVERSITY PRESS

32 Avenue of the Americas, New York NY 10013-2473, USA

Cambridge University Press is part of the University of Cambridge.

It furthers the University's mission by disseminating knowledge in the pursuit of education, learning and research at the highest international levels of excellence.

www.cambridge.org
Information on this title: www.cambridge.org/9780521548663

© Cambridge University Press 2005

First published 2005
Reprinted 2010
First paperback edition 2010

A catalogue record for this publication is available from the British Library

ISBN 978-0-521-83971-6 Hardback
ISBN 978-0-521-54866-3 Paperback

Cover image: Analysis of z-values for gene expression data. *Solid line, f(z); dotted line,* theoretical null; dashed line, empirical null. See Figure 3.7.

Contents

Preface

This book aims to provide a concise but comprehensive account of the essential elements of statistical inference and theory. It is designed to be used as a text for courses on statistical theory for students of mathematics or statistics at the advanced undergraduate or Masters level (UK) or the first-year graduate level (US), or as a reference for researchers in other fields seeking a concise treatment of the key concepts of and approaches to statistical inference. It is intended to give a contemporary and accessible account of procedures used to draw formal inference from data.

The book focusses on a clear presentation of the main concepts and results underlying different frameworks of inference, with particular emphasis on the contrasts among frequentist, Fisherian and Bayesian approaches. It provides a description of basic material on these main approaches to inference, as well as more advanced material on recent developments in statistical theory, including higher-order likelihood inference, bootstrap methods, conditional inference and predictive inference. It places particular emphasis on contemporary computational ideas, such as applied in bootstrap methodology and Markov chain Monte Carlo techniques of Bayesian inference. Throughout, the text concentrates on concepts, rather than mathematical detail, but every effort has been made to present the key theoretical results in as precise and rigorous a manner as possible, consistent with the overall mathematical level of the book. The book contains numerous extended examples of application of contrasting inference techniques to real data, as well as selected historical commentaries. Each chapter concludes with an accessible set of problems and exercises.

Prerequisites for the book are calculus, linear algebra and some knowledge of basic probability (including ideas such as conditional probability, transformations of densities etc., though not measure theory). Some previous familiarity with the objectives of and main approaches to statistical inference is helpful, but not essential. Key mathematical and probabilistic ideas are reviewed in the text where appropriate.

The book arose from material used in teaching of statistical inference to students, both undergraduate and graduate, at the University of Cambridge. We thank the many colleagues at Cambridge who have contributed to that material, especially David Kendall, Elizabeth Thompson, Pat Altham, James Norris and Chris Rogers, and to the many students who have, over many years, contributed hugely by their enthusiastic feedback. Particular thanks go to Richard Samworth, who provided detailed and valuable comments on the whole text. Errors and inconsistencies that remain, however, are our responsibility, not his. David Tranah and Diana Gillooly of Cambridge University Press deserve special praise, for their encouragement over a long period, and for exerting just the right amount of pressure, at just the right time. But it is our families who deserve the biggest 'thank you', and who have suffered most during completion of the book.

1

Introduction

What is statistical inference?

In statistical inference experimental or observational data are modelled as the observed values of random variables, to provide a framework from which inductive conclusions may be drawn about the mechanism giving rise to the data.

We wish to analyse observations $x = (x_1, \ldots, x_n)$ by:

1 Regarding x as the observed value of a random variable $X = (X_1, \ldots, X_n)$ having an (unknown) probability distribution, conveniently specified by a probability density, or probability mass function, $f(x)$.
2 Restricting the unknown density to a suitable family or set $\mathcal{F}$. In *parametric statistical inference*, $f(x)$ is of known analytic form, but involves a finite number of real unknown parameters $\theta = (\theta_1, \ldots, \theta_d)$. We specify the region $\Theta \subseteq \mathbb{R}^d$ of possible values of θ, the *parameter space*. To denote the dependency of $f(x)$ on θ, we write $f(x; \theta)$ and refer to this as the *model function*. Alternatively, the data could be modelled non-parametrically, a non-parametric model simply being one which does not admit a parametric representation. We will be concerned almost entirely in this book with parametric statistical inference.

The objective that we then assume is that of assessing, on the basis of the observed data x, some aspect of θ, which for the purpose of the discussion in this paragraph we take to be the value of a particular component, θ_i say. In that regard, we identify three main types of inference: *point estimation*, *confidence set estimation* and *hypothesis testing*. In point estimation, a single value is computed from the data x and used as an estimate of θ_i. In confidence set estimation we provide a set of values, which, it is hoped, has a predetermined high probability of including the true, but unknown, value of θ_i. Hypothesis testing sets up specific hypotheses regarding θ_i and assesses the plausibility of any such hypothesis by assessing whether or not the data x support that hypothesis.

Of course, other objectives might be considered, such as: (a) prediction of the value of some as yet unobserved random variable whose distribution depends on θ, or (b) examination of the adequacy of the model specified by $\mathcal{F}$ and Θ. These are important problems, but are not the main focus of the present book, though we will say a little on predictive inference in Chapter 10.

How do we approach statistical inference?

Following Efron (1998), we identify three main paradigms of statistical inference: the *Bayesian, Fisherian* and *frequentist.* A key objective of this book is to develop in detail the essential features of all three schools of thought and to highlight, we hope in an interesting way, the potential conflicts between them. The basic differences that emerge relate to interpretation of probability and to the objectives of statistical inference. To set the scene, it is of some value to sketch straight away the main characteristics of the three paradigms. To do so, it is instructive to look a little at the historical development of the subject.

The Bayesian paradigm goes back to Bayes and Laplace, in the late eighteenth century. The fundamental idea behind this approach is that the unknown parameter, θ, should itself be treated as a random variable. Key to the Bayesian viewpoint, therefore, is the specification of a *prior probability distribution* on θ, before the data analysis. We will describe in some detail in Chapter 3 the main approaches to specification of prior distributions, but this can basically be done either in some objective way, or in a subjective way, which reflects the statistician's own prior state of belief. To the Bayesian, inference is the formalisation of how the prior distribution changes, to the *posterior distribution*, in the light of the evidence presented by the available data x, through Bayes' formula. Central to the Bayesian perspective, therefore, is a use of probability distributions as expressing opinion.

In the early 1920s, R.A. Fisher put forward an opposing viewpoint, that statistical inference must be based entirely on probabilities with direct experimental interpretation. As Efron (1998) notes, Fisher's primary concern was the development of a logic of inductive inference, which would release the statistician from the a priori assumptions of the Bayesian school. Central to the Fisherian viewpoint is the *repeated sampling principle.* This dictates that the inference we draw from x should be founded on an analysis of how the conclusions change with variations in the data samples, which would be obtained through hypothetical repetitions, under exactly the same conditions, of the experiment which generated the data x in the first place. In a Fisherian approach to inference, a central role is played by the concept of *likelihood*, and the associated principle of *maximum likelihood*. In essence, the likelihood measures the probability that different values of the parameter θ assign, under a hypothetical repetition of the experiment, to re-observation of the actual data x. More formally, the ratio of the likelihood at two different values of θ compares the relative plausibilities of observing the data x under the models defined by the two θ values. A further fundamental element of Fisher's viewpoint is that inference, in order to be as relevant as possible to the data x, must be carried out *conditional* on everything that is known and uninformative about θ.

Fisher's greatest contribution was to provide for the first time an optimality yardstick for statistical estimation, a description of the optimum that it is possible to do in a given estimation problem, and the technique of maximum likelihood, which produces estimators of θ that are close to ideal in terms of that yardstick. As described by Pace and Salvan (1997), spurred on by Fisher's introduction of optimality ideas in the 1930s and 1940s, Neyman, E.S. Pearson and, later, Wald and Lehmann offered the third of the three paradigms, the frequentist approach. The origins of this approach lay in a detailed mathematical analysis of some of the fundamental concepts developed by Fisher, in particular likelihood and *sufficiency*. With this focus, emphasis shifted from inference as a summary of data, as favoured by Fisher, to inferential procedures viewed as decision problems. Key elements of the frequentist approach are the need for clarity in mathematical formulation, and that optimum inference procedures should be identified *before* the observations x are available, optimality being defined explicitly in terms of the repeated sampling principle.

The plan of the book is as follows. In Chapter 2, we describe the main elements of the *decision theory* approach to frequentist inference, where a strict mathematical statement of the inference problem is made, followed by formal identification of the optimal solution. Chapter 3 develops the key ideas of Bayesian inference, before we consider, in Chapter 4, central optimality results for hypothesis testing from a frequentist perspective. There a comparison is made between frequentist and Bayesian approaches to hypothesis testing. Chapter 5 introduces two special classes of model function of particular importance to later chapters, exponential families and transformation models. Chapter 6 is concerned primarily with point estimation of a parameter θ and provides a formal introduction to a number of key concepts in statistical inference, in particular the notion of sufficiency. In Chapter 7, we revisit the topic of hypothesis testing, to extend some of the ideas of Chapter 4 to more complicated settings. In the former chapter we consider also the conditionality ideas that are central to the Fisherian perspective, and highlight conflicts with the frequentist approach to inference. There we describe also key frequentist optimality ideas in confidence set estimation. The subject of Chapter 8 is maximum likelihood and associated inference procedures. The remaining chapters contain more advanced material. In Chapter 9 we present a description of some recent innovations in statistical inference, concentrating on ideas which draw their inspiration primarily from the Fisherian viewpoint. Chapter 10 provides a discussion of various approaches to *predictive inference*. Chapter 11, reflecting the personal interests of one of us (GAY), provides a description of the *bootstrap* approach to inference. This approach, made possible by the recent availability of cheap computing power, offers the prospect of techniques of statistical inference which avoid the need for awkward mathematical analysis, but retain the key operational properties of methods of inference studied elsewhere in the book, in particular in relation to the repeated sampling principle.

2

Decision theory

In this chapter we give an account of the main ideas of decision theory. Our motivation for beginning our account of statistical inference here is simple. As we have noted, decision theory requires formal specification of all elements of an inference problem, so starting with a discussion of decision theory allows us to set up notation and basic ideas that run through the remainder of the book in a formal but easy manner. In later chapters, we will develop the specific techniques of statistical inference that are central to the three paradigms of inference. In many cases these techniques can be seen as involving the removal of certain elements of the decision theory structure, or focus on particular elements of that structure.

Central to decision theory is the notion of a set of *decision rules* for an inference problem. Comparison of different decision rules is based on examination of the *risk functions* of the rules. The risk function describes the expected *loss* in use of the rule, under hypothetical repetition of the sampling experiment giving rise to the data x, as a function of the *parameter* of interest. Identification of an optimal rule requires introduction of fundamental principles for discrimination between rules, in particular the *minimax* and *Bayes* principles.

2.1 Formulation

A full description of a statistical decision problem involves the following formal elements:

1. A *parameter space* Θ, which will usually be a subset of $\mathbb{R}^d$ for some $d \geq 1$, so that we have a vector of d unknown parameters. This represents the set of possible unknown states of nature. The unknown parameter value $\theta \in \Theta$ is the quantity we wish to make inference about.

2. A *sample space* $\mathcal{X}$, the space in which the data x lie. Typically we have n observations, so the data, a generic element of the sample space, are of the form $x = (x_1, \ldots, x_n) \in \mathbb{R}^n$.

3. A *family of probability distributions* on the sample space $\mathcal{X}$, indexed by values $\theta \in \Theta$, $\{\mathbb{P}_\theta(x), \ x \in \mathcal{X}, \ \theta \in \Theta\}$. In nearly all practical cases this will consist of an assumed parametric family $f(x; \theta)$, of probability mass functions for x (in the discrete case), or probability density functions for x (in the continuous case).

4. An *action space* $\mathcal{A}$. This represents the set of all actions or decisions available to the experimenter.

   Examples of action spaces include the following:

   (a) In a hypothesis testing problem, where it is necessary to decide between two hypotheses H_0 and H_1, there are two possible actions corresponding to 'accept H_0' and 'accept H_1'. So here $\mathcal{A} = \{a_0, a_1\}$, where a_0 represents accepting H_0 and a_1 represents accepting H_1.

(b) In an estimation problem, where we want to estimate the unknown parameter value θ by some function of $x = (x_1, \ldots, x_n)$, such as $\bar{x} = \frac{1}{n} \sum x_i$ or $s^2 = \frac{1}{n-1} \sum (x_i - \bar{x})^2$ or $x_1^3 + 27 \sin(\sqrt{x_2})$, etc., we should allow ourselves the possibility of estimating θ by any point in Θ. So, in this context we typically have $\mathcal{A} \equiv \Theta$.

(c) However, the scope of decision theory also includes things such as 'approve Mr Jones' loan application' (if you are a bank manager) or 'raise interest rates by 0.5%' (if you are the Bank of England or the Federal Reserve), since both of these can be thought of as actions whose outcome depends on some unknown state of nature.

5 A *loss function* $L : \Theta \times \mathcal{A} \to \mathbb{R}$ links the action to the unknown parameter. If we take action $a \in \mathcal{A}$ when the true state of nature is $\theta \in \Theta$, then we incur a loss $L(\theta, a)$.

Note that losses can be positive or negative, a negative loss corresponding to a gain. It is a convention that we formulate the theory in terms of trying to minimise our losses rather than trying to maximise our gains, but obviously the two come to the same thing.

6 A set $\mathcal{D}$ of *decision rules*. An element $d : \mathcal{X} \to \mathcal{A}$ of $\mathcal{D}$ is such that each point x in $\mathcal{X}$ is associated with a specific action $d(x) \in \mathcal{A}$.

For example, with hypothesis testing, we might adopt the rule: 'Accept H_0 if $\bar{x} \leq 5.7$, otherwise accept H_1.' This corresponds to a decision rule,

$$d(x) = \begin{cases} a_0 & \text{if } \bar{x} \leq 5.7, \\ a_1 & \text{if } \bar{x} > 5.7. \end{cases}$$

2.2 The risk function

For parameter value $\theta \in \Theta$, the risk associated with a decision rule d based on random data X is defined by

$$\begin{aligned} R(\theta, d) &= \mathbb{E}_\theta L(\theta, d(X)) \\ &= \begin{cases} \int_{\mathcal{X}} L(\theta, d(x)) f(x; \theta) \, dx & \text{for continuous } X, \\ \sum_{x \in \mathcal{X}} L(\theta, d(x)) f(x; \theta) & \text{for discrete } X. \end{cases} \end{aligned}$$

So, we are treating the observed data x as the realised value of a random variable X with density or mass function $f(x; \theta)$, and defining the risk to be the expected loss, the expectation being with respect to the distribution of X for the particular parameter value θ.

The key notion of decision theory is that different decision rules should be compared by comparing their risk functions, as functions of θ. Note that we are explicitly invoking the repeated sampling principle here, the definition of risk involving hypothetical repetitions of the sampling mechanism that generated x, through the assumed distribution of X.

When a loss function represents the real loss in some practical problem (as opposed to some artificial loss function being set up in order to make the statistical decision problem well defined) then it should really be measured in units of 'utility' rather than actual money. For example, the expected return on a UK lottery ticket is less than the £1 cost of the ticket; if everyone played so as to maximise their expected gain, nobody would ever buy a lottery ticket! The reason that people still buy lottery tickets, translated into the language of statistical decision theory, is that they subjectively evaluate the very small chance of winning, say, £1 000 000 as worth more than a fixed sum of £1, even though the chance of actually winning the £1 000 000 is appreciably less than 1/1 000 000. There is a formal theory, known as utility theory, which asserts that, provided people behave rationally (a considerable assumption in its own right!), then they will always act *as if* they were maximising the expected value of a function known as the utility function. In the lottery example, this implies that we subjectively evaluate the possibility of a massive prize, such as £1 000 000, to be worth more than 1 000 000 times as much as the relatively paltry sum of £1. However in situations involving monetary sums of the same order of magnitude, most people tend to be risk averse. For example, faced with a choice between:

Offer 1: Receive £10 000 with probability 1;

and

Offer 2: Receive £20 000 with probability $\frac{1}{2}$, otherwise receive £0,

most of us would choose Offer 1. This means that, in utility terms, we consider £20 000 as worth less than twice as much as £10 000. Either amount seems like a very large sum of money, and we may not be able to distinguish the two easily in our minds, so that the lack of risk involved in Offer 1 makes it appealing. Of course, if there was a specific reason why we really needed £20 000, for example because this was the cost of a necessary medical operation, we might be more inclined to take the gamble of Offer 2.

Utility theory is a fascinating subject in its own right, but we do not have time to go into the mathematical details here. Detailed accounts are given by Ferguson (1967) or Berger (1985), for example. Instead, in most of the problems we will be considering, we will use various artificial loss functions. A typical example is use of the loss function

$$L(\theta, a) = (\theta - a)^2,$$

the squared error loss function, in a point estimation problem. Then the risk $R(\theta, d)$ of a decision rule is just the mean squared error of $d(X)$ as an estimator of θ, $\mathbb{E}_\theta\{d(X) - \theta\}^2$. In this context, we seek a decision rule d that minimises this mean squared error.

Other commonly used loss functions, in point estimation problems, are

$$L(\theta, a) = |\theta - a|,$$

the absolute error loss function, and

$$L(\theta, a) = \begin{cases} 0 & \text{if } |\theta - a| \leq \delta, \\ 1 & \text{if } |\theta - a| > \delta, \end{cases}$$

where δ is some prescribed tolerance limit. This latter loss function is useful in a Bayesian formulation of interval estimation, as we shall discuss in Chapter 3.

In hypothesis testing, where we have two hypotheses H_0, H_1, identified with subsets of Θ, and corresponding action space $\mathcal{A} = \{a_0, a_1\}$ in which action a_j corresponds to selecting the hypothesis H_j, $j = 0, 1$, the most familiar loss function is

$$L(\theta, a) = \begin{cases} 1 & \text{if } \theta \in H_0 \text{ and } a = a_1, \\ 1 & \text{if } \theta \in H_1 \text{ and } a = a_0, \\ 0 & \text{otherwise.} \end{cases}$$

In this case the risk is the probability of making a wrong decision:

$$R(\theta, d) = \begin{cases} \Pr_\theta\{d(X) = a_1\} & \text{if } \theta \in H_0, \\ \Pr_\theta\{d(X) = a_0\} & \text{if } \theta \in H_1. \end{cases}$$

In the classical language of hypothesis testing, these two risks are called, respectively, the type I error and the type II error: see Chapter 4.

2.3 Criteria for a good decision rule

In almost any case of practical interest, there will be no way to find a decision rule $d \in \mathcal{D}$ which makes the risk function $R(\theta, d)$ uniformly smallest for all values of θ. Instead, it is necessary to consider a number of criteria, which help to narrow down the class of decision rules we consider. The notion is to start with a large class of decision rules d, such as the set of *all* functions from $\mathcal{X}$ to $\mathcal{A}$, and then reduce the number of candidate decision rules by application of the various criteria, in the hope of being left with some unique best decision rule for the given inference problem.

2.3.1 Admissibility

Given two decision rules d and d', we say that d *strictly dominates* d' if $R(\theta, d) \le R(\theta, d')$ for all values of θ, and $R(\theta, d) < R(\theta, d')$ for at least one value θ.

Given a choice between d and d', we would always prefer to use d.

Any decision rule which is strictly dominated by another decision rule (as d' is in the definition) is said to be *inadmissible*. Correspondingly, if a decision rule d is not strictly dominated by any other decision rule, then it is *admissible*.

Admissibility looks like a very weak requirement: it seems obvious that we should always restrict ourselves to admissible decision rules. Admissibility really represents absence of a negative attribute, rather than possession of a positive attribute. In practice, it may not be so easy to decide whether a given decision rule is admissible or not, and there are some surprising examples of natural-looking estimators which are inadmissible. In Chapter 3, we consider an example of an inadmissible estimator, Stein's paradox, which has been described (Efron, 1992) as 'the most striking theorem of post-war mathematical statistics'!

2.3.2 Minimax decision rules

The maximum risk of a decision rule $d \in \mathcal{D}$ is defined by

$$\text{MR}(d) = \sup_{\theta \in \Theta} R(\theta, d).$$

A decision rule d is *minimax* if it minimises the maximum risk:

$$\text{MR}(d) \leq \text{MR}(d') \text{ for all decision rules } d' \in \mathcal{D}.$$

Another way of writing this is to say that d must satisfy

$$\sup_{\theta} R(\theta, d) = \inf_{d' \in \mathcal{D}} \sup_{\theta \in \Theta} R(\theta, d'). \tag{2.1}$$

In most of the problems we will encounter, the supremum and infimum are actually attained, so that we can rewrite (2.1) as

$$\max_{\theta \in \Theta} R(\theta, d) = \min_{d' \in \mathcal{D}} \max_{\theta \in \Theta} R(\theta, d').$$

(Recall that the difference between $\sup_{\theta}$ and $\max_{\theta}$ is that the maximum must actually be attained for some $\theta \in \Theta$, whereas a supremum represents a least upper bound that may not actually be attained for any single value of θ. Similarly for infimum and minimum.)

The *minimax principle* says we should use a minimax decision rule.

A few comments about minimaxity are appropriate.

(a) The motivation may be roughly stated as follows: we do not know anything about the true value of θ, therefore we ought to insure ourselves against the worst possible case. There is also an analogy with game theory. In that context, $L(\theta, a)$ represents the penalty suffered by you (as one player in a game) when you choose the action a and your opponent (the other player) chooses θ. If this $L(\theta, a)$ is also the amount gained by your opponent, then this is called a two-person zero-sum game. In game theory, the minimax principle is well established because, in that context, you know that your opponent is trying to choose θ to maximise your loss. See Ferguson (1967) or Berger (1985) for a detailed exposition of the connections between statistical decision theory and game theory.

(b) There are a number of situations in which minimaxity may lead to a counterintuitive result. One situation is when a decision rule d_1 is better than d_2 for all values of θ except in a very small neighbourhood of a particular value, θ_0 say, where d_2 is much better: see Figure 2.1. In this context one might prefer d_1 unless one had particular reason to think that θ_0, or something near it, was the true parameter value. From a slightly broader perspective, it might seem illogical that the minimax criterion's preference for d_2 is based entirely in its behaviour in a small region of Θ, while the rest of the parameter space is ignored.

(c) The minimax procedure may be likened to an arms race in which both sides spend the maximum sum available on military fortification in order to protect themselves against the worst possible outcome, of being defeated in a war, an instance of a non-zero-sum game!

(d) Minimax rules may not be unique, and may not be admissible. Figure 2.2 is intended to illustrate a situation in which d_1 and d_2 achieve the same minimax risk, but one would obviously prefer d_1 in practice.

2.3.3 Unbiasedness

A decision rule d is said to be *unbiased* if

$$\mathbb{E}_{\theta}\{L(\theta', d(X))\} \geq \mathbb{E}_{\theta}\{L(\theta, d(X))\} \text{ for all } \theta, \theta' \in \Theta.$$

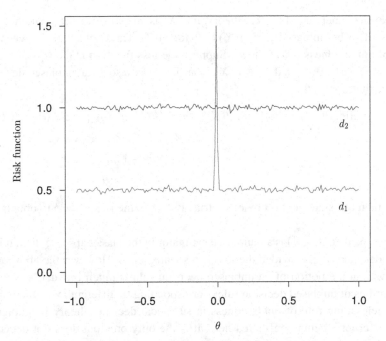

Figure 2.1 Risk functions for two decision rules

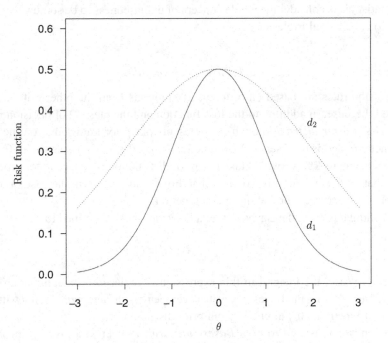

Figure 2.2 Minimax rules may not be admissible

Recall that in elementary statistical theory, if $d(X)$ is an estimator for a parameter θ, then $d(X)$ is said to be unbiased if $\mathbb{E}_\theta d(X) = \theta$ for all θ. The connection between the two notions of unbiasedness is as follows. Suppose the loss function is the squared error loss, $L(\theta, d) = (\theta - d)^2$. Fix θ and let $\mathbb{E}_\theta d(X) = \phi$. Then, for d to be an unbiased decision rule, we require that, for all θ',

$$
\begin{aligned}
0 \leq \mathbb{E}_\theta\{L(\theta', d(X))\} - \mathbb{E}_\theta\{L(\theta, d(X))\} &= \mathbb{E}_\theta\{(\theta' - d(X))^2\} - \mathbb{E}_\theta\{(\theta - d(X))^2\} \\
&= (\theta')^2 - 2\theta'\phi + \mathbb{E}_\theta d^2(X) - \theta^2 \\
&\quad + 2\theta\phi - \mathbb{E}_\theta d^2(X) \\
&= (\theta' - \phi)^2 - (\theta - \phi)^2.
\end{aligned}
$$

If $\phi = \theta$, then this statement is obviously true. If $\phi \neq \theta$, then set $\theta' = \phi$ to obtain a contradiction.

Thus we see that, if $d(X)$ is an unbiased estimator in the classical sense, then it is also an unbiased decision rule, provided the loss is a squared error. However the above argument also shows that the notion of an unbiased decision rule is much broader: we could have whole families of unbiased decision rules corresponding to different loss functions.

Nevertheless, the role of unbiasedness in statistical decision theory is ambiguous. Of the various criteria being considered here, it is the only one that does not depend solely on the risk function. Often we find that biased estimators perform better than unbiased ones from the point of view of, say, minimising mean squared error. For this reason, many modern statisticians consider the whole concept of unbiasedness to be somewhere between a distraction and a total irrelevance.

2.3.4 Bayes decision rules

Bayes decision rules are based on different assumptions from the other criteria we have considered, because, in addition to the loss function and the class $\mathcal{D}$ of decision rules, we must specify a *prior distribution*, which represents our prior knowledge on the value of the parameter θ, and is represented by a function $\pi(\theta)$, $\theta \in \Theta$. In cases where Θ contains an open rectangle in $\mathbb{R}^d$, we would take our prior distribution to be absolutely continuous, meaning that $\pi(\theta)$ is taken to be some probability density on Θ. In the case of a discrete parameter space, $\pi(\theta)$ is a probability mass function.

In the continuous case, the Bayes risk of a decision rule d is defined to be

$$
r(\pi, d) = \int_{\theta \in \Theta} R(\theta, d)\pi(\theta)d\theta.
$$

In the discrete case, the integral in this expression is replaced by a summation over the possible values of θ. So, the Bayes risk is just average risk, the averaging being with respect to the weight function $\pi(\theta)$ implied by our prior distribution.

A decision rule d is said to be a *Bayes rule*, with respect to a given prior $\pi(\cdot)$, if it minimises the Bayes risk, so that

$$
r(\pi, d) = \inf_{d' \in \mathcal{D}} r(\pi, d') = m_\pi, \text{ say.} \tag{2.2}
$$

The *Bayes principle* says we should use a Bayes decision rule.

2.3.5 Some other definitions

Sometimes the Bayes rule is not defined because the infimum in (2.2) is not attained for any decision rule d. However, in such cases, for any $\epsilon > 0$ we can find a decision rule d_ϵ for which

$$r(\pi, d_\epsilon) < m_\pi + \epsilon$$

and in this case d_ϵ is said to be ϵ-*Bayes* with respect to the prior distribution $\pi(\cdot)$.

Finally, a decision rule d is said to be *extended Bayes* if, for every $\epsilon > 0$, we have that d is ϵ-Bayes with respect to *some* prior, which need not be the same for different ϵ. As we shall see in Theorem 2.2, it is often possible to derive a minimax rule through the property of being extended Bayes. A particular example of an extended Bayes rule is discussed in Problem 3.11.

2.4 Randomised decision rules

Suppose we have a collection of I decision rules $d_1, \ldots, d_I$ and an associated set of probability weights $p_1, \ldots, p_I$, so that $p_i \geq 0$ for $1 \leq i \leq I$, and $\sum_i p_i = 1$.

Define the decision rule $d^* = \sum_i p_i d_i$ to be the rule 'select d_i with probability p_i'. Then d^* is a *randomised decision rule*. We can imagine that we first use some randomisation mechanism, such as tossing coins or using a computer random number generator, to select, independently of the observed data x, one of the decision rules $d_1, \ldots, d_I$, with respective probabilities $p_1, \ldots, p_I$. Then, having decided in favour of use of the particular rule d_i, under d^* we carry out the action $d_i(x)$.

For a randomised decision rule d^*, the risk function is defined by averaging across possible risks associated with the component decision rules:

$$R(\theta, d^*) = \sum_{i=1}^{I} p_i R(\theta, d_i).$$

Randomised decision rules may appear to be artificial, but minimax solutions may well be of this form. It is easy to contruct examples in which d^* is formed by randomising the rules $d_1, \ldots, d_I$ but

$$\sup_\theta R(\theta, d^*) < \sup_\theta R(\theta, d_i) \text{ for each } i,$$

so that d^* may be a candidate for the minimax procedure, but none of $d_1, \ldots, d_I$. An example of a decision problem, where the minimax rule indeed turns out to be a randomised rule, is presented in Section 2.5.1, and illustrated in Figure 2.9.

2.5 Finite decision problems

A finite decision problem is one in which the *parameter space* is a finite set: $\Theta = \{\theta_1, \ldots, \theta_t\}$ for some finite t, with $\theta_1, \ldots, \theta_t$ specified values. In such cases the notions of admissible, minimax and Bayes decision rules can be given a geometric interpretation, which leads to some interesting problems in their own right, and which also serves to motivate some properties of decision rules in more general problems.

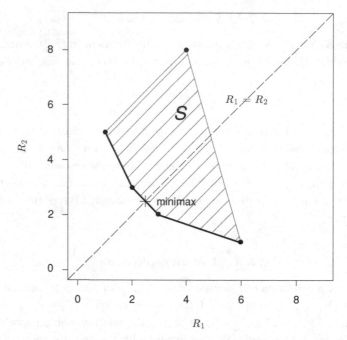

Figure 2.3 An example of a risk set

For a finite decision problem, define the *risk set* to be a subset S of $\mathbb{R}^t$, in which a generic point consists of the t-vector $(R(\theta_1, d), \ldots, R(\theta_t, d))$ associated with a decision rule d. An important point to note is that we assume in our subsequent discussion that the space of decision rules $\mathcal{D}$ includes all randomised rules.

A set A is said to be *convex* if whenever $x_1 \in A$ and $x_2 \in A$ then $\lambda x_1 + (1 - \lambda)x_2 \in A$ for any $\lambda \in (0, 1)$.

Lemma 2.1 *The risk set S is a convex set.*

Proof   Suppose $x_1 = (R(\theta_1, d_1), \ldots, R(\theta_t, d_1))$ and $x_2 = (R(\theta_1, d_2), \ldots, R(\theta_t, d_2))$ are two elements of S, and suppose $\lambda \in (0, 1)$. Form a new randomised decision rule $d = \lambda d_1 + (1 - \lambda)d_2$. Then for every θ, by definition of the risk of a randomised rule,

$$R(\theta, d) = \lambda R(\theta, d_1) + (1 - \lambda)R(\theta, d_2).$$

Then we see that $\lambda x_1 + (1 - \lambda)x_2$ is associated with the decision rule d, and hence is itself a member of S. This proves the result.   □

In the case $t = 2$, it is particularly easy to see what is going on, because we can draw the risk set as a subset of $\mathbb{R}^2$, with coordinate axes $R_1 = R(\theta_1, d)$, $R_2 = R(\theta_2, d)$. An example is shown in Figure 2.3.

The extreme points of S (shown by the dots) correspond to non-randomised decision rules, and points on the lower left-hand boundary (represented by thicker lines) correspond to the admissible decision rules.

For the example shown in Figure 2.3, the minimax decision rule corresponds to the point at the lower intersection of S with the line $R_1 = R_2$ (the point shown by the cross). Note

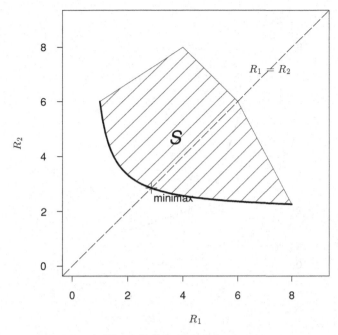

Figure 2.4 Another risk set

that this is a randomised decision rule. However, we shall see below that not all minimax rules are formed in this way.

In cases where the boundary of the risk set is a smooth curve, as in Figure 2.4, the properties are similar, except that we now have an infinite number of non-randomised admissible decision rules and the minimax point is non-randomised.

To find the Bayes rules, suppose we have prior probabilities (π_1, π_2), where $\pi_1 \geq 0$, $\pi_2 \geq 0$, $\pi_1 + \pi_2 = 1$, so that π_j, for $j = 1$ or 2, represents the prior probability that θ_j is the true parameter value. For any c, the straight line $\pi_1 R_1 + \pi_2 R_2 = c$ represents a class of decision rules with the same Bayes risk. By varying c, we get a family of parallel straight lines. Of course, if the line $\pi_1 R_1 + \pi_2 R_2 = c$ does not intersect S, then the Bayes risk c is unattainable. Then the Bayes risk for the decision problem is c' if the line $\pi_1 R_1 + \pi_2 R_2 = c'$ just hits the set S on its lower left-hand boundary: see Figure 2.5. Provided S is a closed set, which we shall assume, the Bayes decision rule corresponds to the point at which this line intersects S.

In many cases of interest, the Bayes rule is unique, and is then automatically both admissible and non-randomised. However, it is possible that the line $\pi_1 R_1 + \pi_2 R_2 = c'$ hits S along a line segment rather than at a single point, as in Figure 2.6. In that case, any point along the segment identifies a Bayes decision rule with respect to this prior. Also, in this case the interior points of the segment will identify randomised decision rules, but the endpoints of the segment also yield Bayes rules, which are non-randomised. Thus we can always find a Bayes rule which is non-randomised. Also, it is an easy exercise to see that, provided $\pi_1 > 0$, $\pi_2 > 0$, the Bayes rule is admissible.

It can easily happen (see Figure 2.7) that the same decision rule is Bayes with respect to a whole family of different prior distributions.

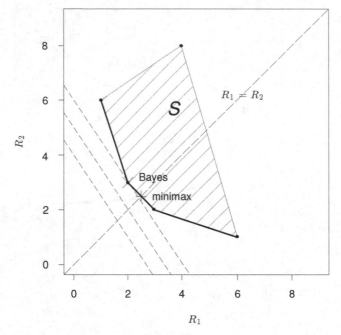

Figure 2.5 Finding the Bayes rule

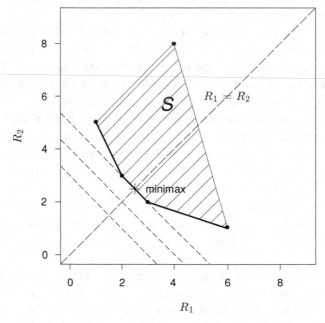

Figure 2.6 Non-uniqueness of Bayes rule

Not all minimax rules satisfy $R_1 = R_2$. See Figures 2.8 for several examples. By contrast, Figure 2.3 is an example where the minimax rule does satisfy $R_1 = R_2$. In Figures 2.8(a) and (b), S lies entirely to the left of the line $R_1 = R_2$, so that $R_1 < R_2$ for every point in S, and therefore the minimax rule is simply that which minimises R_2. This is the

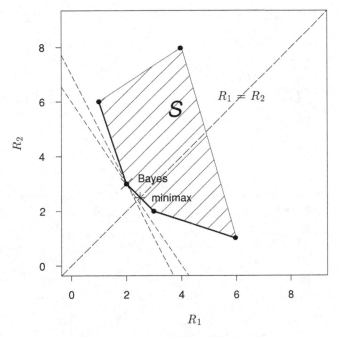

Figure 2.7 Bayes with respect to family of priors

Bayes rule for the prior $\pi_1 = 0, \pi_2 = 1$, and may be attained either at a single point (Figure 2.8(a)) or along a segment (Figure 2.8(b)). In the latter case, only the left-hand element of the segment (shown by the cross) corresponds to an admissible rule. Figures 2.8(c) and (d) are the mirror images of Figures 2.8(a) and (b) in which every point of the risk set satisfies $R_1 > R_2$ so that the minimax rule is Bayes for the prior $\pi_1 = 1$, $\pi_2 = 0$.

2.5.1 A story

The Palliser emerald necklace has returned from the cleaners, together with a valueless imitation which you, as Duchess of Omnium, wear on the less important State occasions. The tag identifying the imitation has fallen off, and so you have two apparently identical necklaces in the left- and right-hand drawers of the jewelcase. You consult your Great Aunt, who inspects them both (left-hand necklace first, and then right-hand necklace), and then from her long experience pronounces one of them to be the true necklace. *But is she right?* You know that her judgement will be infallible if she happens to inspect the true necklace first and the imitation afterwards, but that if she inspects them in the other order she will in effect select one of them at random, with equal probabilities $\frac{1}{2}$ on the two possibilities. With a loss of £0 being attributed to a correct decision (choosing the real necklace), you know that a mistaken one (choosing the imitation) will imply a loss of £1 million. You will wear the necklace tonight at an important banquet, where the guest of honour is not only the Head of State of a country with important business contracts with Omnium, but also an expert on emerald jewellery, certain to be able to spot an imitation.

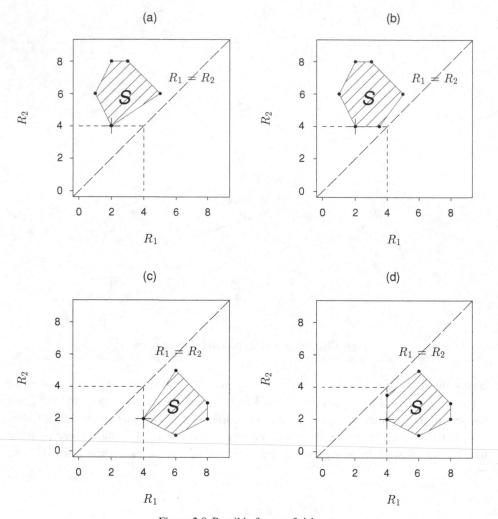

Figure 2.8 Possible forms of risk set

Our data x are your Great Aunt's judgement. Consider the following exhaustive set of four non-randomised decision rules based on x:

(d_1): accept the left-hand necklace, irrespective of Great Aunt's judgement;
(d_2): accept the right-hand necklace, irrespective of Great Aunt's judgement;
(d_3): accept your Great Aunt's judgement;
(d_4): accept the reverse of your Great Aunt's judgement.

Code the states of nature 'left-hand necklace is the true one' as $\theta = 1$, and 'right-hand necklace is the true one' as $\theta = 2$. We compute the risk functions of the decision rules as follows:

$$R_1 = R(\theta = 1, d_1) = 0, R_2 = R(\theta = 2, d_1) = 1;$$
$$R_1 = R(\theta = 1, d_2) = 1, R_2 = R(\theta = 2, d_2) = 0;$$
$$R_1 = R(\theta = 1, d_3) = 0, R_2 = R(\theta = 2, d_3) = \tfrac{1}{2};$$
$$R_1 = R(\theta = 1, d_4) = 1, R_2 = R(\theta = 2, d_4) = \tfrac{1}{2}.$$

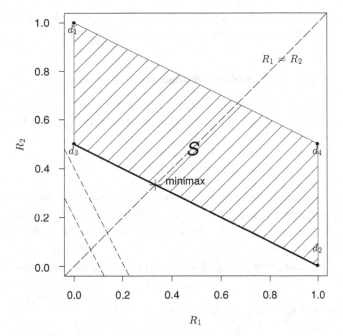

Figure 2.9 Risk set, Palliser necklace

To understand these risks, note that when $\theta = 1$ your Great Aunt chooses correctly, so that d_3 is certain to make the correct choice, while d_4 is certain to make the wrong choice. When $\theta = 2$, Great Aunt chooses incorrectly with probability $\frac{1}{2}$, so our expected loss is $\frac{1}{2}$ with both rules d_3 and d_4.

The risks associated with these decision rules and the associated risk set are shown in Figure 2.9. The only admissible rules are d_2 and d_3, together with randomised rules formed as convex combinations of d_2 and d_3.

The minimax rule d^* is such a randomised rule, of the form $d^* = \lambda d_3 + (1 - \lambda)d_2$. Then setting $R(\theta = 1, d^*) = R(\theta = 2, d^*)$ gives $\lambda = \frac{2}{3}$.

Suppose that the Duke now returns from hunting and points out that the jewel cleaner will have placed the true necklace in the left-hand drawer with some probability ψ, which we know from knowledge of the way the jewelcase was arranged on its return from previous trips to the cleaners.

The Bayes risk of a rule d is then $\psi R(\theta = 1, d) + (1 - \psi)R(\theta = 2, d)$. There are three groups of Bayes rules according to the value of ψ:

(i) If $\psi = \frac{1}{3}$, d_2 and d_3, together with all convex combinations of the two, give the same Bayes risk ($= \frac{1}{3}$) and are Bayes rules. In this case the minimax rule is also a Bayes rule.

(ii) If $\psi > \frac{1}{3}$ the unique Bayes rule is d_3, with Bayes risk $(1 - \psi)/2$. This is the situation with the lines of constant Bayes risk illustrated in Figure 2.9, and makes intuitive sense. If our prior belief that the true necklace is in the left-hand drawer is strong, then we attach a high probability to Great Aunt inspecting the necklaces in the order true necklace first, then imitation, in which circumstances she is certain to identify them correctly. Then following her judgement is sensible.

(iii) If $\psi < \frac{1}{3}$ the unique Bayes rule is d_2, with Bayes risk ψ. Now the prior belief is that the true necklace is unlikely to be in the left-hand drawer, so we are most likely to be in the situation where Great Aunt basically guesses which is the true necklace, and it is better to go with our prior hunch of the right-hand drawer.

2.6 Finding minimax rules in general

Although the formula $R_1 = R_2$ to define the minimax rule is satisfied in situations such as that shown in Figure 2.3, all four situations illustrated in Figure 2.8. satisfy a more general formula:

$$\max_\theta R(\theta, d) \le r(\pi, d), \tag{2.3}$$

where π is *a* prior distribution with respect to which the minimax rule d is Bayes. In the cases of Figure 2.8(a) and Figure 2.8(b), for example, this is true for the minimax rules, for the prior $\pi_1 = 0$, $\pi_2 = 1$, though we note that in Figure 2.8(a) the unique minimax rule is Bayes for other priors as well.

These geometric arguments suggest that, in general, a minimax rule is one which satisfies:

(a) it is Bayes with respect to some prior $\pi(\cdot)$,
(b) it satisfies (2.3).

A complete classification of minimax decision rules in general problems lies outside the scope of this text, but the following two theorems give simple sufficient conditions for a decision rule to be minimax. One generalisation that is needed in passing from the finite to the infinite case is that the class of Bayes rules must be extended to include sequences of either Bayes rules, or extended Bayes rules.

Theorem 2.1 *If δ_n is Bayes with respect to prior $\pi_n(\cdot)$, and $r(\pi_n, \delta_n) \to C$ as $n \to \infty$, and if $R(\theta, \delta_0) \le C$ for all $\theta \in \Theta$, then δ_0 is minimax.*

Of course this includes the case where $\delta_n = \delta_0$ for all n and the Bayes risk of δ_0 is exactly C.

To see the infinite-dimensional generalisation of the condition $R_1 = R_2$, we make the following definition.

Definition *A decision rule d is an* equaliser decision rule *if $R(\theta, d)$ is the same for every value of θ.*

Theorem 2.2 *An equaliser decision rule δ_0 which is extended Bayes must be minimax.*

Proof of Theorem 2.1 Suppose δ_0 satisfies the conditions of the theorem but is not minimax. Then there must exist some decision rule δ' for which $\sup_\theta R(\theta, \delta') < C$: the inequality must be strict, because, if the maximum risk of δ' was the same as that of δ_0, that would not contradict minimaxity of δ_0. So there is an $\epsilon > 0$ for which $R(\theta, \delta') < C - \epsilon$ for every θ. Now, since $r(\pi_n, \delta_n) \to C$, we can find an n for which $r(\pi_n, \delta_n) > C - \epsilon/2$. But $r(\pi_n, \delta') \le C - \epsilon$. Therefore, δ_n cannot be the Bayes rule with respect to π_n. This creates a contradiction, and hence proves the theorem. □

Proof of Theorem 2.2. The proof here is almost the same. If we suppose δ_0 is not minimax, then there exists a δ' for which $\sup_\theta R(\theta, \delta') < C$, where C is the common value of $R(\theta, \delta_0)$.

So let $\sup_\theta R(\theta, \delta') = C - \epsilon$, for some $\epsilon > 0$. By the extended Bayes property of δ_0, we can find a prior π for which

$$r(\pi, \delta_0) = C < \inf_\delta r(\pi, \delta) + \frac{\epsilon}{2}.$$

But $r(\pi, \delta') \leq C - \epsilon$, so this gives another contradiction, and hence proves the theorem. $\qquad\square$

2.7 Admissibility of Bayes rules

In Chapter 3 we will present a general result that allows us to characterise the Bayes decision rule for any given inference problem. An immediate question that then arises concerns admissibility. In that regard, the rule of thumb is that Bayes rules are nearly always admissible. We complete this chapter with some specific theorems on this point. Proofs are left as exercises.

Theorem 2.3 *Assume that $\Theta = \{\theta_1, \ldots, \theta_t\}$ is finite, and that the prior $\pi(\cdot)$ gives positive probability to each θ_i. Then a Bayes rule with respect to π is admissible.*

Theorem 2.4 *If a Bayes rule is unique, it is admissible.*

Theorem 2.5 *Let Θ be a subset of the real line. Assume that the risk functions $R(\theta, d)$ are continuous in θ for all decision rules d. Suppose that for any $\epsilon > 0$ and any θ the interval $(\theta - \epsilon, \theta + \epsilon)$ has positive probability under the prior $\pi(\cdot)$. Then a Bayes rule with respect to π is admissible.*

2.8 Problems

2.1 Let X be uniformly distributed on $[0, \theta]$, where $\theta \in (0, \infty)$ is an unknown parameter. Let the action space be $[0, \infty)$ and the loss function $L(\theta, d) = (\theta - d)^2$, where d is the action chosen. Consider the decision rules $d_\mu(x) = \mu x$, $\mu \geq 0$. For what value of μ is d_μ unbiased? Show that $\mu = 3/2$ is a necessary condition for d_μ to be admissible.

2.2 Dashing late into King's Cross, I discover that Harry Potter must have already boarded the Hogwart's Express. I must therefore make my own way on to platform nine and three-quarters. Unusually, there are two guards on duty, and I will ask one of them for directions. It is safe to assume that one guard is a Wizard, who will certainly be able to direct me, and the other a Muggle, who will certainly not. But which is which? Before choosing one of them to ask for directions to platform nine and three-quarters, I have just enough time to ask one of them 'Are you a Wizard?', and on the basis of their reply I must make my choice of which guard to ask for directions. I know that a Wizard will answer this question truthfully, but that a Muggle will, with probability $1/3$, answer it untruthfully.

Failure to catch the Hogwart's Express results in a loss that I measure as 1000 galleons, there being no loss associated with catching up with Harry on the train.

Write down an exhaustive set of non-randomised decision rules for my problem and, by drawing the associated risk set, determine my minimax decision rule.

My prior probability is 2/3 that the guard I ask 'Are you a Wizard?' is indeed a Wizard. What is my Bayes decision rule?

2.3 Each winter evening between Sunday and Thursday, the superintendent of the Chapel Hill School District has to decide whether to call off the next day's school because of snow conditions. If he fails to call off school and there is snow, there are various possible consequences, including children and teachers failing to show up for school, the possibility of traffic accidents etc. If he calls off school, then regardless of whether there actually is snow that day there will have to be a make-up day later in the year. After weighing up all the possible outcomes he decides that the costs of failing to close school when there is snow are twice the costs incurred by closing school, so he assigns two units of loss to the first outcome and one to the second. If he does not call off school and there is no snow, then of course there is no loss.

Two local radio stations give independent and identically distributed weather forecasts. If there is to be snow, each station will forecast this with probability 3/4, but predict no snow with probability 1/4. If there is to be no snow, each station predicts snow with probability 1/2.

The superintendent will listen to the two forecasts this evening, and then make his decision on the basis of the data x, the number of stations forecasting snow.

Write down an exhaustive set of non-randomised decision rules based on x.

Find the superintendent's admissible decision rules, and his minimax rule. Before listening to the forecasts, he believes there will be snow with probability 1/2; find the Bayes rule with respect to this prior.

(Again, include randomised rules in your analysis when determining admissible, minimax and Bayes rules.)

2.4 An unmanned rocket is being launched in order to place in orbit an important new communications satellite. At the time of launching, a certain crucial electronic component is either functioning or not functioning. In the control centre there is a warning light that is not completely reliable. If the crucial component is not functioning, the warning light goes on with probability 2/3; if the component is functioning, it goes on with probability 1/4. At the time of launching, an observer notes whether the warning light is on or off. It must then be decided immediately whether or not to launch the rocket. There is no loss associated with launching the rocket with the component functioning, or aborting the launch when the component is not functioning. However, if the rocket is launched when the component is not functioning, the satellite will fail to reach the desired orbit. The Space Shuttle mission required to rescue the satellite and place it in the correct orbit will cost 10 billion dollars. Delays caused by the decision not to launch when the component is functioning result, through lost revenue, in a loss of 5 billion dollars.

Suppose that the prior probability that the component is not functioning is $\psi = 2/5$. If the warning light does not go on, what is the decision according to the Bayes rule?

For what values of the prior probability ψ is the Bayes decision to launch the rocket, even if the warning light comes on?

2.5 Bacteria are distributed at random in a fluid, with mean density θ per unit volume, for some $\theta \in H \subseteq [0, \infty)$. This means that

$$\Pr_\theta(\text{no bacteria in volume } v) = e^{-\theta v}.$$

We remove a sample of volume v from the fluid and test it for the presence or absence of bacteria. On the basis of this information we have to decide whether there are any bacteria in the fluid at all. An incorrect decision will result in a loss of 1, a correct decision in no loss.

(i) Suppose $H = [0, \infty)$. Describe all the non-randomised decision rules for this problem and calculate their risk functions. Which of these rules are admissible?

(ii) Suppose $H = \{0, 1\}$. Identify the risk set

$$S = \{(R(0, d), R(1, d)): d \text{ a randomised rule}\} \subseteq \mathbb{R}^2,$$

where $R(\theta, d)$ is the expected loss in applying d under Pr_θ. Determine the minimax rule.

(iii) Suppose again that $H = [0, \infty)$.

Determine the Bayes decision rules and Bayes risk for prior

$$\pi(\{0\}) = 1/3,$$
$$\pi(A) = 2/3 \int_A e^{-\theta} d\theta, A \subseteq (0, \infty).$$

(So the prior probability that $\theta = 0$ is 1/3, while the prior probability that $\theta \in A \subseteq (0, \infty)$ is $2/3 \int_A e^{-\theta} d\theta$.)

(iv) If it costs $v/24$ to test a sample of volume v, what is the optimal volume to test? What if the cost is $1/6$ per unit volume?

2.6 Prove Theorems 2.3, 2.4 and 2.5, concerning admissibility of Bayes rules.

2.7 In the context of a finite decision problem, decide whether each of the following statements is true, providing a proof or counterexample as appropriate.

(i) The Bayes risk of a minimax rule is never greater than the minimax risk.

(ii) If a Bayes rule is not unique, then it is inadmissible.

2.8 In a Bayes decision problem, a prior distribution π is said to be *least favourable* if $r_\pi \geq r_{\pi'}$, for all prior distributions π', where r_π denotes the Bayes risk of the Bayes rule d_π with respect to π.

Suppose that π is a prior distribution, such that

$$\int R(\theta, d_\pi)\pi(\theta)d\theta = \sup_\theta R(\theta, d_\pi).$$

Show that (i) d_π is minimax, (ii) π is least favourable.

3

Bayesian methods

This chapter develops the key ideas in the Bayesian approach to inference. Fundamental ideas are described in Section 3.1. The key conceptual point is the way that the *prior* distribution on the unknown parameter θ is updated, on observing the realised value of the data x, to the *posterior* distribution, via Bayes' law. Inference about θ is then extracted from this posterior. In Section 3.2 we revisit decision theory, to provide a characterisation of the Bayes decision rule in terms of the posterior distribution. The remainder of the chapter discusses various issues of importance in the implementation of Bayesian ideas. Key issues that emerge, in particular in realistic data analytic examples, include the question of choice of prior distribution and computational difficulties in summarising the posterior distribution. Of particular importance, therefore, in practice are ideas of *empirical Bayes inference* (Section 3.5), *Monte Carlo techniques* for application of Bayesian inference (Section 3.7) and *hierarchical modelling* (Section 3.8). Elsewhere in the chapter we provide discussion of *Stein's paradox* and the notion of *shrinkage* (Section 3.4). Though not primarily a Bayesian problem, we shall see that the *James–Stein estimator* may be justified (Section 3.5.1) as an empirical Bayes procedure, and the concept of shrinkage is central to practical application of Bayesian thinking. We also provide here a discussion of *predictive inference* (Section 3.9) from a Bayesian perspective, as well as a historical description of the development of the Bayesian paradigm (Section 3.6).

3.1 Fundamental elements

In non-Bayesian, or classical, statistics X is random, with a density or probability mass function given by $f(x; \theta)$, but θ is treated as a *fixed* unknown parameter value.

Instead, in Bayesian statistics X and θ are *both* regarded as random variables, with joint density (or probability mass function) given by $\pi(\theta)f(x; \theta)$, where $\pi(\cdot)$ represent the prior density of θ, and $f(\cdot; \theta)$ is the conditional density of X, given θ.

The *posterior density* of θ, given observed value $X = x$, is given by applying Bayes' law of conditional probabilities:

$$\pi(\theta|x) = \frac{\pi(\theta)f(x; \theta)}{\int_{\Theta} \pi(\theta')f(x; \theta')d\theta'}.$$

Commonly we write

$$\pi(\theta|x) \propto \pi(\theta)f(x; \theta),$$

where the constant of proportionality is allowed to depend on x but not on θ. This may be written in words as

$$\text{posterior} \propto \text{prior} \times \text{likelihood}$$

since $f(x; \theta)$, treated as a function of θ for fixed x, is called the *likelihood function* – for example, maximum likelihood estimation (which is not a Bayesian procedure) proceeds by maximising this expression with respect to θ: see Chapter 8.

Example 3.1 Consider a binomial experiment in which $X \sim \text{Bin}(n, \theta)$ for known n and unknown θ. Suppose the prior density is a Beta density on (0,1),

$$\pi(\theta) = \frac{\theta^{a-1}(1-\theta)^{b-1}}{B(a, b)}, \quad 0 < \theta < 1,$$

where $a > 0, b > 0$ and $B(\cdot, \cdot)$ is the beta function ($B(a, b) = \Gamma(a)\Gamma(b)/\Gamma(a+b)$, where Γ is the gamma function, $\Gamma(t) = \int_0^\infty x^{t-1}e^{-x}dx$). For the density of X, here interpreted as a probability mass function, we have

$$f(x; \theta) = \binom{n}{x}\theta^x(1-\theta)^{n-x}.$$

Ignoring all components of π and f which do not depend on θ, we have

$$\pi(\theta|x) \propto \theta^{a+x-1}(1-\theta)^{n-x+b-1}.$$

This is also of Beta form, with the parameters a and b replaced by $a + x$ and $b + n - x$, so the full posterior density is

$$\pi(\theta|x) = \frac{\theta^{a+x-1}(1-\theta)^{n-x+b-1}}{B(a+x, b+n-x)}.$$

Recall (or easily verify for yourself!) that for the Beta distribution with parameters a and b, we have

$$\text{mean} = \frac{a}{a+b}, \quad \text{variance} = \frac{ab}{(a+b)^2(a+b+1)}.$$

Thus the mean and variance of the posterior distribution are respectively

$$\frac{a+x}{a+b+n}$$

and

$$\frac{(a+x)(b+n-x)}{(a+b+n)^2(a+b+n+1)}.$$

For large n, the influence of a and b will be negligible and we can write

$$\text{posterior mean} \approx \frac{x}{n}, \quad \text{posterior variance} \approx \frac{x(n-x)}{n^3}.$$

In classical statistics, we very often take $\widehat{\theta} = X/n$ as our estimator of θ, based on a binomial observation X. Its variance is $\theta(1-\theta)/n$, but, when we do not know θ, we usually substitute its estimated value and quote the approximate variance as $X(n-X)/n^3$. (More commonly, in practice, we use the square root of this quantity, which is called the *standard error* of $\hat{\theta}$.) This illustrates a very general property of Bayesian statistical procedures. In large samples, they give answers which are very similar to the answers provided by classical statistics: the data swamp the information in the prior. For this reason, many statisticians feel that the distinction between Bayesian and classical methods is not so important in practice. Nevertheless, in small samples the procedures do lead to different answers, and the Bayesian solution does depend on the prior adopted, which can be viewed as either an advantage or a disadvantage, depending on how much you regard the prior density as representing real prior information and the aim of the investigation.

This example illustrates another important property of some Bayesian procedures: by adopting a prior density of Beta form we obtained a posterior density that was also a member of the Beta family, but with different parameters. When this happens, the common parametric form of the prior and posterior are called a *conjugate prior family* for the problem. There is no universal law that says we must use a conjugate prior. Indeed, if it really was the case that we had genuine prior information about θ, there would be no reason to assume that it took the form of a Beta distribution. However, the conjugate prior property is often a very convenient one, because it avoids having to integrate numerically to find the normalising constant in the posterior density. In non-conjugate cases, where we have to do everything numerically, this is the hardest computational problem associated with Bayesian inference. Therefore, in cases where we can find a conjugate family, it is very common to use it.

Example 3.2 Suppose $X_1, \ldots, X_n$ are independent, identically distributed from the normal distribution $N(\theta, \sigma^2)$, where the mean θ is unknown and the variance σ^2 is *known*. Let us also assume that the prior density for θ is $N(\mu_0, \sigma_0^2)$, with μ_0, σ_0^2 known. We denote by X the vector $(X_1, \ldots, X_n)$ and let its observed value be $x = (x_1, \ldots, x_n)$. Ignoring all quantities that do not depend on θ, the prior $\times$ likelihood can be written in the form

$$\pi(\theta)f(x;\theta) \propto \exp\left\{-\frac{(\theta - \mu_0)^2}{2\sigma_0^2} - \sum_{i=1}^{n}\frac{(x_i - \theta)^2}{2\sigma^2}\right\}.$$

Completing the square shows that

$$\frac{(\theta - \mu_0)^2}{\sigma_0^2} + \sum_{i=1}^{n}\frac{(x_i - \theta)^2}{\sigma^2} = \theta^2\left(\frac{1}{\sigma_0^2} + \frac{n}{\sigma^2}\right) - 2\theta\left(\frac{\mu_0}{\sigma_0^2} + \frac{n\bar{x}}{\sigma^2}\right) + C_1$$

$$= \frac{1}{\sigma_1^2}(\theta - \mu_1)^2 + C_2,$$

where $\bar{x} = \sum x_i/n$, and where C_1 and C_2 denote quantities which do not depend on θ (though they do depend on $x_1, \ldots, x_n$), and μ_1 and σ_1^2 are defined by

$$\frac{1}{\sigma_1^2} = \frac{1}{\sigma_0^2} + \frac{n}{\sigma^2}, \quad \mu_1 = \sigma_1^2\left(\frac{\mu_0}{\sigma_0^2} + \frac{n\bar{x}}{\sigma^2}\right).$$

Thus we see that

$$\pi(\theta|x) \propto \exp\left\{-\frac{1}{2\sigma_1^2}(\theta - \mu_1)^2\right\},$$

allowing us to observe that the posterior density is the normal density with mean μ_1 and variance σ_1^2. This, therefore, is another example of a conjugate prior family. Note that as

$n \to \infty$, $\sigma_1^2 \approx \frac{\sigma^2}{n}$ and hence $\mu_1 \approx \bar{x}$, so that again the Bayesian estimates of the mean and variance become indistinguishable from their classical, frequentist counterparts as n increases.

Example 3.3 Here is an extension of the previous example in which the normal variance as well as the normal mean is unknown. It is convenient to write τ in place of $1/\sigma^2$ and μ in place of θ, so that $\theta = (\tau, \mu)$ may be reserved for the two-dimensional pair. Consider the prior in which τ has a Gamma distribution with parameters $\alpha > 0$, $\beta > 0$ and, conditionally on τ, μ has distribution $N(\nu, 1/(k\tau))$ for some constants $k > 0$, $\nu \in \mathbb{R}$. The full prior density is $\pi(\tau, \mu) = \pi(\tau)\pi(\mu \mid \tau)$, which may be written as

$$\pi(\tau, \mu) = \frac{\beta^\alpha}{\Gamma(\alpha)} \tau^{\alpha-1} e^{-\beta\tau} \cdot (2\pi)^{-1/2} (k\tau)^{1/2} \exp\left\{-\frac{k\tau}{2}(\mu - \nu)^2\right\},$$

or more simply

$$\pi(\tau, \mu) \propto \tau^{\alpha-1/2} \exp\left[-\tau\left\{\beta + \frac{k}{2}(\mu - \nu)^2\right\}\right].$$

We have $X_1, \ldots, X_n$ independent, identically distributed from $N(\mu, 1/\tau)$, so the likelihood is

$$f(x; \mu, \tau) = (2\pi)^{-n/2} \tau^{n/2} \exp\left\{-\frac{\tau}{2}\sum(x_i - \mu)^2\right\}.$$

Thus

$$\pi(\tau, \mu | x) \propto \tau^{\alpha+n/2-1/2} \exp\left[-\tau\left\{\beta + \frac{k}{2}(\mu - \nu)^2 + \frac{1}{2}\sum(x_i - \mu)^2\right\}\right].$$

Complete the square to see that

$$k(\mu - \nu)^2 + \sum(x_i - \mu)^2$$
$$= (k + n)\left(\mu - \frac{k\nu + n\bar{x}}{k + n}\right)^2 + \frac{nk}{n + k}(\bar{x} - \nu)^2 + \sum(x_i - \bar{x})^2.$$

Hence the posterior satisfies

$$\pi(\tau, \mu \mid x) \propto \tau^{\alpha'-1/2} \exp\left[-\tau\left\{\beta' + \frac{k'}{2}(\mu - \nu')^2\right\}\right],$$

where

$$\alpha' = \alpha + \frac{n}{2},$$
$$\beta' = \beta + \frac{1}{2}\frac{nk}{n + k}(\bar{x} - \nu)^2 + \frac{1}{2}\sum(x_i - \bar{x})^2,$$
$$k' = k + n,$$
$$\nu' = \frac{k\nu + n\bar{x}}{k + n}.$$

Thus the posterior distribution is of the same parametric form as the prior (the above form of prior is a conjugate family), but with (α, β, k, ν) replaced by $(\alpha', \beta', k', \nu')$.

Sometimes we are particularly interested in the posterior distribution of μ alone. This may be simplified if we assume $\alpha = m/2$ for integer m. Then we may write the prior distribution, equivalently to the above, as

$$\tau = \frac{W}{2\beta}, \quad \mu = v + \frac{Z}{\sqrt{k\tau}},$$

where W and Z are independent random variables with the distributions χ_m^2 (the chi-squared distribution on m degrees of freedom) and $N(0, 1)$ respectively. Recalling that $Z\sqrt{m/W}$ has a t_m distribution (the t distribution on m degrees of freedom), we see that under the prior distribution,

$$\sqrt{\frac{km}{2\beta}}(\mu - v) \sim t_m.$$

For the posterior distribution of μ, we replace m by $m' = m + n$, etc., to obtain

$$\sqrt{\frac{k'm'}{2\beta'}}(\mu - v') \sim t_{m'}.$$

In general, the marginal posterior for a parameter μ of interest is obtained by integrating a joint posterior of μ and τ with respect to τ:

$$\pi(\mu \mid x) = \int \pi(\tau, \mu \mid x)d\tau.$$

Example 3.4 * (The asterisk denotes that this is a more advanced section, and optional reading.) If you are willing to take a few things on trust about multivariate generalisations of the normal and χ^2 distribution, we can do all of the above for multivariate data as well.

With some abuse of notation, the model of Example 3.3 may be presented in the form

$$\tau \sim \text{Gamma}(\alpha, \beta),$$

$$\mu|\tau \sim N\left(v, \frac{1}{k\tau}\right),$$

$$X_i|\tau, \mu \sim N\left(\mu, \frac{1}{\tau}\right).$$

A few definitions follow. A p-dimensional random vector X with mean vector μ and non-singular covariance matrix Σ is said to have a *multivariate normal distribution* if its p-variate probability density function is

$$(2\pi)^{-p/2}|\Sigma|^{-1/2} \exp\left\{-\frac{1}{2}(x - \mu)^T\Sigma^{-1}(x - \mu)\right\}.$$

We use the notation $N_p(\mu, \Sigma)$ to describe this. The distribution is also defined if Σ is singular, but then it is concentrated on some subspace of $\mathbb{R}^p$ and does not have a density in the usual sense. We shall not consider that case here. A result important to much of applied statistics is that, if $X \sim N_p(\mu, \Sigma)$ with Σ non-singular, the quadratic form $(X - \mu)^T\Sigma^{-1}(X - \mu) \sim \chi_p^2$.

The multivariate generalisation of a χ^2 random variable is called the *Wishart distribution*. A $p \times p$ symmetric random matrix D has the Wishart distribution $W_p(A, m)$ if its density is given by

$$\frac{c_{p,m}|D|^{(m-p-1)/2}}{|A|^{m/2}} \exp\left\{-\frac{1}{2}\text{tr}(DA^{-1})\right\},$$

where $\text{tr}(A)$ denotes the trace of the matrix A and $c_{p,m}$ is the constant

$$c_{p,m} = \left[2^{mp/2} \pi^{p(p-1)/4} \prod_{j=1}^{p} \Gamma\left(\frac{m+1-j}{2} \right) \right]^{-1}.$$

For m integer, this is the distribution of $D = \sum_{j=1}^{m} Z_j Z_j^T$ when $Z_1, \ldots, Z_m$ are independent $N_p(0, A)$. The density is proper provided $m > p - 1$.

We shall not prove these statements, or even try to establish that the Wishart density is a valid density function, but they are proved in standard texts on multivariate analysis, such as Anderson (1984) or Mardia, Kent and Bibby (1979). You do not need to know these proofs to be able to understand the computations which follow.

Consider the following scheme:

$$V \sim W_p(\Psi^{-1}, m),$$

$$\mu|V \sim N_p(v, (kV)^{-1}),$$

$$X_i|\mu, V \sim N_p(\mu, V^{-1}) \quad (1 \le i \le n).$$

Here $X_1, \ldots, X_n$ are conditionally independent p-dimensional random vectors, given (μ, V), k and m are known positive constants, v a known vector and Ψ a known positive definite matrix.

The prior density of $\theta = (V, \mu)$ is proportional to

$$|V|^{(m-p)/2} \exp\left[-\frac{1}{2} \text{tr}\{ V(\Psi + k(\mu - v)(\mu - v)^T) \} \right].$$

In deriving this, we have used the elementary relation $\text{tr}(AB) = \text{tr}(BA)$ (for any pair of matrices A and B for which both AB and BA are defined) to write $(\mu - v)^T V(\mu - v) = \text{tr}\{ V(\mu - v)(\mu - v)^T \}$.

Multiplying by the joint density of $X_1, \ldots, X_n$, the prior $\times$ likelihood is proportional to

$$|V|^{(m+n-p)/2} \exp\left[-\frac{1}{2} \text{tr}\left\{ V\left(\Psi + k(\mu - v)(\mu - v)^T + \sum_i (x_i - \mu)(x_i - \mu)^T \right) \right\} \right].$$

However, again we may simplify this by completing a (multivariate) square:

$$k(\mu - v)(\mu - v)^T + \sum_i (x_i - \mu)(x_i - \mu)^T$$

$$= (k + n)\left(\mu - \frac{kv + n\bar{x}}{k + n} \right)\left(\mu - \frac{kv + n\bar{x}}{k + n} \right)^T$$

$$+ \frac{nk}{k + n}(\bar{x} - v)(\bar{x} - v)^T + \sum_i (x_i - \bar{x})(x_i - \bar{x})^T.$$

Thus we see that the posterior density of (V, μ) is proportional to

$$|V|^{(m'-p)/2} \exp\left[-\frac{1}{2} \text{tr}\{ V(\Psi' + k'(\mu - v')(\mu - v')^T) \} \right],$$

which is of the same form as the prior density but with the parameters (m, k, Ψ, v)

replaced by

$$m' = m + n,$$
$$k' = k + n,$$
$$\Psi' = \Psi + \frac{nk}{k+n}(\bar{x} - v)(\bar{x} - v)^T + \sum_i (x_i - \bar{x})(x_i - \bar{x})^T,$$
$$v' = \frac{kv + n\bar{x}}{k+n}.$$

The joint prior for (V, μ) in this example is sometimes called the normal-inverse Wishart prior. The final result shows that the joint posterior density is of the same form, but with the four parameters m, k, Ψ and v updated as shown. Therefore, the normal-inverse Wishart prior is a conjugate prior in this case. The example shows that exact Bayesian analysis, of both the mean and covariance matrix in a multivariate normal problem, is feasible and elegant.

3.2 The general form of Bayes rules

We now return to our general discussion of how to solve Bayesian decision problems. For notational convenience, we shall write formulae assuming both X and θ have continuous densities, though the concepts are exactly the same in the discrete case.

Recall that the risk function of a decision rule d is given by

$$R(\theta, d) = \int_{\mathcal{X}} L(\theta, d(x)) f(x; \theta) dx$$

and the Bayes risk of d by

$$r(\pi, d) = \int_{\Theta} R(\theta, d)\pi(\theta)d\theta$$
$$= \int_{\Theta} \int_{\mathcal{X}} L(\theta, d(x)) f(x; \theta)\pi(\theta) dx d\theta$$
$$= \int_{\Theta} \int_{\mathcal{X}} L(\theta, d(x)) f(x)\pi(\theta|x) dx d\theta$$
$$= \int_{\mathcal{X}} f(x) \left\{ \int_{\Theta} L(\theta, d(x))\pi(\theta|x)d\theta \right\} dx.$$

In the third line here, we have written the joint density $f(x; \theta)\pi(\theta)$ in a different way as $f(x)\pi(\theta|x)$, where $f(x) = \int f(x; \theta)\pi(\theta)d\theta$ is the marginal density of X. The change of order of integration in the fourth line is trivially justified because the integrand is non-negative.

From the final form of this expression, we can see that, to find the action $d(x)$ specified by the Bayes rule for any x, it suffices to minimise the expression inside the curly brackets. In other words, for each x we choose $d(x)$ to minimise

$$\int_{\Theta} L(\theta, d(x))\pi(\theta|x)d\theta,$$

the *expected posterior loss* associated with the observed x. This greatly simplifies the calculation of the Bayes rule in a particular case. It also illustrates what many people feel is an intuitively natural property of Bayesian procedures: in order to decide what to do, based on a particular observed X, it is only necessary to think about the losses that follow from one value $d(X)$. There is no need to worry (as would be the case with a minimax procedure) about all the other values of X that might have occurred, but did not. This property, a simplified form of the *likelihood principle* (Chapter 8), illustrates just one of the features that have propelled many modern statisticians towards Bayesian methods.

We consider a number of specific cases relating to hypothesis testing, as well as point and interval estimation. Further Bayesian approaches to hypothesis testing are considered in Section 4.4.

Case 1: Hypothesis testing Consider testing the hypothesis $H_0 : \theta \in \Theta_0$ against the hypothesis $H_1 : \theta \in \Theta_1 \equiv \Theta \setminus \Theta_0$, the complement of Θ_0. Now the action space $\mathcal{A} = \{a_0, a_1\}$, where a_0 denotes 'accept H_0' and a_1 denotes 'accept H_1'. Assume the following form of loss function:

$$L(\theta, a_0) = \begin{cases} 0 & \text{if } \theta \in \Theta_0, \\ 1 & \text{if } \theta \in \Theta_1, \end{cases}$$

and

$$L(\theta, a_1) = \begin{cases} 1 & \text{if } \theta \in \Theta_0, \\ 0 & \text{if } \theta \in \Theta_1. \end{cases}$$

The Bayes decision rule is: accept H_0 if

$$\Pr(\theta \in \Theta_0 | x) < \Pr(\theta \in \Theta_1 | x).$$

Since $\Pr(\theta \in \Theta_1 | x) = 1 - \Pr(\theta \in \Theta_0 | x)$, this is equivalent to accepting H_0 if $\Pr(\theta \in \Theta_0 | x) > 1/2$. We leave the reader to consider what happens when $\Pr(\theta \in \Theta_0 | x) = 1/2$.

Case 2: Point estimation Suppose loss is squared error: $L(\theta, d) = (\theta - d)^2$. For observed $X = x$, the Bayes estimator chooses $d = d(x)$ to minimise

$$\int_\Theta (\theta - d)^2 \pi(\theta | x) d\theta.$$

Differentiating with respect to d, we find

$$\int_\Theta (\theta - d) \pi(\theta | x) d\theta = 0.$$

Taking into account that the posterior density integrates to 1, this becomes

$$d = \int_\Theta \theta \pi(\theta | x) d\theta,$$

the *posterior mean* of θ. In words, for a squared error loss function, the Bayes estimator is the mean of the posterior distribution.

Case 3: Point estimation Suppose $L(\theta, d) = |\theta - d|$. The Bayes rule will minimise

$$\int_{-\infty}^d (d - \theta) \pi(\theta | x) d\theta + \int_d^\infty (\theta - d) \pi(\theta | x) d\theta.$$

Differentiating with respect to d, we must have

$$\int_{-\infty}^{d} \pi(\theta|x)d\theta - \int_{d}^{\infty} \pi(\theta|x)d\theta = 0$$

or in other words

$$\int_{-\infty}^{d} \pi(\theta|x)d\theta = \int_{d}^{\infty} \pi(\theta|x)d\theta = \frac{1}{2}.$$

The Bayes rule is the *posterior median* of θ.

Case 4: Interval estimation Suppose

$$L(\theta, d) = \begin{cases} 0 & \text{if } |\theta - d| \leq \delta, \\ 1 & \text{if } |\theta - d| > \delta, \end{cases}$$

for prescribed $\delta > 0$. The expected posterior loss in this case is the posterior probability that $|\theta - d| > \delta$. This can be most easily motivated as a Bayesian version of interval estimation: we want to find the 'best' interval of the form $(d - \delta, d + \delta)$, of predetermined length 2δ. 'Best' here means the interval that maximises the posterior probability that θ is within the interval specified.

The resulting interval is often called the HPD (for *highest posterior density*) interval. The resulting posterior probability that $\theta \in (d - \delta, d + \delta)$ is often written $1 - \alpha$, by analogy with the notation used for classical confidence intervals. Of course, the interpretation here is quite different from that of confidence intervals, where $1 - \alpha$ is the probability, under repeated sampling of X, that the (random) confidence interval covers θ *whatever the true value of θ may be*. The Bayesian interpretation, as a probability statement about θ, is the one that many beginners wrongly attribute to a confidence interval!

The formulation given here – first decide δ, then choose the interval to minimise α – is the one most naturally represented in terms of loss functions and the like, but in practice we very often go in the opposite direction, that is first specify a suitably small α such as 0.05, then find the interval of smallest length subject to posterior coverage probability $1 - \alpha$. In the case of a *unimodal* posterior density (one which achieves its maximum at a unique point, and decreases monotonically to either side) the solution to both problems is the same, and is achieved by an interval of the form

$$\{\theta : \pi(\theta|x) \geq c\}$$

for some suitable c, as illustrated in the example shown in Figure 3.1.

Although from this point of view the best Bayesian interval is always of HPD form, in practice it is not universally used. The reason is that it may be rather difficult to compute in cases where the posterior density has to be evaluated numerically. Two alternatives are:

(a) the equal-tailed interval, in which the interval is constructed so that $\Pr\{\theta > d + \delta|x\} = \Pr\{\theta < d - \delta|x\} = \alpha/2$, and

(b) the normal interval $(\mu - z_{\alpha/2}\sigma, \mu + z_{\alpha/2}\sigma)$, where μ and σ are the posterior mean and posterior standard deviation and z_α is the upper-α point of the standard normal distribution. This will be a very good approximation to either the equal-tailed or the HPD interval if the posterior density is of approximately normal shape (which it very often is in practice).

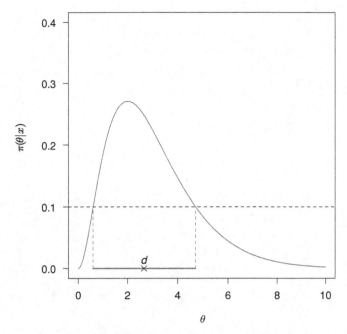

Figure 3.1 A unimodal posterior distribution

The second of these possibilities is an example of application of *Bayesian asymptotics*, the primary conclusion of which is that the posterior distribution $\pi(\theta|x)$ is, quite generally, asymptotically normal: see Chapter 9.

In cases where the posterior density is not unimodal, there are two possible strategies. One is simply to ignore the lack of unimodality and proceed as if the posterior density *were* unimodal. The alternative solution, which is more appropriate if the modes are far apart relative to the scale of the posterior distribution, is to abandon the concept of an 'HPD interval' and instead find an 'HPD set', which may be the union of several intervals, as in the example shown in Figure 3.2.

Example 3.2 (continued) Consider Example 3.2 above, in which $X_1, \ldots, X_n$ are independent, identically distributed from $N(\theta, \sigma^2)$, with known σ^2, and θ has a normal prior distribution. Then we saw that the posterior distribution is also normal of the form $N(\mu_1, \sigma_1^2)$, where μ_1 (depending on $x_1, \ldots, x_n$) and σ_1^2 were computed in the example. In this case the HPD, equal-tailed and normal posterior intervals all coincide and are of the form

$$(\mu_1 - \sigma_1 z_{\alpha/2}, \mu_1 + \sigma_1 z_{\alpha/2}).$$

In the limit as the prior variance $\sigma_0^2 \to \infty$ (a diffuse or *improper prior* for θ, which is not a proper density in the sense of integrating to 1 over the whole range of θ), we have

$$\mu_1 \to \bar{x}, \quad \sigma_1^2 \to \frac{\sigma^2}{n},$$

so the Bayesian interval agrees exactly with the usual, frequentist confidence interval, though the interpretations of the intervals are completely different. This is therefore another case

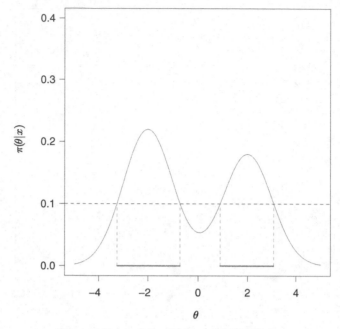

Figure 3.2 A multimodal posterior distribution

in which the Bayesian and classical situations ultimately coincide. In this case, though, the limit is as the prior variance tends to ∞, not the sample size, though the same limits apply as $n \to \infty$, provided $\bar{x}$ and $\frac{\sigma^2}{n}$ are interpreted as asymptotic expressions, rather than limits in the strict sense.

Exercise Work out the corresponding procedure for Example 3.3 (normal mean with unknown variance), to calculate a Bayesian interval estimate for the mean μ. The boundary points in this case are defined by the percentage points of a t distribution; does there exist a prior for which this coincides with the standard (non-Bayesian) confidence interval?

3.3 Back to minimax...

We now give an example to show how some of the ideas we have developed may be applied to solve a non-trivial problem in minimax decision theory.

The problem is: find a minimax estimator of θ based on a single observation $X \sim \text{Bin}(n, \theta)$ with n known, under squared error loss $L(\theta, d) = (\theta - d)^2$.

We know by Theorem 2.2 of Chapter 2 that, if we can find a Bayes (or extended Bayes) estimator that has constant mean squared error (that is, risk), this will also be a minimax rule.

We do not know all the possible Bayes estimators for this problem, but we do know a very large class of them, namely all those that arise from the conjugate prior, a Beta prior with parameters $a > 0$, $b > 0$. For such a prior, the posterior distribution is Beta with the parameters a and b replaced by $a + X$, $b + n - X$. We also know that, with squared error loss, the Bayes estimator is the mean of the posterior distribution,

$$d(X) = \frac{a + X}{a + b + n}.$$

The question therefore arises: is there any estimator in this class which has constant mean squared error? If there is, then it is necessarily the minimax estimator.

Recall $\mathbb{E}X = n\theta$, $\mathbb{E}X^2 = n\theta(1 - \theta) + n^2\theta^2$. Writing $c = a + b + n$, we have

$$\mathbb{E}\left\{\left(\frac{a + X}{c} - \theta\right)^2\right\} = \frac{1}{c^2}\mathbb{E}\{(X + a - c\theta)^2\}$$

$$= \frac{1}{c^2}\{n\theta(1 - \theta) + n^2\theta^2 + 2n\theta(a - c\theta) + (a - c\theta)^2\}.$$

This is a quadratic function of θ, and will be constant if the coefficients of θ and θ^2 are 0. This requires

$$n + 2na - 2ac = 0, \tag{3.1}$$

$$-n + n^2 - 2nc + c^2 = 0. \tag{3.2}$$

Equation (3.2) has roots $c = n \pm \sqrt{n}$, but we need $c > n$ for a proper Bayes rule, so take $c = n + \sqrt{n}$. Then (3.1) gives $a = \sqrt{n}/2$ so the final result is that

$$d(X) = \frac{X + \sqrt{n}/2}{n + \sqrt{n}}$$

is the minimax decision rule with respect to squared error loss. A prior with respect to which the Bayes rule is minimax is called a *least favourable prior*.

Remark Problem 3.4 considers a different loss function, with respect to which the usual estimator $d(X) = X/n$ *is* minimax. It can be shown that this is not a Bayes rule with respect to any proper prior, but it arises in the limit of the above scheme as $a \to 0$, $b \to 0$, that is it is extended Bayes.

3.4 Shrinkage and the James–Stein estimator

We now move on to some broader aspects of the interplay between Bayesian methods and decision theory. Recall from Section 2.7 that, subject to certain restrictions on the prior, Bayes decision rules are admissible. However, minimax rules may not be admissible, and, more generally, statistical estimators that are derived from other criteria, such as maximum likelihood (Chapter 8), may not be admissible. In situations like this, it may be possible to use Bayesian ideas to improve upon classical estimators, even when they are assessed by frequentist criteria. The earliest and most famous example of this is *Stein's paradox*, which we now describe.

Example 3.5: Stein's paradox Let X have a p-dimensional ($p \geq 3$) normal distribution with mean vector μ and known covariance matrix equal to the identity I, meaning that $X_i \sim N(\mu_i, 1)$, independently, $i = 1, \ldots, p$.

Consider estimation of μ, with loss function $L(\mu, d) = \|\mu - d\|^2 = \sum_{i=1}^{p}(\mu_i - d_i)^2$ equal to the sum of squared errors.

If we had just one $X \sim N(\mu, 1)$, $p = 1$, we would certainly estimate μ by X. In the general case $p > 1$, if, as we have assumed, the X_i are independent and we use as loss the sum of squared error losses for the individual components, it seems obvious that the X_i have nothing to do with one another and that we should therefore use X as the multivariate estimator of μ. Stein's paradox is so called because what seems obvious turns out not to be true.

Consider the class of 'James–Stein estimators' of the form

$$d^a(X) = \left(1 - \frac{a}{\|X\|^2}\right) X,$$

indexed by $a \geq 0$, which (at least if $\|X\|^2 > a$) shrink X towards 0.

Now $X \equiv d^0(X)$ has risk

$$R\big(\mu, d^0(X)\big) = \mathbb{E}\|\mu - X\|^2 = \sum_{i=1}^{p} \mathbb{E}(\mu_i - X_i)^2 = \sum_{i=1}^{p} \text{var } X_i$$
$$= p, \quad \text{irrespective of } \mu.$$

Integration by parts shows that, for each i, for suitably behaved real-valued functions h,

$$\mathbb{E}\{(X_i - \mu_i)h(X)\} = \mathbb{E}\left\{\frac{\partial h(X)}{\partial X_i}\right\}.$$

Verification is left to the reader. This result, known as *Stein's Lemma*, enables us to compute the risk of the estimator $d^a(X)$:

$$R\big(\mu, d^a(X)\big) = \mathbb{E}\|\mu - d^a(X)\|^2$$
$$= \mathbb{E}\|\mu - X\|^2 - 2a\mathbb{E}\left[\frac{X^T(X - \mu)}{\|X\|^2}\right] + a^2\mathbb{E}\left[\frac{1}{\|X\|^2}\right].$$

We have

$$\mathbb{E}\left[\frac{X^T(X - \mu)}{\|X\|^2}\right] = \mathbb{E}\left[\sum_{i=1}^{p} \frac{X_i(X_i - \mu_i)}{\Sigma_j X_j^2}\right]$$
$$= \sum_{i=1}^{p} \mathbb{E}\left[\frac{\partial}{\partial X_i}\left\{\frac{X_i}{\Sigma_j X_j^2}\right\}\right]$$
$$= \sum_{i=1}^{p} \mathbb{E}\left[\frac{\Sigma_j X_j^2 - 2X_i^2}{(\Sigma_j X_j^2)^2}\right]$$
$$= \mathbb{E}\left[\frac{p - 2}{\|X\|^2}\right],$$

so

$$R\big(\mu, d^a(X)\big) = p - [2a(p - 2) - a^2]\mathbb{E}\left(\frac{1}{\|X\|^2}\right). \tag{3.3}$$

Remember that here $\mathbb{E}$ denotes expectation with respect to the distribution of X for the given μ. We then note immediately that $R(\mu, d^a(X)) < p \equiv R(\mu, d^0(X))$, provided $2a(p - 2) - a^2 > 0$, that is $0 < a < 2(p - 2)$. For such a, $d^a(X)$ strictly dominates $d^0(X)$, so that the obvious estimator X is inadmissible!

Note also that the risk of $d^a(X)$ is minimised for $a = p - 2$. When $\mu = 0$, $X^T X \sim \mathcal{X}_p^2$, so that $\mathbb{E}[1/(\|X\|^2)] = \frac{1}{p-2}$, by a straightforward direct calculation. Hence, when $\mu = 0$, $d^{p-2}(X)$ has risk $p - [(p-2)^2/(p-2)] = 2$, which is substantially less than the risk of X if p is large.

This inadmissibility result was first pointed out by Charles Stein in 1956, but then proved in more detail by James and Stein (1961). Stein (1981) presented a simpler proof, on which the above analysis is essentially based. At first sight, the result seems incredible: there is no apparent 'tying together' of the losses in different components, yet the obvious estimator, the sample 'mean' X, is not admissible. It is now known that this is a very general phenomenon when comparing three or more populations – the present setting of normal means with known common variance is just the simplest case.

There are, however, many well-documented examples which give intuitive justification for not using the sample mean in practice. The most famous of these concerns an analysis of baseball batting data, by Efron and Morris (1975): see also Efron and Morris (1977). Here we consider a more contemporary data analytic example of the James–Stein estimator, using the same baseball context.

3.4.1 Data example: Home run race, 1998 baseball season

In 1998, American sports fans became gripped by a race between two leading baseball players, Mark McGwire of St Louis Cardinals and Sammy Sosa of Chicago Cubs, to beat the record, set by Roger Maris in 1961, for the number of 'home runs' hit during a major league baseball season. Maris' record of 61 home runs was beaten by McGwire on 8 September, with Sosa hitting his 62nd home run of the season on 13 September. The two players finished the season having hit 70 and 66 home runs respectively. McGwire's new record for the number of home runs in a single season actually lasted just until 2001, when Barry Bonds of San Francisco Giants hit 73 home runs in the season. Could the breaking of Maris' record have been predicted at the beginning of the 1998 season?

Consider the following exercise, based on data obtained from the USAToday baseball statistical archive website. We examine the batting records in pre-season exhibition matches for a set of 17 players, including McGwire and Sosa, those who had hit the most home runs in the previous season, 1997. By considering the proportion of times at bat in those exhibition matches on which each player hit a home run, we attempt to estimate the proportion of times at bat that each will score a home run over the actual competitive season. Of course, to give a precise prediction of the number of home runs over the season we would also need to know the number of at bats in a season, which cannot be predicted in advance; therefore, our formal focus is on prediction of the home run strike rate of each player, rather than prediction of the number of home runs.

The pre-season statistics show the ith player to have achieved Y_i home runs, in n_i times at bat: these figures are shown in the second and third columns of Table 3.1. Then assume that the home runs occur according to a binomial distribution (so that player i has probability p_i of hitting a home run each time at bat, independently of other at bats),

$$Y_i \sim \text{Bin}(n_i, p_i), \quad i = 1, \ldots, 17.$$

Bayesian methods

Table 3.1 *Home run race 1998: James–Stein and maximum likelihood estimators*

	Y_i	n_i	p_i	AB	X_i	JS_i	μ_i	HR	$\hat{HR}$	$\hat{HR}_s$
McGwire	7	58	0.138	509	−6.56	−7.12	−6.18	70	61	50
Sosa	9	59	0.103	643	−5.90	−6.71	−7.06	66	98	75
Griffey	4	74	0.089	633	−9.48	−8.95	−8.32	56	34	43
Castilla	7	84	0.071	645	−9.03	−8.67	−9.44	46	54	61
Gonzalez	3	69	0.074	606	−9.56	−9.01	−8.46	45	26	35
Galaragga	6	63	0.079	555	−7.49	−7.71	−7.94	44	53	48
Palmeiro	2	60	0.070	619	−9.32	−8.86	−8.04	43	21	28
Vaughn	10	54	0.066	609	−5.01	−6.15	−7.73	40	113	78
Bonds	2	53	0.067	552	−8.59	−8.40	−7.62	37	21	24
Bagwell	2	60	0.063	540	−9.32	−8.86	−8.23	34	18	24
Piazza	4	66	0.057	561	−8.72	−8.48	−8.84	32	34	38
Thome	3	66	0.068	440	−9.27	−8.83	−8.47	30	20	25
Thomas	2	72	0.050	585	−10.49	−9.59	−9.52	29	16	28
T. Martinez	5	64	0.053	531	−8.03	−8.05	−8.86	28	41	41
Walker	3	42	0.051	454	−6.67	−7.19	−7.24	23	32	24
Burks	2	38	0.042	504	−6.83	−7.29	−7.15	21	27	19
Buhner	6	58	0.062	244	−6.98	−7.38	−8.15	15	25	21

Here p_i is the true, full-season strike rate, and Y_i/n_i is the strike rate in the pre-season. Table 3.1 shows these values, and the actual number of at bats of each player over the season, in columns 4 and 5, as collected from full-season statistics. Note that the actual season involves roughly ten times as many at bats as the pre-season.

At first sight, this problem may not seem to have very much to do with the rather abstract nature of Stein's paradox. However, the central issue is the same: we are trying to estimate many means simultaneously, and Stein's paradox suggests there may be a better way to do it than simply estimate each player's home run probability as $\hat{p}_i = Y_i/n_i$. We now show how we can first translate this problem into one where a modified version of the James–Stein procedure is applicable, and then we demonstrate how this actually improves the estimates.

Following Efron and Morris (1975), suppose we define $X_i = f_{n_i}(Y_i/n_i)$, $i = 1, \ldots, 17$, with

$$f_n(y) = n^{1/2} \sin^{-1}(2y - 1).$$

This is an example of a *variance stabilising transformation*. Then (Efron and Morris, 1975) X_i has mean approximately equal to $\mu_i = f_{n_i}(p_i)$ and unit variance. The values of X_i and μ_i are shown in columns 6 and 8 respectively of Table 3.1. We have, in fact, that the X_i are approximately, independently distributed as

$$X_i \sim N(\mu_i, 1), \quad i = 1, \ldots, 17.$$

The specific inference problem we consider is that of estimating the μ_i from the observed X_i. To do so, we consider a slight extension of the James–Stein estimator $d^{p-2}(X)$, by which the estimate of μ_i is

$$JS_i = \bar{X} + \{1 - (p - 3)/V\}(X_i - \bar{X}),$$

with $V = \sum(X_i - \bar{X})^2$ and $\bar{X} = \sum X_i/p$, with $p = 17$ the number of component estimation problems. Properties of this estimator are considered in Problems 3.6 and 3.7. Now, the individual X_i are shrunk towards their mean $\bar{X}$, rather than 0. The intuition behind such an estimator is that the true μ_i are likely to be more closely clustered than the raw X_i, which are subject to substantial random variability arising from the small amount of pre-season data. It therefore seems sensible to shrink the individual X_i towards the overall mean $\bar{X}$, to obtain a more clustered set of estimates of the μ_i.

Analogous to the result shown for the simple James–Stein estimator, the risk function of the James–Stein estimator $JS = (JS_1, \ldots, JS_p)$ is bounded above (Problem 3.6) by the (constant) risk p of the naive estimator $X = (X_1, \ldots, X_p)$, which estimates μ_i by X_i, a direct extrapolation of the ith player's pre-season strike rate. The components JS_i of the James–Stein estimator are shown in column 7 of Table 3.1.

The total squared error $\sum(X_i - \mu_i)^2$ of X is 19.68, while the James–Stein estimator has total squared error $\sum(JS_i - \mu_i)^2 = 8.07$. Further, for 14 of the 17 players JS_i is closer to μ_i than is X_i. So, the James–Stein estimator has reduced the total squared error by more than half.

The final columns of Table 3.1 present the hypothetical estimates of the number of home runs of each player over the season, together with the actual home runs totals HR. (These estimates are constructed using information on the number of at bats over the full season: this information is, of course, not known in advance.) The simple estimates $\hat{HR}$ based on direct extrapolation include some wild estimates, such as the estimate of 113 home runs for Mo Vaughn, but even the James–Stein-based estimates $\hat{HR}_s$ suggest that two players, Vaughn and Sosa, would beat the Maris record.

The figures in Table 3.1 provide a stark illustration of an important point. The James–Stein estimator JS achieves a uniformly lower *aggregate risk* than X, but allows increased risk in estimation of individual components μ_i. If we were interested, as with hindsight we might well have been, in estimation of μ_i for McGwire *alone*, then the simple estimator X_i is preferable over the corresponding component JS_i of the James–Stein estimator. Direct extrapolation from his pre-season batting would have suggested that McGwire *would* just have reached Maris' record of 61 home runs, while the James–Stein procedure estimates just 50 home runs, further away from McGwire's actual tally of 70.

3.4.2 Some discussion

Admissibility of $d(X) = X$ in dimension $p = 1$ was established by Blyth (1951). A simple, direct proof is possible: see, for example Casella and Berger (1990: Section 10.4). Admissibility is more awkward to prove in the case $p = 2$, but was established by Stein (1956). Berger (1985: Chapter 8) gives the admissibility results a Bayesian interpretation, using the notion of a *generalised Bayes rule*. Though the formal definition of a generalised Bayes rule is mathematically awkward, the rough idea is that of a rule which minimises the expected posterior loss, obtained from an improper prior. In the estimation problem at hand, any admissible estimator is a generalised Bayes rule, and results are available which determine whether or not a generalised Bayes estimator is admissible. Since X is a generalised Bayes estimator in any dimension p, these latter results lead immediately to the conclusions that X is admissible if $p = 1$ or 2, but not if $p \geq 3$.

A point of clarification should be noted here: although the estimator $d^a(X)$ defined in Example 3.5 dominates $d^0(X) = X$ for certain values of a, this does not mean we would actually want to use the estimator $d^a(X)$ in applications. Once the idea is presented, that we might not want to use X as our estimator, then there are many so-called shrinkage estimators which potentially improve on X, and the task of deciding which of them to adopt is an important focus of practical discussion. A key point to note is that the estimator $d^{p-2}(X)$ is actually inadmissible: it is strictly dominated by the estimator $d_+^{p-2}(X)$, which replaces the factor $(1 - \frac{p-2}{X^T X})$ by zero whenever it is negative.

From a modern viewpoint, there are many applications involving the comparison of large numbers of populations in which ideas related to shrinkage have an important role to play. Topical applications include the comparison of the success rates of operations in different hospitals, and the comparison of examination results over many schools (school 'league tables'): see, for example, Goldstein and Spiegelhalter (1996). Although such modern applications seem rather far removed from the original theoretical result discovered by Stein, there is a sense in which they derive ultimately from it.

3.5 Empirical Bayes

In a standard Bayesian analysis, there will usually be parameters in the prior distribution that have to be specified.

For example, consider the simple normal model in which $X \mid \theta \sim N(\theta, 1)$ and θ has the prior distribution $\theta \mid \tau^2 \sim N(0, \tau^2)$. If a value is specified for the parameter τ^2 of the prior, a standard Bayesian analysis can be carried out. Noting that $f(x) = \int f(x; \theta) \pi(\theta) d\theta$, it is readily shown that the marginal distribution of X is $N(0, \tau^2 + 1)$, and can therefore be used to estimate τ^2, in circumstances where a value is not specified.

Empirical Bayes analysis is characterised by the estimation of prior parameter values from marginal distributions of data. Having estimated the prior parameter values, we proceed as before, as if these values had been fixed at the beginning.

3.5.1 James–Stein estimator, revisited

In the Stein's paradox Example 3.5 above, the estimator $d^{p-2}(X)$ may be viewed as an empirical Bayes estimator of μ, the Bayes rule with prior parameter values replaced by estimates constructed from the marginal distribution of the X_i.

Specifically, let $X_i \mid \mu_i$ be distributed as $N(\mu_i, 1)$, independently, $i = 1, \ldots, p$, and suppose $\mu_1, \ldots, \mu_p$ are independent, identically distributed $N(0, \tau^2)$.

If τ^2 is known, the Bayes estimator $\delta^\tau(X)$, for the given sum of squared errors loss, of $\mu = (\mu_1, \ldots, \mu_p)^T$ is the posterior mean $\delta^\tau(X) = \frac{\tau^2}{\tau^2+1} X$, on observing that the posterior distribution of μ_i is $N(\frac{\tau^2}{\tau^2+1} X_i, \frac{\tau^2}{\tau^2+1})$, independently for $i = 1, \ldots, p$. Straightforward calculations then show that the Bayes risk of $\delta^\tau(X)$, $r(\tau, \delta^\tau(X))$, say, in an obvious notation, is given by

$$r(\tau, \delta^{\tau}(X)) = \sum_{i=1}^{p} \mathrm{var}(\mu_i | X_i) = \sum_{i=1}^{p} \frac{\tau^2}{\tau^2 + 1} = \frac{p\tau^2}{\tau^2 + 1}.$$

Marginally the X_i are independent, identically distributed $N(0, \tau^2 + 1)$, so that $X_i/\sqrt{\tau^2 + 1} \sim N(0, 1)$ and marginally $\|X\|^2/(\tau^2 + 1) \sim \chi_p^2$. Since we know that $E(1/Z) = 1/(p - 2)$ if $Z \sim \chi_p^2$ and $p \geq 3$, we see that taking the expectation with respect to this marginal distribution of X gives

$$\mathbb{E}\left[1 - \frac{(p - 2)}{\|X\|^2}\right] = \frac{\tau^2}{\tau^2 + 1}, \qquad (3.4)$$

if $p \geq 3$.

In the case when τ^2 is unknown, estimating $\tau^2/(\tau^2 + 1)$ by $1 - (p - 2)/(\|X\|^2)$ yields the James–Stein estimator $d^{p-2}(X)$.

Under our assumed model, the Bayes risk of the James–Stein estimator $d^{p-2}(X)$ is

$$r(\tau, d^{p-2}(X)) = \int R(\mu, d^{p-2}(X))\pi(\mu)d\mu$$

$$= \int_{\mathbb{R}^p} \int_{\mathcal{X}} \left[p - \frac{(p - 2)^2}{\|x\|^2}\right] f(x|\mu)\pi(\mu)dxd\mu$$

$$= \int_{\mathcal{X}} \left\{\int_{\mathbb{R}^p} \left[p - \frac{(p - 2)^2}{\|x\|^2}\right] \pi(\mu|x)d\mu\right\} f(x)dx,$$

where we have used (3.3) and then changed the order of integration. Now, the integrand in the inner integral is independent of μ, and $\int \pi(\mu|x)d\mu$ is trivially equal to 1, and therefore

$$r(\tau, d^{p-2}(X)) = p - (p - 2)^2 \mathbb{E}\left(\frac{1}{\|X\|^2}\right).$$

Now the expectation is, as in (3.4), with respect to the marginal distribution of X, so that (3.4) immediately gives

$$r(\tau, d^{p-2}(X)) = p - \frac{p - 2}{\tau^2 + 1} = r(\tau, \delta^{\tau}(X)) + \frac{2}{\tau^2 + 1}.$$

The second term represents the increase in Bayes risk associated with the need to estimate τ^2: the increase tends to 0 as $\tau^2 \to \infty$.

3.6 Choice of prior distributions

To understand some of the controversies about Bayesian statistics, including various ways of thinking about the choice of prior distributions, it is helpful to know something more of the history of the subject.

Bayesian statistics takes its name from an eighteenth-century English clergyman, the Reverend Thomas Bayes. Bayes died in 1761 but his most famous work, 'An essay towards solving a problem in the doctrine of chances', was published posthumously in the *Philosophical Transactions of the Royal Society* (Bayes, 1763). The problem considered by Bayes was, in modern terminology, the problem of estimating θ in a binomial (n, θ) distribution and he worked out what we now call the Bayesian solution, under the assumption that θ has a uniform prior density on $(0,1)$, equivalent to $a = b = 1$ in our Beta prior formulation of Example 3.1. This assumption, sometimes called *Bayes' postulate*, is the controversial assumption in the paper (not Bayes' Theorem itself, which is just an elementary statement about conditional probabilities). Some authors have held, though modern scholars dispute this, that Bayes' dissatisfaction with this assumption is the reason that he did not publish his paper during his lifetime. Whether this is correct or not, it is the case that much of the paper is devoted to justifying this assumption, for which Bayes gave an ingenious physical argument. However, Bayes' argument is difficult to generalise to other situations in which one might want to apply Bayesian statistics.

At the time, Bayes' paper had very little influence and much of what we now call Bayesian statistics was developed, independently of Bayes, by the French mathematician Laplace (resulting in this *Théorie Analytique des Probabilités*, published in 1812, though the bulk of the work was done in the 1770s and 1780s). Laplace widely used the 'principle of insufficient reason' to justify uniform prior densities: we do not have any reason to think that one value of θ is more likely than any other, therefore we should use a uniform prior distribution. One disadvantage of that argument is that if we apply the principle of insufficient reason to θ^2, say, this results in a different prior from the same principle applied to θ. The argument used by Bayes was more subtle than that, and did lead to a uniform prior on θ itself rather than some transformation of θ, but only for a specific physical model.

By the time more-modern theories of statistical inference were being developed, starting with the work of Francis Galton and Karl Pearson in the late nineteenth century, Bayesian ideas were under a cloud, and R.A. Fisher, arguably the greatest contributor of all to modern statistical methods, was vehemently anti-Bayesian throughout his career. (Fisher never held an academic post in statistics or mathematics, but for many years was Professor of Genetics in Cambridge, and a Fellow of Gonville and Caius College.) However, the tide began to swing back towards Bayesian statistics beginning with the publication of Jeffreys' book *Theory of Probability* in 1939. Jeffreys was also a Cambridge professor, most famous for his contributions to applied mathematics, geophysics and astronomy, but he also thought deeply about the foundations of scientific inference, and his book, despite its title, is a treatise on Bayesian methods. Following in the tradition of Laplace, Jeffreys believed that the prior distribution should be as uninformative as possible, and proposed a general formula, now known as the Jeffreys prior, for achieving this. However, his arguments did not convince the sceptics; Fisher, in a review of his book, stated that there was a mistake on page 1 (that is the use of a Bayesian formulation) and this invalidated the whole book!

One feature of the arguments of Laplace and Jeffreys is that they often result in what we have termed improper priors. Suppose we use the principle of insufficient reason to argue in favour of a uniform prior for a parameter θ. When the range of θ is the whole real line (for instance, if θ is the unknown mean of a normal distribution) then this would lead to a prior which cannot be normalised to form a proper density. The limit in Example 3.2 above, where $\sigma_0^2 \to \infty$, is a case in point. However, in many such cases the *posterior* density is still proper, and can be thought of as a limit of posterior densities based on proper priors. Alternatively, a decision rule of this form is extended Bayes. Most modern Bayesians do not have a problem with improper prior distributions, though with very complicated problems there is a danger that an improper prior density will result in an improper posterior density, and this must of course be avoided!

While Jeffreys was developing his theory, Neyman and Egon Pearson (son of Karl) had published their theory of hypothesis testing (Neyman and Pearson, 1933), which also avoided any reference to Bayesian ideas. (Fisher also disagreed with Neyman's approach, but the source of their disagreement is too complicated to summarise in a couple of sentences. The one thing they agreed on was that Bayesian ideas were no good.) The ideas started by Neyman and Pearson were taken up in the United States, in particular by Abraham Wald, whose book *Statistical Decision Functions* (1950) developed much of the abstract theory of statistical decisions which we see in this text.

However at about the same time B. de Finetti (in Italy) and L.J. Savage (in the USA) were developing an alternative approach to Bayesian statistics based on subjective probability. In the UK, the leading exponent of this approach was D.V. Lindley. According to de Finetti, Savage and Lindley, the only logically consistent theory of probability, and therefore of statistics, is one based on personal probability, in which each individual behaves in such a way as to maximise his/her expected utility according to his/her own judgement of the probabilities of various outcomes. Thus they rejected not only the whole of classical (non-Bayesian) statistics, but also the 'uninformative prior' approach of Laplace and Jeffreys. They believed that the only way to choose a prior distribution was subjectively, and they had no problem with the fact that this would mean different statisticians reaching different conclusions from the same set of data.

There are many situations where subjective judgement of probability is essential. The most familiar situation is at a racetrack! When a bookmaker quotes the odds on a horse race, he is using his subjective judgement, but a bookmaker who did not consistently get the odds right (or very nearly right) would soon go out of business. American weather forecasters also make widespread use of subjective probabilities, because their forecasts always include statements like 'the chance of rain is 40%'. Although they have all the modern tools of computer-based weather forecasting to help them, the actual probability quoted is a subjective judgement by the person making the forecast, and much research has been done on assessing and improving the skills of forecasters in making these subjectively based forecasts.

Thus there are many situations where subjective probability methods are highly appropriate; the controversial part about the theories of de Finetti and Savage is the assertion that *all* probabilistic and statistical statements should be based on subjective probability.

From the perspective of present-day statistics, Bayesian and non-Bayesian methods happily co-exist most of the time. Some modern theoreticians have taken a strongly pro-Bayesian approach (see, for example, the introduction to the 1985 second edition of Berger's book) but much of the modern interest in Bayesian methods for applied statistics has resulted from more pragmatic considerations: in the very complicated models analysed in present-day statistics, often involving thousands of observations and hundreds of parameters, Bayesian methods can be implemented computationally using devices such as the Gibbs sampler (see Section 3.7), whereas the calculation of, for instance, a minimax decision rule, is too complicated to apply in practice. Nevertheless, the arguments are very far from being resolved. Consider, for example, the problem of estimating a density $f(x)$, when we have independent, identically distributed observations $X_1, \ldots, X_n$ from that density, but where we do not make any parametric assumption, such as normal, gamma, etc. This kind of problem can be thought of as one with an infinite-dimensional unknown parameter, but in that case it is a hard problem (conceptually, not just practically) to formulate the kind of prior distribution necessary to apply Bayesian methods. Meanwhile, comparisons of different estimators by means of a criterion such as mean squared error are relatively straightforward, and some modern theoreticians have developed ingenious minimax solutions to problems of this nature, which have no counterpart in the Bayesian literature.

Thus, the main approaches to the selection of prior distributions may be summarised as:

(a) physical reasoning (Bayes) – too restrictive for most practical purposes;
(b) flat or uniform priors, including improper priors (Laplace, Jeffreys) – the most widely used method in practice, but the theoretical justification for this approach is still a source of argument;
(c) subjective priors (de Finetti, Savage) – used in certain specific situations such as weather forecasting (though even there it does not tend to be as part of a formal Bayesian analysis with likelihoods and posterior distributions) and for certain kinds of business applications where prior information is very important and it is worthwhile to go to the trouble of trying to establish ('elicit' is the word most commonly used for this) the client's true subjective opinions, but hardly used at all for routine statistical analysis;
(d) prior distributions for convenience, for example conjugate priors – in practice these are very often used just to simplify the calculations.

3.7 Computational techniques

As mentioned previously, one of the main practical advantages of Bayesian methods is that they may often be applied in very complicated situations where both X and θ are very high dimensional. In such a situation, the main computational problem is to compute numerically the normalising constant that is required to make the posterior density a proper density function.

Direct numerical integration is usually impracticable in more than four or five dimensions. Instead, *Monte Carlo methods* – in which random numbers are drawn to simulate a sample from the posterior distribution – have become very widely used. These methods use computational algorithms known as *pseudorandom number generators* to obtain streams of numbers, which look like independent, identically distributed uniform random numbers over $(0,1)$, and then a variety of transformation techniques to convert these uniform random numbers to any desired distribution.

3.7.1 Gibbs sampler

One computational technique in common use is the *Gibbs sampler*. Suppose θ is d-dimensional: $\theta = (\theta_1, \ldots, \theta_d) \in \Theta \subseteq \mathbb{R}^d$. We know that

$$\pi(\theta|X = x) \propto \pi(\theta) f(x; \theta),$$

but we have no practical method of computing the normalising constant needed to make this into a proper density function. So, instead of doing that, we try to generate a pseudorandom sample of observations from $\pi(\cdot|x)$, sampling from the distribution of θ, holding x fixed. If we can do that, then we can easily approximate probabilities of interest (for example what is $\Pr\{\theta_1 > 27.15|X = x\}$?) from the empirical distribution of the simulated sample.

Start off with an arbitrary initial vector, say $\theta^{(0)} = (\theta_1^{(0)}, \ldots, \theta_d^{(0)})$. Now carry out the following procedure:

Step 1: Holding $\theta_2^{(0)}, \ldots, \theta_d^{(0)}$ fixed, generate a new value of θ_1 conditional on $\theta_2 = \theta_2^{(0)}, \ldots, \theta_d = \theta_d^{(0)}$ and of course $X = x$, to obtain a new value $\theta_1^{(1)}$.

Step 2: Generate a new value $\theta_2 = \theta_2^{(1)}$ from the conditional distribution given $\theta_1 = \theta_1^{(1)}, \theta_3 = \theta_3^{(0)}, \ldots, \theta_d = \theta_d^{(0)}, X = x$.

Step 3: Generate a new value $\theta_3 = \theta_3^{(1)}$ from the conditional distribution given $\theta_1 = \theta_1^{(1)}, \theta_2 = \theta_2^{(1)}, \theta_4 = \theta_4^{(0)}, \ldots, \theta_d = \theta_d^{(0)}, X = x$.

. . .

Step d: Generate a new value $\theta_d = \theta_d^{(1)}$ from the conditional distribution given $\theta_1 = \theta_1^{(1)}, \ldots, \theta_{d-1} = \theta_{d-1}^{(1)}, X = x$.

This completes one iteration of the Gibbs sampler, and generates a new vector $\theta^{(1)}$. Notice that at each step we only have to simulate from a *one*-dimensional (conditional) distribution. We then repeat this process to get $\theta^{(2)}, \theta^{(3)}, \ldots$. After many such iterations (usually several hundred or even several thousand are required) the sampling distribution of θ will approximate the posterior distribution we are trying to calculate, and the Monte Carlo sample can then be used directly to approximate the probabilities of interest.

This method still requires that we have some efficient way to generate the individual $\theta_1, \ldots, \theta_d$ values. However, very often we can simplify this problem by using conjugate priors for them. It is a familiar situation, when Bayesian analysis is applied to very high-dimensional problems, where we can find conjugate prior families for individual components of θ, but where there is no way to find a single conjugate prior family for the entire vector θ. In problems where we cannot use conjugate priors at all, there are other efficient methods of generating Monte Carlo samples for a single component of θ, which may then be used in conjunction with the Gibbs sampler to create samples from the whole vector.

Consecutive observations $\theta^{(j)}$ and $\theta^{(j+1)}$ will not be independent, so if a random sample of approximately independent observations is required we must sample intermittently from the simulated sequence. For instance, we might take observations $\theta^{(s)}, \theta^{(s+t)}, \ldots, \theta^{(s+(n-1)t)}$, for suitable s and large t, as our sample of size n from $\pi(\theta \mid x)$. An alternative is to repeat the whole process n times, using n different starting values $\theta^{(0)}$, and obtain $\theta^{(s)}$ from each of the n runs of the iterative procedure as our sample of observations, or to carry out n independent parallel runs from the same starting value.

The Gibbs sampler is an example of a Markov chain Monte Carlo (MCMC) method. The iterative procedure is simulating a Markov chain which, under suitable regularity conditions, has equilibrium distribution the posterior distribution $\pi(\theta \mid x)$.

Suppose that we have simulated a random sample of observations $\theta^{[1]}, \theta^{[2]}, \ldots, \theta^{[n]}$ from $\pi(\theta \mid x)$, and that we wish to make inference about one component of θ, say θ_i. Let $\theta_i^{[j]}$ denote the ith component of $\theta^{[j]}$ and let $\theta_{-i}^{[j]}$ denote $\theta^{[j]}$ with this ith component deleted, $j = 1, \ldots, n$.

Inferences about θ_i may be based on either $\theta_i^{[1]}, \theta_i^{[2]}, \ldots, \theta_i^{[n]}$ or $\theta_{-i}^{[1]}, \theta_{-i}^{[2]}, \ldots, \theta_{-i}^{[n]}$. An estimate of the posterior mean of θ_i, $E(\theta_i \mid x)$, for example, might be $n^{-1} \sum_{j=1}^{n} \theta_i^{[j]}$. However, in general, it is more efficient to base inferences on $\theta_{-i}^{[1]}, \ldots, \theta_{-i}^{[n]}$. The conditional density $\pi_i(\theta_i \mid \theta_{-i}, x)$, where θ_{-i} is θ with the ith component deleted, is known (we have drawn samples from it at Step i of the iterative procedure) and $\pi(\theta_i \mid x)$ may be estimated by

$$\hat{\pi}(\theta_i \mid x) = n^{-1} \sum_{j=1}^{n} \pi_i\left(\theta_i \mid \theta_{-i} = \theta_{-i}^{[j]}, x\right). \tag{3.5}$$

This process of obtaining estimates is usually referred to as Rao–Blackwellisation, after the Rao–Blackwell Theorem (Chapter 6) which is the basis of proof that such estimates are to be preferred to those based on $\theta_i^{[1]}, \theta_i^{[2]}, \ldots, \theta_i^{[n]}$.

3.7.2 The Metropolis–Hastings sampler

Next to the Gibbs sampler, the most popular method used in Bayesian statistics is the *Metropolis–Hastings sampler*, originally developed by Metropolis *et al.* (1953) for the numerical solution of certain problems in statistical physics, then reformulated and extended, in language more familiar to statisticians, by Hastings (1970). Very often it is combined with the Gibbs sampler: in cases where it is not possible to draw a random variable directly from the conditional distribution of one component of θ given the rest, an alternative approach is to perform one step of a Metropolis–Hastings sampler on each component in turn, iterating among the components in the same manner as the Gibbs sampler. However, the basic idea of the Metropolis–Hastings algorithm works in any number of dimensions, so it is also possible to update the entire vector θ in a single step. The main practical advantage of updating one component at a time is in cases where some rescaling is needed to achieve an efficient sampler, because it is easier to find an optimal scaling in one component at a time than over all components simultaneously. We consider scaling issues in Section 3.7.6.

In the next two sections, we assume some knowledge of the theory of Markov chains. However the exposition is at a fairly rudimentary level, and the reader who has not taken a course in Markov chains should still be able to follow the basic concepts. Although the Metropolis–Hastings algorithm is usually applied to settings with continuous distributions, the detailed discussion will be only for discrete distributions; we argue largely by analogy that the same formulae apply in the continuous case. Tierney (1994) gave a detailed discussion of continuous-state Markov chains as they are used in the context of Monte Carlo sampling algorithms.

3.7.3 Metropolis–Hastings algorithm for a discrete state space

Let $\mathcal{X} = \{x_1, x_2, \ldots\}$ be a discrete (countable) state space and let X be a random variable for which $\Pr\{X = x_i\} = f_i$ for known $\{f_i\}$ satisfying $f_i > 0$, $\sum_i f_i = 1$. In principle it is easy to sample from X directly. For example, let U be a pseudorandom variable which is uniform on $[0, 1]$; let I be the first index for which $\sum_{i=1}^{I} f_i > U$; set $X = x_I$. In practice that may not be so easy for two reasons: (i) if f_i converges only very slowly to 0, it may take a lot of searching to find the right I; (ii) in some cases the f_i may only be known up to an unspecified constant of proportionality and in that case it would be impossible to apply the direct method without first evaluating the constant of proportionality. As we have already noted, the latter situation is particularly common in the case of Bayesian statistics, where exact evaluation of the normalising constant requires summation over a large state space (in the discrete case) or integration over a multidimensional space (in the continuous case), and the whole point of using Markov chain simulation methods is to avoid that calculation. As we shall see, the Metropolis–Hastings algorithm is readily adapted to the case where each f_i is known only up to a constant of proportionality.

The steps of the algorithm are as follows:

Step 1: Start from an arbitrary $X^{(0)}$. In Bayesian statistics, where X is a parameter of a model (more usually denoted θ), the starting value is often taken as the maximum likelihood estimator (Chapter 8), though this is by no means necessary for the algorithm to work.

Step 2: Given $X^{(n)} = x_i$, generate a random variable $Y = x_J$, where the index J is chosen by $\Pr\{J = j \mid X^{(n)} = x_i\} = q_{ij}$ for some family $\{q_{ij}\}$ such that $\sum_j q_{ij} = 1$ for each i. The family $\{q_{ij}\}$ is arbitrary except for an irreducibility condition which we explain below; it is generally chosen for convenience and the ease with which the random variable J may be generated. The value Y is called a *trial value* for the next step of the Markov chain.

Step 3: Define $\alpha = \min\left(\frac{f_J q_{Ji}}{f_i q_{ij}}, 1\right)$. If $\alpha = 1$, then set $X^{(n+1)} = Y$ (in this case Y is *accepted*). If $0 < \alpha < 1$ perform an auxiliary randomisation to accept Y with probability α (for example, let U be a uniform $[0, 1]$ random variable, independent of all previous random variables, and accept Y if $U < \alpha$). If Y is accepted then $X^{(n+1)} = Y$; else $X^{(n+1)} = X^{(n)}$.

Step 4: Replace n by $n + 1$ and return to Step 2.

The irreducibility condition in Step 2 essentially says that it is possible to get from any state to any other state in a finite number of steps, with positive probability. For example, a sufficient condition for that would be $q_{i,i+1} > 0$ for all i, $q_{i,i-1} > 0$ for all $i > 1$. However, subject to that, the choice of $\{q_{ij}\}$ really is arbitrary, though it can have quite a drastic effect on the rate of convergence. It is even possible to use a different $\{q_{ij}\}$ array for each sampling step n of the process, though this complicates considerably the proofs of convergence.

Note also that Step 3 of the algorithm depends on the individual f_i values only through *ratios* f_j/f_i, so, if a normalising constant is undefined, this does not invalidate the algorithm.

3.7.4 Proof of convergence of Metropolis–Hastings in the discrete case

The process $X^{(n)}$ described in Section 3.7.3 is a discrete state Markov chain; from the theory of such chains it is known that if a Markov chain with transition probabilities $p_{ij} = \Pr\{X^{(n+1)} = x_j \mid X^{(n)} = x_i\}$ is (a) irreducible, (b) aperiodic and (c) has a stationary measure $\{\pi_i\}$ satisfying

$$\sum_i \pi_i p_{ij} = \pi_j \text{ for all } j, \tag{3.6}$$

then the Markov chain converges, that is $\lim_{n\to\infty} \Pr\{X^{(n)} = x_i\} = \pi_i$ for all i. In the present case, irreducibility is guaranteed as a condition on the array $\{q_{ij}\}$, and aperiodicity is automatic from the property that $\Pr\{X^{(n+1)} = X^{(n)}\} > 0$, so it will suffice to show that (3.6) holds with $\pi_i = f_i$.

First note that (3.6) will follow at once if we can verify the simpler condition

$$\pi_i p_{ij} = \pi_j p_{ji} \text{ for all } j \neq i, \qquad (3.7)$$

which if often called the *detailed balance* condition. If (3.7) holds, then the Markov chain is called *reversible*.

We now show that (3.7) holds for the Metropolis–Hastings sampler, with $\pi_i = f_i$.

In this case, combining the trial selection and the accept/reject step into a single operation, we have, for all $i \neq j$,

$$p_{ij} = q_{ij} \min \left(\frac{f_j q_{ji}}{f_i q_{ij}}, 1 \right).$$

There are two cases. If $f_j q_{ji} \geq f_i q_{ij}$ then $p_{ij} = q_{ij}$, $p_{ji} = f_i q_{ij}/f_j$. So $f_i p_{ij} = f_i q_{ij} = f_j p_{ji}$. If $f_j q_{ji} < f_i q_{ij}$ then exactly the same argument holds after switching i and j. Therefore, in either case, $f_i p_{ij} = f_j p_{ji}$. However this is the detailed balance condition we were trying to show, so convergence of the Markov chain is established.

3.7.5 Metropolis–Hastings algorithm for a continuous state space

As we have already intimated, rigorous proof of the convergence of the Metropolis–Hastings algorithm is significantly harder in the case of a continuous state space, but the basic concept is identical, so we content ourselves with describing the method without rigorously attempting to justify it.

In this case we assume $\mathcal{X}$ is a continuous state space, usually a subspace of $\mathbb{R}^d$ for some d. The true density of X is $f(x)$, which may be known only up to a constant of proportionality. In Bayesian statistics, we usually write θ instead of X; we know that the posterior density of θ is proportional to the product of the prior density and likelihood, but the constant of proportionality is usually intractable except in cases involving a conjugate prior. As in the discrete case, we require a trial density $q(x, y) \geq 0$, which is some probability density satisfying $\int_{\mathcal{X}} q(x, y)dy = 1$ for all x, and an irreducibility condition that we shall not attempt to make precise, but the basic idea is that, from any starting point $X^{(0)} = x$, it should be possible to get arbitrarily close to any other point y, for which $f(y) > 0$, in a finite number of steps, with positive probability.

The algorithm mimics Steps 1–4 in the discrete case, as follows:

Step 1:   Start from an arbitrary $X^{(0)}$.

Step 2:   Given $X^{(n)} = x$, generate a trial value $Y = y$ from the probability density $q(x, y)$.

Step 3:   Define $\alpha = \min \left(\frac{f(y)q(y,x)}{f(x)q(x,y)}, 1 \right)$. If $\alpha = 1$ then set $X^{(n+1)} = Y$. If $0 < \alpha < 1$ perform an auxiliary randomisation to accept Y with probability α. If Y is accepted then $X^{(n+1)} = Y$; else $X^{(n+1)} = X^{(n)}$.

Step 4:   Replace n by $n + 1$ and return to Step 2.

In both the discrete and continuous cases, the algorithm continues for a large number of steps until it is judged to have converged. It is quite common to delete a number of iterations at the beginning to allow *burn-in* – that is, a period during which the state probabilities or probability density are assumed to be settling down to the true $\{f_i\}$ or $f(\cdot)$. For example, a typical use of the algorithm might treat the first 10 000 iterations as a burn-in sample which is discarded, then continue for another 100 000 iterations to provide what is then treated as a random sample from the desired distribution. Often, it is a subjective decision (affected as much by available computer time as by formal convergence considerations) how many iterations are needed for each of these stages, but there are also diagnostics for convergence, and, in a few cases, rigorous mathematical upper bounds for the number of iterations required for any desired precision.

In the symmetric case, where $q(x, y) = q(y, x)$ for all x, y, Step 3 of the algorithm may be simplified to $\alpha = \min\left(\frac{f(y)}{f(x)}, 1\right)$. This was the original form given by Metropolis *et al.* (1953); the fact that the same idea would work when q is not symmetric was noted by Hastings (1970). Thus in principle the Hastings algorithm is more general than the Metropolis algorithm, though in practice it is very common to use a symmetric q function as in the original Metropolis procedure.

3.7.6 Scaling

The key quantity that has to be specified in the Metropolis–Hastings algorithm is the trial density function $q(x, y)$. This is often simplified by writing $q(x, y) = \frac{1}{h} g \left(\frac{y-x}{h}\right)$ for some density g symmetric about 0 and a *scaling constant* h; note that in this case we automatically have $q(x, y) = q(y, x)$. We may take g to be some standard form such as normal or uniform; the critical issue then becomes how to choose h so that the algorithm converges in reasonable time. We must not take h to be too large because, in that case, a single trial step will often lead to a value Y very far from the centre of the distribution, so that the probability of rejection is very high. However, it can be equally disastrous to choose h too small; in that case, the algorithm will have a high probability of accepting the trial Y, but it will proceed with very small steps and therefore take a long time to cover the whole sample space.

In a remarkable result, Roberts, Gelman and Gilks (1997) argued that the correct scaling constant is one that leads to an overall acceptance rate (average value of α) of about 0.23. To be fair, the mathematical derivation of this result involves a number of steps that take us rather far from the original Hastings–Metropolis algorithm, for example assuming the dimension of the sampling space tends to ∞ and approximating the Markov chain in Langevin diffusion. Nevertheless, it has been found in practical examples that this rule gives good guidance to the optimal scaling even in cases not formally covered by their theorem. However, it can be hard in practice to find h to achieve some predetermined acceptance rate. Gilks *et al.* (1996) recommended, as a rule of thumb, trying to achieve an acceptance rate between 15% and 50%, and this seems good enough for most practical purposes.

In the case that Metropolis–Hastings sampling is used to sample from the posterior distribution of a d-dimensional parameter vector θ, there are still two basic ways of doing it: either the algorithm is applied directly to sampling in $\mathbb{R}^d$, or it could be applied one component at a time, writing $\theta^{(n)} = (\theta_1^{(n)}, \ldots, \theta_d^{(n)})$ and updating $\theta_j^{(n)}$, for $1 \le j \le d$, by applying one step of a Metropolis–Hastings sampler to the conditional distribution of θ_j given $\theta_1^{(n+1)}, \ldots, \theta_{j-1}^{(n+1)}, \theta_{j+1}^{(n)}, \ldots, \theta_d^{(n)}$. This algorithm (sometimes called Gibbs–Metropolis) therefore combines the essential features of the Gibbs sampler and the Metropolis–Hastings sampler. Although it does not appear to be theoretically superior to the direct Metropolis–Hastings sampler, it does have the practical advantage that the appropriate scaling factors

$h_1, \ldots, h_d$ can be chosen and optimised separately for each component of θ. Therefore, for practical purposes, this is often the recommended method.

3.7.7 Further reading

The description here covers only the two best-known forms of Markov chain Monte Carlo algorithm; there now are many others. For example, in recent years much attention has been given to variants on the concept of *perfect sampling*, in which a Markov chain algorithm is adapted to generate a sample *directly* from the stationary distribution, without an initial convergence step. However, this method requires a much more complicated set-up than the Gibbs sampler or Metropolis algorithm, and it is not clear that it is superior in practice. There are also many issues that we have not covered regarding convergence diagnostics, the effects of reparametrisation of θ, etc., that are important in practice. There are by now a number of excellent books that specialise in these topics, and we refer to them for further study. Gamerman (1997), Gelman *et al.* (2003), Gilks *et al.* (1996) and Robert and Casella (1999) are recommended.

3.8 Hierarchical modelling

Another way of dealing with the specification of prior parameter values in Bayesian inference is with a hierarchical specification, in which the prior parameter values are themselves given a (second-stage) prior.

For example, in the simple normal model considered previously we might specify $X \mid \theta \sim N(\theta, 1)$, $\theta \mid \tau^2 \sim N(0, \tau^2)$ and $\tau^2 \sim$ uniform $(0, \infty)$, another example of an improper, diffuse prior. Inference on θ is based on the marginal posterior of θ, obtained by integrating out τ^2 from the joint posterior of θ and τ^2:

$$\pi(\theta \mid x) = \int \pi(\theta, \tau^2 \mid x)d\tau^2,$$

where the joint posterior $\pi(\theta, \tau^2 \mid x) \propto f(x;\theta)\pi(\theta \mid \tau^2)\pi(\tau^2)$.

Hierarchical modelling is a very effective practical tool and usually yields answers that are reasonably robust to misspecification of the model. Often, answers from a hierarchical analysis are quite similar to those obtained from an empirical Bayes analysis. In particular, when the second-stage prior is relatively flat compared with the first-stage prior and the density of the observable X, answers from the two approaches are close to one another. We now give a detailed numerical example of hierarchical modelling. A further example is given later, in Section 3.10.

3.8.1 Normal empirical Bayes model rewritten as a hierarchical model

Consider, as a model for the baseball example of Section 3.4.1, the hierarchy

$$X_i \mid \mu_i \sim N(\mu_i, 1), i = 1, \ldots, p,$$
$$\mu_i \mid \theta, \tau \sim N(\theta, \tau^2), \tag{3.8}$$
$$\pi(\theta, \tau) \propto \tau^{-1-2\alpha^*}e^{-\beta^*/\tau^2},$$

where $\theta \in \mathbb{R}$, $\tau \in (0, \infty)$ and where α^* and β^* are set to some small positive number (we use 0.001). The roles of α^* and β^* are explained further below. The first two equations here are standard as in empirical Bayes analysis, but the third indicates that, rather than treat θ and τ^2 as constants to be estimated, we are treating them as random parameters with a prior density π.

Under this model (3.8), the joint density of $\theta, \tau, \mu_1, \ldots, \mu_p, X_1, \ldots, X_p$ is (ignoring irrelevant constants)

$$\frac{1}{\tau^{p+1+2\alpha^*}} \exp\left[-\frac{1}{2} \sum_{i=1}^{p} \left\{ (X_i - \mu_i)^2 + \frac{(\mu_i - \theta)^2}{\tau^2} \right\} - \frac{\beta^*}{\tau^2} \right]. \tag{3.9}$$

Bayesian analysis proceeds by calculating the joint conditional distribution of $\theta, \tau, \mu_1, \ldots, \mu_p$, given $X_1, \ldots, X_p$. However, there is no way to calculate (or even to construct a Monte Carlo sample from) this conditional density directly from the joint density (3.9) – the normalising constants are too intractable. However, we can characterise the conditional density of any one of these unknowns given the rest. Specifically:

1 *Conditional distribution of μ_i, given everything else.* By completing the square in the exponent of (3.9), we see that the conditional distribution of μ_i given X_i, θ, τ is $N\left(\frac{\theta + \tau^2 X_i}{1+\tau^2}, \frac{\tau^2}{1+\tau^2} \right)$. We generate a random value μ_i from this distribution, successively for $i = 1, \ldots, p$.

2 *Conditional distribution of θ, given everything else.* Completing the square with respect to θ in the exponent of (3.9), this has the distribution $N\left(\bar{\mu}, \frac{\tau^2}{p} \right)$, where $\bar{\mu} = \frac{1}{p} \sum \mu_i$. Hence we generate a random value of θ from this distribution.

3 *Conditional distribution of τ, given everything else.* As a function of τ, the density (3.9) is proportional to $\tau^{-p-1-2\alpha^*} e^{-A/\tau^2}$, where $A = \frac{1}{2} \sum (\mu_i - \theta)^2 + \beta^*$. This is equivalent, on performing an elementary change of variables, to the statement that $1/\tau^2$ has the distribution Gamma $\left(\frac{p}{2} + \alpha^*, A \right)$. (Here we continue to follow the convention that the Gamma(α, β) distribution has density $\propto y^{\alpha-1} e^{-\beta y}$, $0 < y < \infty$.) Hence we generate Y from the distribution Gamma $\left(\frac{p}{2} + \alpha^*, A \right)$, and set $\tau = \frac{1}{\sqrt{Y}}$.

The prior distribution described by (3.8) is equivalent to the statement that $1/\tau^2$ has the Gamma(α^*, β^*) distribution. If we had genuine prior information about τ (as we could gain, for example, by the analysis of previous baseball seasons in our baseball example, Section 3.4.1), we could use this to select suitable informative values of α^* and β^*. In this case we choose not to use such information (and, in many situations where similar analyses might be applied, no such information is available), so the obvious thing to do would be to set $\alpha^* = \beta^* = 0$, an uninformative prior. There is a difficulty with this, however: in hierarchical models, there is often no guarantee that an improper prior leads to a proper posterior. In this specific example, if the Monte Carlo simulation is run with $\alpha^* = \beta^* = 0$, we find that the value of τ eventually collapses to 0 (typically, only after several thousand iterations). By setting α^* and β^* to small positive values, we ensure that the prior density is proper and therefore so is the posterior. (Note, however, that (3.8) assumes the prior density of θ to be uniform over the whole real line, which is improper, but this does not appear to cause the same difficulty.)

The simulation we report here followed steps 1–3 above for 10 000 iterations, storing every tenth value of θ and $\tau_2 = \frac{\tau^2}{1+\tau^2}$. (There is no need to store every individual value;

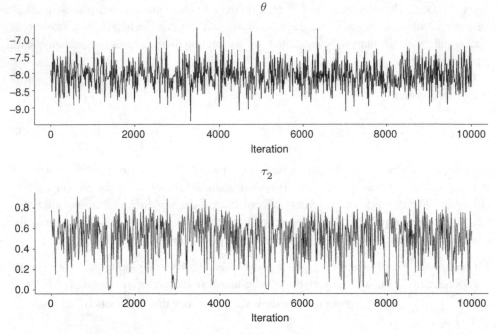

Figure 3.3 Baseball data example. Time series plot of 1000 values (one value for every tenth iteration) from the simulated posterior densities of θ and τ_2

by taking every tenth, we ensure that successive values of θ and τ_2 correspond to nearly independent draws from the posterior distribution.) Figure 3.3 shows, for the baseball data of Section 3.4.1, a time series plot of the 1000 stored values; the appearance of the plots supports our contention that the values are roughly independent and also that there is no apparent drift in the distribution (which might indicate non-convergence).

We are interested in computing posterior densities of the individual μ_i. Two methods immediately come to mind:

1 Store the individual μ_i values generated by the above simulation (in other words, for each $i \in \{1, 2, \ldots, 17\}$, store the 1000 values generated at step 1 of every tenth iteration of the sampler) and construct an estimate of the posterior density of each μ_i by using a numerical density estimation procedure, such as the 'density' command in S-PLUS;

2 Store only the θ and τ_2 values from the simulation. For each such θ and τ_2, the conditional density of each μ_i, given X_i, θ and τ_2, is the $N(\theta + \tau_2(X_i - \theta), \tau_2)$ density. The estimated unconditional posterior density of μ_i is therefore the average of these normal densities over all the simulated posterior values of θ and τ_2.

Method 2 has two advantages. First, it avoids the numerical estimation of a density function. Second, it is theoretically more efficient – in fact, this is an example of the Rao–Blackwellisation procedure mentioned in Section 3.7.1.

Figure 3.4 shows the Rao–Blackwellised estimates of the posterior densities of μ_i (dashed curves) for each of the 17 baseball players. Also shown for comparison (solid curves) are the corresponding conditional normal curves, where we estimate θ and τ_2 respectively as

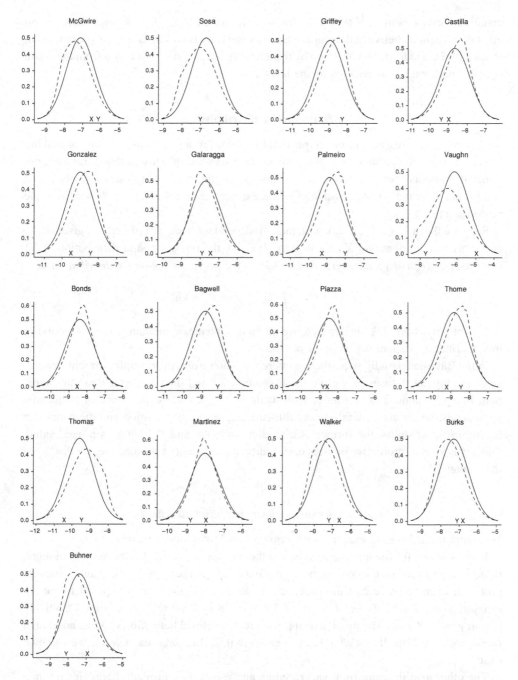

Figure 3.4 Baseball data example. Comparison of posterior densities of μ_i by the empirical Bayes method (solid curve) and the fully (hierarchical) Bayes method (dashed curve)

$\bar{X}$ and $1 + 14/\sum(X_i - \bar{X})^2$, which are the values corresponding to the empirical Bayes procedure. Most statisticians consider the fully Bayesian estimates (dashed curves) to be more reliable characterisations of the uncertainty in μ_i because they take account of the fact that θ and τ_2 are unknown, whereas the empirical Bayes procedure, having estimated these quantities, treats them as if they were known. Nevertheless, in this example there are no great discrepancies between the empirical Bayes and fully Bayes densities. For comparison, in each plot we indicate the X_i value by the letter X, and the true μ_i derived from the entire season's performance is denoted by the letter Y.

3.9 Predictive distributions

So far, we have stressed use of the posterior distribution $\pi(\theta \mid x)$ as a means of making inference about the parameter θ. We may not be interested directly in that parameter, but rather in some independent future observation depending on θ. It is possible to obtain the conditional distribution of the value of a future observation $X^\dagger$, given the data x, from the posterior $\pi(\theta \mid x)$.

Suppose that $x = (x_1, \ldots, x_n)$, with the x_i independent from $f(x; \theta)$. Since, given θ, $X^\dagger$ and x are independent and $X^\dagger$ has density $f(x^\dagger; \theta)$, the posterior joint distribution of $X^\dagger$ and θ is $f(x^\dagger; \theta)\pi(\theta \mid x)$. Integrating out θ gives the posterior *predictive distribution* as

$$g(X^\dagger \mid x) = \int f(X^\dagger; \theta)\pi(\theta \mid x)d\theta.$$

If a point prediction of $X^\dagger$ is required, we might use the mean, median or other function of this distribution, depending on our loss function.

In the Bayesian paradigm, predictive inference is therefore, in principle, straightforward, since the logical statuses of the future observation $X^\dagger$ and the parameter θ are the same, both being random. This contrasts with methods for predictive inference in frequentist approaches, which are generally more difficult, due to the observation and the parameter having *different* status, the former as a random variable, and the latter as a fixed value. Further aspects of predictive inference, including the frequentist perspective, are discussed in Chapter 10.

3.9.1 An example involving baseball data

We return to the baseball example of Section 3.4 with two modifications.

If we write m_i for the number of at bats in the full season and Z_i for the number of home runs, it seems reasonable to assume $Z_i \sim \text{Bin}(m_i, p_i)$, where p_i is the batter's success probability, and then re-cast the problem as one of *predicting* Z_i as opposed to merely *estimating* p_i. The analyses of Sections 3.4 and 3.8 ignored this distinction, effectively defining $p_i = Z_i/m_i$. The predictive approach to be outlined here allows for the additional randomness in Z_i itself that would be present even if the binomial success rate were known exactly.

The other modification from our preceding analysis is that, instead of applying a transformation and treating the data as normal, we use the binomial distribution directly (for the pre-season home runs Y_i as well as for Z_i). The penalty for this is that we can no longer use the Gibbs sampler, so instead we apply the Metropolis–Hastings sampler as described in Section 3.7.5.

As in our earlier analyses, we assume $Y_i \sim \text{Bin}(n_i, p_i)$, where n_i is the number of pre-season at bats. Write $\mu_i = \log\{p_i/(1 - p_i)\}$: this is known as the *logit transformation* and translates the parameter range from $(0, 1)$ for p_i to $(-\infty, \infty)$ for μ_i. As in Section 3.8.1, we assume

$$\mu_i \mid \theta, \tau \sim N(\theta, \tau^2),$$
$$\pi(\theta, \tau) \sim \tau^{-1-2\alpha^*} e^{-\beta^*/\tau^2},$$

where α^*, β^* are positive (we take $\alpha^* = \beta^* = 0.001$ to get a proper but highly diffuse prior for τ).

This defines a hierarchical model in which the individual batters' success rates are translated into parameters μ_i, and θ and τ are hyperparameters. As in Section 3.8.1, the analysis proceeds by simulation, in which the values of $\mu_1, \ldots, \mu_p, \theta, \tau$ are successively updated conditionally on all the other parameters. The updating rules for θ and τ are exactly the same as earlier, but there is no direct algorithm for sampling μ_i from its conditional posterior distribution. To see this, write the conditional density of μ_i as the product of the densities of $(Y_i \mid \mu_i)$ and $(\mu_i \mid \theta, \tau)$:

$$p(\mu_i \mid Y_i, \theta, \tau) \propto \frac{e^{Y_i \mu_i}}{(1 + e^{\mu_i})^{n_i}} \exp\left\{-\frac{1}{2}\left(\frac{\mu_i - \theta}{\tau}\right)^2\right\}. \tag{3.10}$$

The right-hand side of (3.10) cannot be integrated analytically and therefore there is no direct Gibbs sampler to sample from this distribution.

We therefore proceed by a Metropolis–Hastings update, as follows:

1 Evaluate (3.10) for the current value of μ_i; call the answer f_0.
2 Define a new trial value μ_i^*, which is sampled uniformly from the interval $(\mu_i - \delta, \mu_i + \delta)$; choice of δ will be considered below.
3 Evaluate (3.10) for the new value μ_i^*; call the answer f_1.
4 Let U be an independent random variable, uniform on $(0, 1)$. If $U < \frac{f_1}{f_0}$ accept μ_i^*; otherwise reject (in which case the value μ_i is kept unchanged for this iteration).

This sequence of steps applies for a single value μ_i at a single iteration; the full algorithm will do this for each of the batters in turn $(i = 1, 2, \ldots, p)$, followed by an update of $\theta \sim N(\bar{\mu}, \frac{\tau^2}{p})$ and then $\frac{1}{\tau^2} \sim \text{Gamma}(\frac{p}{2} + \alpha^*, \frac{1}{2}\sum(\mu_i - \theta)^2 + \beta^*)$ as derived in Section 3.8.1.

For the analysis reported here, 10 000 cycles of this complete sequence were run and discarded as a warm-up sample; then, 100 000 cycles were run and the values of $\mu_1, \ldots, \mu_p$ from every tenth cycle were written to a file for subsequent post processing. (The whole algorithm was programmed in FORTRAN and ran in about 5 seconds on a IBM laptop.)

We still have to choose δ, the step length for the Metropolis update. As noted in Section 3.7.6, the value of δ is often chosen so that there is an acceptance probability between 15% and 50%. Table 3.2 shows the acceptance rate for five different values of δ. (For brevity, only

Table 3.2 *Acceptance rates for the Metropolis algorithm as a function of the step length δ*

δ	0.1	0.5	1	2	10
Acceptance Probability	0.87	0.55	0.35	0.17	0.04

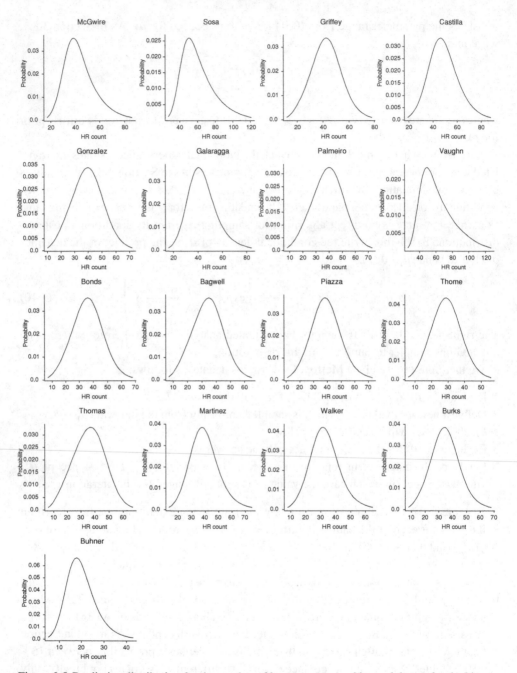

Figure 3.5 Predictive distribution for the number of home runs scored by each batter, by the hierarchical analysis of Section 3.9

the overall acceptance rate is shown, averaged over all 17 batters; however, the acceptance rates for the individual μ_i did not differ much from these values. If there were substantial differences among the acceptance rates for the individual μ_is, we might consider using a different δ_i for each batter.) Based on these results, $\delta = 1$ was selected for the subsequent results.

We now consider how to calculate predictive distributions. As noted already, the output of the sampling algorithm is a sample of 10 000 stored values from the posterior distribution of each μ_i. For any given μ_i, the predictive probability that $Z_i = k$ is

$$\binom{m_i}{k} \frac{e^{k\mu_i}}{(1 + e^{\mu_i})^{m_i}}. \tag{3.11}$$

The Bayesian predictive probability that $Z_i = k$ is therefore the average of (3.11) over the posterior distribution. In practice, this is calculated by averaging (3.11) over the stored values of each μ_i from the Monte Carlo sampling algorithm.

Figure 3.5 shows the $p = 17$ predictive distributions that are obtained by this procedure. This is much more informative than a simple table of mean predictions, as in the columns $\hat{HR}$ and $\hat{HR}_s$ of Table 3.1. If we wanted to make a bet about the home run total of a particular batter or group of batters, we would much prefer to base it on the full predictive distribution rather than just the mean prediction!

However, it is also worth looking at the mean predictions. For the 17 batters in Table 3.1, these are, in order: 44, 62, 43, 49, 40, 44, 39, 64, 36, 34, 39, 29, 36, 40, 33, 35, 20. Comparing these with the actual number of home runs scored, we find that the mean squared prediction error is 2052, compared with 3201 for the James–Stein estimator $\hat{HR}_s$ and 9079 for the simple predictor $\hat{HR}$.

In conclusion, this analysis improves on the preceding analyses of this data set by calculating a full predictive distribution for the number of home runs by each batter, but even when just used to calculate a predictive mean for each batter, improves substantially on the James–Stein estimator. However, the results for individual batters are not necessarily better: in particular, the new analysis notably fails to predict McGwire's actual home run count. This reinforces a general point about shrinkage methods, that, while they improve the overall prediction in a group of individuals, they do not necessarily improve the prediction on any specific individual.

3.10 Data example: Coal-mining disasters

Carlin, Gelfand and Smith (1992) analyse data giving the number of British coal-mining disasters each year between 1851 and 1962. The data are shown in Figure 3.6(a).

It would appear that there is a 'changepoint' in the data around 1890–1895, after which time the yearly number of disasters tends to be less. Carlin, Gelfand and Smith (1992) analyse this hypothesis using a three-stage Bayesian hierarchical model and Gibbs sampling to make inference on the position of the changepoint, k, in the series.

Specifically, we model the observed data $X_1, X_2, \ldots, X_m, m = 112$, as

$$X_i \sim \text{Poisson}(\mu), i = 1, \ldots, k; X_i \sim \text{Poisson}(\lambda), i = k+1, \ldots, m.$$

Then $k = m$ would be interpreted as 'no change' in the yearly mean number of disasters.

At the second stage of the hierarchical model we place independent priors on k, μ and λ. We take k uniform on $\{1, \ldots, m\}$, $\mu \sim \text{Gamma}(a_1, b_1)$ and $\lambda \sim \text{Gamma}(a_2, b_2)$, where the Gamma (a, b) distribution has density $\propto t^{a-1} e^{-t/b}$. (Note that this is slightly different from the parametrisation we used earlier, for example in Example 3.3.)

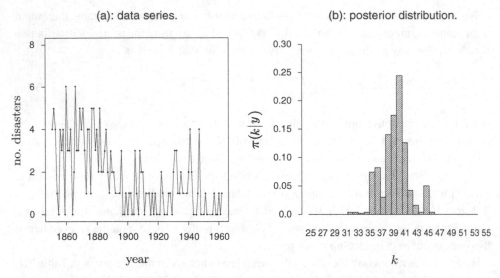

Figure 3.6 Coal mining disasters: (a) data series and (b) simulated posterior

At the third stage, we take $b_1 \sim$ inverse Gamma (c_1, d_1) independent of $b_2 \sim$ inverse Gamma (c_2, d_2). (A random variable Z is said to have an inverse Gamma (a, b) distribution if $1/Z$ is distributed as Gamma (a, b).) We assume that a_1, a_2, c_1, c_2, d_1 and d_2 are known: in our analysis, following that of Carlin, Gelfand and Smith (1992), we set $a_1 = a_2 = 0.5$, $c_1 = c_2 = 0, d_1 = d_2 = 1$.

Interest lies specifically in the marginal posterior distribution of k. The parameter vector is five-dimensional: $\theta = (\mu, \lambda, b_1, b_2, k)$. The conditional distributions required by the Gibbs sampler are easily checked to be:

$$\mu \mid X, \lambda, b_1, b_2, k \sim \text{Gamma}\left(a_1 + \sum_1^k X_i, (k + b_1^{-1})^{-1}\right);$$

$$\lambda \mid X, \mu, b_1, b_2, k \sim \text{Gamma}\left(a_2 + \sum_{k+1}^m X_i, (m - k + b_2^{-1})^{-1}\right);$$

$$b_1 \mid X, \mu, \lambda, b_2, k \sim \text{inverse Gamma}\left(a_1 + c_1, (\mu + d_1^{-1})^{-1}\right);$$

$$b_2 \mid X, \mu, \lambda, b_1, k \sim \text{inverse Gamma}\left(a_2 + c_2, (\lambda + d_2^{-1})^{-1}\right)$$

and

$$f(k \mid X, \mu, \lambda, b_1, b_2) = \frac{L(X; k, \mu, \lambda)}{\sum_{k'} L(X; k', \mu, \lambda)},$$

where

$$L(X; k, \mu, \lambda) = \exp\{(\lambda - \mu)k\} \left(\frac{\mu}{\lambda}\right)^{\sum_1^k X_i}.$$

We initialised the Gibbs sampler by setting $\theta^{(0)} = (3, 1, 0.5, 0.5, 40)$. Here $k = 40$

corresponds to the year 1890. A series of 200 independent runs from this starting value was carried out, and the corresponding values $\theta^{[1]}, \theta^{[2]}, \ldots, \theta^{[200]}$ after 100 iterations obtained. Figure 3.6(b) displays the estimate obtained using (3.5) of the marginal posterior for $k \mid Y$.

The posterior mode is at $k = 41$, the three largest spikes being at $k = 39, 40, 41$, suggesting strongly that a changepoint occurred between 1889 and 1892.

3.11 Data example: Gene expression data

Microarrays are a biogenetic technology for measuring gene 'expression levels', how active a particular gene is in the workings of a given cell. Background on microarrays is given by Efron, Tibshirani, Storey and Tusher (2001). A characteristic of microarray experiments is that they typically provide expression levels for thousands of genes at once, and therefore pose interesting statistical challenges. The effective analysis of gene expression data provides a sophisticated, contemporary example of application of Bayesian inferential ideas. The following data example is adapted from Efron (2003a), using ideas of Efron (2004).

Our analysis involves data from a microarray experiment concerning stomach cancer, analysed by Efron (2003a). In the experiment, 2638 genes were analysed on each of 48 microarrays, each microarray using cells from a different cancer patient, 24 with less aggressive disease (Type 1) and the other 24 having more aggressive disease (Type 2). The raw gene expression data therefore consist of a 2638×48 matrix, standardised so that in each column the 2638 values have mean 0 and variance 1. The purpose of the study is to identify genes which are more active or less active in Type 2 compared with Type 1 tumours. Table 3.3 shows a small portion of the data.

If we had data on just one gene, gene i, a simple test for a difference between the 24 Type 1 measurements and the 24 Type 2 measurements might be based on a (two-sided) t-test, and the associated t-statistic value, y_i. Our analysis will actually be based on the

Table 3.3 *Gene expression data, cancer study*

Gene	Cancer type								z_i
	1	1	... 1	1	2	2	... 2	2	
1	0.03	0.25	0.98	−1.34	−0.43	0.10	−0.71	−0.10	−1.809
2	0.66	0.68	0.23	0.01	−0.04	−0.09	0.82	−0.69	−2.158
3	−0.64	−0.36	−0.56	−0.75	0.18	0.31	−0.18	0.84	0.354
4	−0.02	−0.15	−0.79	−0.01	−0.55	−0.24	−0.02	0.92	−0.020
5	0.71	−0.29	−0.02	−0.39	−0.55	−0.78	−0.44	0.62	0.457
6	0.16	−0.04	−1.14	0.26	−0.90	−0.41	2.21	0.86	1.867
7	0.78	0.24	−1.36	0.19	−1.50	−0.36	1.90	0.80	0.668
8	0.78	0.09	−1.28	0.17	−1.54	−0.28	2.22	1.51	1.012
⋮	⋮	⋮	⋮	⋮	⋮	⋮	⋮	⋮	⋮

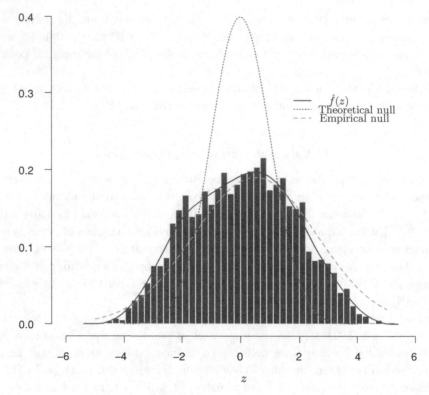

Figure 3.7 Analysis of z-values for gene expression data

transformed z-values z_i defined by

$$z_i = \Phi^{-1}\{F_{46}(y_i)\}, \tag{3.12}$$

where Φ denotes the standard normal cumulative distribution function and F_{46} is the cumulative distribution function of a t-distribution with 46 ($= 48 - 2$) degrees of freedom. Values of z_i are shown in Table 3.3 for the first few genes in our dataset. The histogram in Figure 3.7 displays all 2638 z-values.

Considering again just gene i, if there is no difference in expression between Type 1 and Type 2 tumours we have

$$z_i \sim N(0, 1), \tag{3.13}$$

assuming that the assumptions (in particular that of independence between the 48 measurements) underlying the t-test are met. However, we actually have 2638 *simultaneous* hypotheses to examine, and care is required in the use of (3.13) in the analysis. Performing a series of 2638 separate tests based on (3.13), say reporting all those significant at the 5% level, is inappropriate, and an inference reflecting the multiple hypothesis testing nature of the problem is required.

The $N(0, 1)$ curve shown in Figure 3.7 is much narrower than the histogram of z-values, suggesting that some of the genes show different expression in Type 1 and Type 2 tumours. Efron (2003a, 2004) develops simple (empirical) Bayesian ideas, which enable us to quantify on a gene-by-gene basis evidence for different expression in the two types of tumour.

Suppose that the $N = 2638$ z-values fall into two classes, 'Interesting' and 'Uninteresting', according to whether the gene is differently expressed or not in the two tumour types. Let the prior probabilities for the two classes be p_0 and $p_1 = 1 - p_0$ respectively, with corresponding prior densities $f_0(z)$ and $f_1(z)$ for the z-values (3.12). Finally, let $f(z)$ be the mixture density

$$f(z) = p_0 f_0(z) + p_1 f_1(z).$$

Since it is constructed from the raw collection of all 2638 z-values, the histogram in Figure 3.7 can be thought of as estimating $f(z)$. Smoothing the histogram (in fact by fitting a 'spline smoother' to the histogram counts) yields the solid curve $\hat{f}(z)$ shown in the figure. This $\hat{f}(z)$ can be thought of as an empirical Bayes estimate of the mixture density $f(z)$.

Now, according to Bayes' law the posterior probability of being in the 'Uninteresting' class, given z, is

$$\Pr(\text{Uninteresting}|z) = p_0 f_0(z)/f(z).$$

Efron (2004) defines the *local false discovery rate* to be

$$fdr(z) \equiv f_0(z)/f(z),$$

an upper bound on the probability of 'Uninteresting', given z. The suggestion, therefore, is to estimate $fdr(z)$ by use of $\hat{f}(z)$ (for the denominator) and the null $N(0, 1)$ density (3.13) (for the numerator), and then report as 'Interesting' those genes with $fdr(z_i)$ less than some low threshold value, such as 0.1.

Application of this procedure to the cancer data flags 347 of the 2638 genes as 'Interesting'. This figure seems high, but Efron (2003a) notes that the cancer study actually started with over 10 000 genes, most of which were discarded by a rough screening process, so being left with 347 'Interesting' cases is not implausible.

There is, however, a potential flaw in the analysis, relating to use of the $N(0, 1)$ distribution as the null density $f_0(z)$ in the Bayesian calculation. Efron (2004) comments that, in an observational study of the kind leading to the gene expression data, the 48 measurements for each gene cannot reasonably be expected to be independent. The t-statistic cannot, therefore, reasonably be assumed to have a null distribution which is a t-distribution, so that our analysis, even though it has taken account of the simultaneous nature of the inference, is likely to have been compromised by use of an inappropriate null density $f_0(z)$. Efron (2004) develops a more sophisticated analysis, appropriate to situations such as that here. Extending the idea we have used to estimate the mixture density $f(z)$, Efron suggests using the assemblage of z-values near $z = 0$ to *estimate* an appropriate null density $f_0(z)$, in effect replacing the $N(0, 1)$ 'theoretical null distribution' with an 'empirical null distribution'.

Application of Efron's procedure for construction of an empirical null distribution to the cancer data yields the null density shown by the broken line in Figure 3.7. Recalculation of the local false discovery rate using this empirical null density as $f_0(z)$ has the dramatic effect of leaving *none* of the 2638 genes with $fdr(z_i) < 0.1$. The raw histogram of z-values consists here entirely of a 'central peak' around $z = 0$ (it certainly does not have the multi-modal kind of appearance we would normally associate with a mixture distribution), and it is quite plausible that none of the genes is differentially expressed in the two tumour types.

3.12 Problems

3.1 Let θ be a random variable in $(0, \infty)$ with density

$$\pi(\theta) \propto \theta^{\gamma-1} e^{-\beta\theta},$$

where $\beta, \gamma \in (1, \infty)$. Calculate the mean and mode of θ.

Suppose that $X_1, \ldots, X_n$ are random variables, which, conditional on θ, are independent and each have the Poisson distribution with parameter θ. Find the form of the posterior density of θ given observed values $X_1 = x_1, \ldots, X_n = x_n$. What is the posterior mean?

Suppose now that $T_1, \ldots, T_n$ are random variables, which, conditional on θ, are independent, each exponentially distributed with parameter θ. What is the mode of the posterior distribution of θ, given $T_1 = t_1, \ldots, T_n = t_n$?

3.2 Suppose that Y, the number of heads in n tosses of a coin, is binomially distributed with index n and with parameter θ and that the prior distribution of θ is of Beta form with density

$$\pi(\theta) \propto \theta^{a-1}(1 - \theta)^{b-1}.$$

Find the posterior distribution of θ, given $Y = y$. How would you estimate the probability that the next observation is a head?

Suppose that prior to any tossing the coin seems to be fair, so that we would take $a = b$.

Suppose also that the tossing yields 1 tail and $n - 1$ heads. How large should n be in order that we would just give odds of 2 to 1 in favour of a head occurring at the next toss? Show that for $a = b = 1$ we obtain $n = 4$.

3.3 Find the form of the Bayes rule in an estimation problem with loss function

$$L(\theta, d) = \begin{cases} a(\theta - d) & \text{if } d \le \theta, \\ b(d - \theta) & \text{if } d > \theta, \end{cases}$$

where a and b are given positive constants.

3.4 Suppose that X is distributed as a binomial random variable with index n and parameter θ. Calculate the Bayes rule (based on the single observation X) for estimating θ when the prior distribution is the uniform distribution on $[0, 1]$ and the loss function is

$$L(\theta, d) = (\theta - d)^2 / \{\theta(1 - \theta)\}.$$

Is the rule you obtain minimax?

3.5 At a critical stage in the development of a new aeroplane, a decision must be taken to continue or to abandon the project. The financial viability of the project can be measured by a parameter θ, $0 < \theta < 1$, the project being profitable if $\theta > \frac{1}{2}$. Data x provide information about θ.

If $\theta < \frac{1}{2}$, the cost to the taxpayer of continuing the project is $(\frac{1}{2} - \theta)$ (in units of $\$$ billion), whereas if $\theta > \frac{1}{2}$ it is zero (since the project will be privatised if profitable). If $\theta > \frac{1}{2}$ the cost of abandoning the project is $(\theta - \frac{1}{2})$ (due to contractual arrangements for purchasing the aeroplane from the French), whereas if $\theta < \frac{1}{2}$ it is zero. Derive the Bayes decision rule in terms of the posterior mean of θ given x.

The Minister of Aviation has prior density $6\theta(1 - \theta)$ for θ. The Prime Minister has prior density $4\theta^3$. The prototype aeroplane is subjected to trials, each independently having probability θ of success, and the data x consist of the total number of trials required for the first successful result to be obtained. For what values of x will there be serious ministerial disagreement?

3.6 Consider the following extension of the James–Stein estimator. X has a p-dimensional normal distribution with mean vector μ and covariance matrix I, the $p \times p$ identity matrix. Consider the estimator

$$\bar{d}^a(X) = \bar{X}e_p + \left(1 - \frac{a}{V}\right)(X - \bar{X}e_p),$$

where $\bar{X} = \frac{1}{p}\sum_{i=1}^{p} X_i$, $V = \sum_{i=1}^{p}(X_i - \bar{X})^2$ and e_p is the p-dimensional vector of ones. This modifies the classical James–Stein estimator by shrinking the natural estimator X towards $\bar{X}e_p$, rather than shrinking it towards 0. Assume the loss function is still the classical least-squares loss, $L(\mu, d) = ||\mu - d||^2$. Show that, when $p \geq 3$, the risk function is

$$R(\mu, \bar{d}^a(X)) = p - \left[2a(p - 3) - a^2\right] \mathbb{E}\left(\frac{1}{V}\right).$$

Hence deduce that the modified James–Stein estimator improves on $d(X) = X$ whenever $p \geq 4$, and that the optimal value of a is $p - 3$.

3.7 Consider the situation in Section 3.5.1, where $X_i \mid \mu_i$ are independent $N(\mu_i, 1)$ but with $\mu_i \sim N(\theta, \tau^2)$ with θ and τ^2 both unknown. (In the earlier discussion, we assumed $\theta = 0$.) Compute the Bayes estimator of μ_i for known θ, τ_i and calculate the Bayes risk. Show that, provided $p > 3$, an unbiased estimator of θ is the sample mean $\bar{X}$, and an unbiased estimator of $\tau^2/(\tau^2 + 1)$ is $1 - (p - 3)/V$, where $V = \sum(X_i - \bar{X})^2$. Hence derive the James–Stein estimator with shrinkage to the mean, as considered in Problem 3.6, as an empirical Bayes estimator.

3.8 Suppose $X_1, \ldots, X_n$ are independent, identically distributed random variables, which, given μ, have the normal distribution $N(\mu, \sigma_0^2)$, with σ_0^2 known. Suppose also that the prior distribution of μ is normal with known mean ξ_0 and known variance ν_0.

Let X_{n+1} be a single future observation from the same distribution which is, given μ, independent of $X_1, \ldots, X_n$. Show that, given $(X_1, \ldots, X_n)$, X_{n+1} is normally distributed with mean

$$\left\{\frac{1}{\sigma_0^2/n} + \frac{1}{\nu_0}\right\}^{-1}\left\{\frac{\bar{X}}{\sigma_0^2/n} + \frac{\xi_0}{\nu_0}\right\}$$

and variance

$$\sigma_0^2 + \left\{ \frac{1}{\sigma_0^2/n} + \frac{1}{v_0} \right\}^{-1}.$$

3.9 Let $X_1, \ldots, X_n$ be independent, identically distributed $N(\mu, \sigma^2)$, with both μ and σ^2 unknown. Let $\bar{X} = n^{-1} \sum_{i=1}^n X_i$, and $s^2 = (n-1)^{-1} \sum_{i=1}^n (X_i - \bar{X})^2$. Assume the (improper) prior $\pi(\mu, \sigma)$ with

$$\pi(\mu, \sigma) \propto \sigma^{-1}, (\mu, \sigma) \in \mathbb{R} \times (0, \infty).$$

Show that the marginal posterior distribution of $n^{1/2}(\mu - \bar{X})/s$ is the t distribution with $n-1$ degrees of freedom, and find the marginal posterior distribution of σ.

3.10 Consider a Bayes decision problem with scalar parameter θ. An estimate is required for $\phi \equiv \phi(\theta)$, with loss function

$$L(\theta, d) = (d - \phi)^2.$$

Find the form of the Bayes estimator of ϕ.

Let $X_1, \ldots, X_n$ be independent, identically distributed random variables from the density $\theta e^{-\theta x}$, $x > 0$, where θ is an unknown parameter. Let Z denote some hypothetical future value derived from the same distribution, and suppose we wish to estimate $\phi(\theta) = \Pr(Z > z)$, for given z.

Suppose we assume a gamma prior, $\pi(\theta) \propto \theta^{\alpha-1} e^{-\beta\theta}$ for θ. Find the posterior distribution for θ, and show that the Bayes estimator of ϕ is

$$\widehat{\phi}_B = \left(\frac{\beta + S_n}{\beta + S_n + z} \right)^{\alpha+n},$$

where $S_n = X_1 + \cdots + X_n$.

3.11 Let the distribution of X, given θ, be normal with mean θ and variance 1. Consider estimation of θ with squared error loss $L(\theta, a) = (\theta - a)^2$ and action space $A \equiv \Theta \equiv \mathbb{R}$.

Show that the usual estimate of θ, $d(X) = X$, is not a Bayes rule. (Show that if $d(X)$ were Bayes with respect to a prior distribution π, we should have $r(\pi, d) = 0$.)

Show that X is extended Bayes and minimax. (Consider a family of normal priors for θ. A straightforward proof that X is also admissible is easily constructed.)

3.12 The posterior density of the real parameter θ, given data x, is $\pi(\theta \mid x)$. Assuming that $\pi(\theta \mid x)$ is unimodal, show that if we choose $\theta_1 < \theta_2$ to minimise $\theta_2 - \theta_1$ subject to

$$\int_{\theta_1}^{\theta_2} \pi(\theta \mid x) d\theta = 1 - \alpha,$$

α given, then we have $\pi(\theta_1 \mid x) = \pi(\theta_2 \mid x)$.

3.13 Let $X \sim \text{Bin}(n, \theta)$, and consider a conjugate $\text{Beta}(a, b)$ prior distribution for θ, as in Example 3.1. Show that if we reparametrise from (a, b) to (μ, M), where $\mu = a/(a+b)$ and $M = a + b$, the marginal distribution of X is of *beta-binomial* form:

$$\Pr(X = x \mid \mu, M) = \frac{\Gamma(M)}{\Gamma(M\mu)\Gamma\{M(1-\mu)\}} \binom{n}{x} \frac{\Gamma(x + M\mu)\Gamma\{n - x + M(1-\mu)\}}{\Gamma(n + M)}.$$

Verify that the marginal expectation and variance of X/n are respectively

$$\mathbb{E}(X/n) = \mu,$$

and

$$\text{var}(X/n) = \frac{\mu(1-\mu)}{n}\left[1 + \frac{n-1}{M+1}\right].$$

Consider a model in which (X_i, θ_i), $i = 1, \ldots, k$ are a sequence of independent, identically distributed random variables, for which only the X_i are observable. Suppose that $X_i \sim \text{Bin}(n, \theta_i)$ and $\theta_i \sim \text{Beta}(\mu, M)$, in terms of the above parametrisation of the beta distribution. Find appropriate estimates, based on $X_1, \ldots, X_k$, of μ and M, and an empirical Bayes estimate of the posterior mean of θ_i.

3.14 Let $X_1, \ldots, X_n$ be independent, identically distributed $N(\mu, 1/\tau)$, and suppose that independent priors are placed on μ and τ, with $\mu \sim N(\xi, \kappa^{-1})$ and $\tau \sim \text{Gamma}(\alpha, \beta)$ (see the notation of Example 3.3).

Find the form of the joint posterior distribution $\pi(\mu, \tau \mid X)$, and note that this is not of standard form. Show that the conditional (posterior) distributions *are* of simple forms:

$$\mu \mid \tau, X \sim N\left(\frac{\tau \sum_{i=1}^{n} X_i + \kappa\xi}{\tau n + \kappa}, \frac{1}{\tau n + \kappa}\right),$$

$$\tau \mid \mu, X \sim \text{Gamma}\left(\alpha + \frac{n}{2}, \beta + \frac{\sum_{i=1}^{n}(X_i - \mu)^2}{2}\right).$$

3.15 Suppose $X_1, \ldots, X_n, Z$ are independent $N(\mu, \tau^{-1})$ random variables and the prior density of (τ, μ) is

$$\pi(\tau, \mu; \alpha, \beta, k, \nu) = \frac{\beta^\alpha}{\Gamma(\alpha)} \tau^{\alpha-1} e^{-\beta\tau} \cdot (2\pi)^{-1/2}(k\tau)^{1/2} \exp\left\{-\frac{k\tau}{2}(\mu - \nu)^2\right\}$$

as in Example 3.3. Recall that in Example 3.3 we showed that the posterior density of (τ, μ), given $X_1, \ldots, X_n$, is of the same form with parameters (α, β, k, ν) replaced by $(\alpha', \beta', k', \nu')$.

Suppose now that the objective is prediction about Z, conditionally on $X_1, \ldots, X_n$. Show that, if $\pi(\tau, \mu; \alpha', \beta', k', \nu')$ is the posterior density of (τ, μ) and $Z|(\tau, \mu) \sim N(\mu, \tau^{-1})$, then the joint posterior density of (τ, Z) is of the form $\pi(\tau, Z; \alpha'', \beta'', k'', \nu'')$ and state explicitly what are $\alpha'', \beta'', k'', \nu''$.

Show that the marginal predictive density of Z, given $X_1, \ldots, X_n$, is

$$\frac{(\beta'')^{\alpha''}}{\Gamma(\alpha'')} \cdot \left(\frac{k''}{2\pi}\right)^{1/2} \cdot \frac{\Gamma(\alpha'' + \frac{1}{2})}{\{\beta'' + \frac{1}{2}k''(Z - \nu'')^2\}^{\alpha''+1/2}}$$

and interpret this in terms of the t distribution.

3.16* Suppose $V \sim W_p(\Psi^{-1}, m)$, $\mu|V \sim N_p(\nu, (kV)^{-1})$ and, conditionally on (V, μ), $X_1, \ldots, X_n, Z$ are independent $N_p(\mu, V^{-1})$. If we write the joint prior density of (V, μ) as $\pi(V, \mu; m, k, \Psi, \nu)$ then, as shown in Example 3.4, the joint posterior density of (V, μ), given $X_1, \ldots, X_n$, is of the form $\pi(V, \mu; m', k', \Psi', \nu')$, where we gave explicit formulae for m', k', Ψ', ν' as functions of m, k, Ψ, ν.

Suppose, now, the objective is Bayesian prediction of Z, conditional on $X_1, \ldots, X_n$. Show that the joint conditional density of (V, Z), given $X_1, \ldots, X_n$, is also of the form $\pi(V, Z; m'', k'', \Psi'', \nu'')$, and give explicit expressions for m'', k'', Ψ'', ν''.

Calculate the conditional density of Z alone, given $X_1, \ldots, X_n$.

4

Hypothesis testing

From now on, we consider a variety of specific statistical problems, beginning in this chapter with a re-examination of the theory of hypothesis testing. The concepts and terminology of decision theory will always be present in the background, but inevitably, each method that we consider has developed its own techniques.

In Section 4.1 we introduce the key ideas in the Neyman–Pearson framework for hypothesis testing. The fundamental notion is that of seeking a test which maximises *power*, the probability under repeated sampling of correctly rejecting an incorrect hypothesis, subject to some pre-specified fixed *size*, the probability of incorrectly rejecting a true hypothesis. When the hypotheses under test are *simple*, so that they completely specify the distribution of X, the Neyman–Pearson Theorem (Section 4.2) gives a simple characterisation of the optimal test. We shall see in Section 4.3 that this result may be extended to certain *composite* (non-simple) hypotheses, when the family of distributions under consideration possesses the property of *monotone likelihood ratio*. Other, more elaborate, hypothesis testing problems require the introduction of further structure, and are considered in Chapter 7. The current chapter finishes (Section 4.4) with a description of the Bayesian approach to hypothesis testing based on *Bayes factors*, which may conflict sharply with the Neyman–Pearson frequentist approach.

4.1 Formulation of the hypothesis testing problem

Throughout we have a parameter space Θ, and consider hypotheses of the form

$$H_0 : \theta \in \Theta_0 \quad \text{vs.} \quad H_1 : \theta \in \Theta_1,$$

where Θ_0 and Θ_1 are two disjoint subsets of Θ, possibly, but not necessarily, satisfying $\Theta_0 \cup \Theta_1 = \Theta$.

If a hypothesis consists of a single member of Θ, for example if $\Theta_0 = \{\theta_0\}$ for some $\theta_0 \in \Theta$, then we say that it is a *simple* hypothesis. Otherwise it is called *composite*.

Sometimes hypotheses which at first sight appear to be simple hypotheses are really composite. This is especially common when we have *nuisance parameters*. For example, suppose $X_1, \ldots, X_n$ are independent, identically distributed $N(\mu, \sigma^2)$, with μ and σ^2 both unknown, and we want to test $H_0 : \mu = 0$. This is a composite hypothesis because $\Theta = \{(\mu, \sigma^2) : -\infty < \mu < \infty, 0 < \sigma^2 < \infty\}$, while $\Theta_0 = \{(\mu, \sigma^2) : \mu = 0, 0 < \sigma^2 < \infty\}$. Here σ^2 is a nuisance parameter: it does not enter into the hypothesis we want to test, but nevertheless we have to take it into account in constructing a test.

For most problems we adopt the following criterion: fix a small number α (often fixed to be 0.05, but any value in (0,1) is allowable), and seek a test of *size* α, so that

$$\Pr_\theta\{\text{Reject } H_0\} \le \alpha \quad \text{for all } \theta \in \Theta_0.$$

Thus H_0 and H_1 are treated *asymmetrically*. Usually H_0 is called the *null hypothesis* and H_1 the *alternative hypothesis*.

4.1.1 Test functions

The usual way hypothesis testing is formulated in elementary statistics texts is as follows: choose a test statistic $t(X)$ (some function of the observed data X) and a critical region C_α, then reject H_0 based on $X = x$ if and only if $t(x) \in C_\alpha$. The critical region must be chosen to satisfy

$$\Pr_\theta\{t(X) \in C_\alpha\} \le \alpha \quad \text{for all } \theta \in \Theta_0.$$

For example, consider the problem in which $X = (X_1, \ldots, X_n)$ with $X_1, \ldots, X_n$ independent, identically distributed from $N(\mu, 1)$ and suppose the problem is to test $H_0 : \mu \le 0$ against $H_1 : \mu > 0$. The standard procedure is to reject H_0 if $\bar{X} > z_\alpha \sqrt{n}$, where $\bar{X}$ is the mean of $X_1, \ldots, X_n$ and z_α is the upper-α point of the standard normal distribution. Thus in this case

$$t(X) = \bar{X}, \quad C_\alpha = \{t : t > z_\alpha \sqrt{n}\}.$$

We consider here a slight reformulation of this. Define a *test function* $\phi(x)$ by

$$\phi(x) = \begin{cases} 1 & \text{if } t(x) \in C_\alpha, \\ 0 & \text{otherwise.} \end{cases}$$

So, whenever we observe $\phi(X) = 1$, we reject H_0, while, if $\phi(X) = 0$, we accept.

Recall that in our discussion of abstract decision theory, it was necessary sometimes to adopt a randomised decision rule. The same concept arises in hypothesis testing as well: sometimes we want to use a randomised test. This may be done by generalising the concept of a test function to allow $\phi(x)$ to take on any value in the interval $[0, 1]$. Thus having observed data X and evaluated $\phi(X)$, we use some independent randomisation device to draw a Bernoulli random number W which takes value 1 with probability $\phi(X)$, and 0 otherwise. We then reject H_0 if and only if $W = 1$. Thus we may interpret $\phi(x)$ to be 'the probability that H_0 is rejected when $X = x$'. Of course, in cases where $\phi(x)$ takes on only the values 0 and 1, this is identical with the usual formulation.

Example 4.1 Suppose $X \sim \text{Bin}(10, \theta)$ and we want to test $H_0 : \theta \le \frac{1}{2}$ against $H_1 : \theta > \frac{1}{2}$. The obvious test will be: reject H_0 whenever $X \ge k_\alpha$, where k_α is chosen so that $\Pr_\theta\{X \ge k_\alpha\} \le \alpha$ when $\theta = \frac{1}{2}$. However, if we work out $\Pr_\theta\{X \ge k\}$ with $\theta = \frac{1}{2}$ for several k, we get the answers 0.00098, 0.01074, 0.05469 etc. for $k = 10, 9, 8, \ldots$. So, if $\alpha = 0.05$, the test that takes $k_\alpha = 8$ has size 0.054..., which is no good, so, for a non-randomised test, we would have to take $k_\alpha = 9$, which has actual size only 0.01074 and is therefore very conservative. We can resolve this problem, and achieve size exactly 0.05, by defining

$$\phi(x) = \begin{cases} 1 & \text{if } x = 9 \text{ or } 10, \\ 67/75 & \text{if } x = 8, \\ 0 & \text{if } x \leq 7. \end{cases}$$

In words: if we observe $X = 9$ or 10 we reject H_0, but if we observe $X = 8$ we flip a hypothetical coin, which has probability $67/75$ of coming down heads, and reject H_0 if this coin does come down heads. In all other cases, we accept H_0.

Randomisation looks a very artificial procedure, and it is. If one came across this situation in the course of some consulting, it would probably be sensible to ask the client if she was satisfied with a test of size 0.055 instead of one of size 0.05. However the point of the example is this: there is no test, based on the value of X alone, which achieves size exactly 0.05. Therefore, if we want to construct a theory of hypothesis tests of a *given* size, we have to allow the possibility of randomised tests, regardless of whether we would actually want to use such a test in a practical problem.

4.1.2 Power

Having defined a general randomised test in terms of its test function ϕ, we now need some criterion for deciding whether one test is better than another. We do this by introducing the concept of *power*.

The power function of a test ϕ is defined to be

$$w(\theta) = \text{Pr}_\theta \{\text{Reject } H_0\} = \mathbb{E}_\theta \{\phi(X)\},$$

which is defined for all $\theta \in \Theta$. When testing a simple null hypothesis against a simple alternative hypothesis, the term 'power' is often used to signify the probability of rejecting the null hypothesis when the alternative hypothesis is true.

The idea is this: a good test is one which makes $w(\theta)$ as large as possible on Θ_1, while satisfying the constraint $w(\theta) \leq \alpha$ for all $\theta \in \Theta_0$.

Within this framework, we can consider various classes of problems:

(i) Simple H_0 vs. simple H_1: here there is an elegant and complete theory, which tells us exactly how to construct the best test, given by the Neyman–Pearson Theorem.

(ii) Simple H_0 vs. composite H_1: in this case the obvious approach is to pick out a representative value of Θ_1, say θ_1, and construct the Neyman–Pearson test of H_0 against θ_1. In some cases the test so constructed is the same for every $\theta_1 \in \Theta_1$. When this happens, the test is called 'uniformly most powerful' or UMP. We would obviously like to use a UMP test if we can find one, but there are many problems for which UMP tests do not exist, and then the whole problem is harder.

(iii) Composite H_0 vs. composite H_1: in this case the problem is harder again. It may not be so easy even to find a test which satisfies the size constraint, because of the requirement that $\mathbb{E}_\theta \{\phi(X)\} \leq \alpha$ for all $\theta \in \Theta_0$; if Θ_0 contains a nuisance parameter such as σ^2 in the above $N(\mu, \sigma^2)$ example, we must find a test which satisfies this constraint regardless of the value of σ^2.

4.2 The Neyman–Pearson Theorem

Consider the test of a simple null hypothesis $H_0 : \theta = \theta_0$ against a simple alternative hypothesis $H_1 : \theta = \theta_1$, where θ_0 and θ_1 are specified. Let the probability density function or probability mass function of X be $f(x; \theta)$, specialised to $f_0(x) = f(x; \theta_0)$ and $f_1(x) = f(x; \theta_1)$. Define the likelihood ratio $\Lambda(x)$ by

$$\Lambda(x) = \frac{f_1(x)}{f_0(x)}.$$

According to the Neyman–Pearson Theorem, the best test of size α is of the form: reject H_0 when $\Lambda(X) > k_\alpha$, where k_α is chosen so as to guarantee that the test has size α. However, we have seen above that this method of constructing the test can fail when X has a discrete distribution (or, more precisely, when $\Lambda(X)$ has a discrete distribution under H_0). In the following generalised form of the Neyman–Pearson Theorem, we remove this difficulty by allowing for the possibility of randomised tests.

The (randomised) test with test function ϕ_0 is said to be a *likelihood ratio test* (LRT for short) if it is of the form

$$\phi_0(x) = \begin{cases} 1 & \text{if } f_1(x) > K f_0(x), \\ \gamma(x) & \text{if } f_1(x) = K f_0(x), \\ 0 & \text{if } f_1(x) < K f_0(x), \end{cases}$$

where $K \geq 0$ is a constant and $\gamma(x)$ an arbitrary function satisfying $0 \leq \gamma(x) \leq 1$ for all x.

Theorem 4.1 (Neyman–Pearson)

(a) *(Optimality). For any K and $\gamma(x)$, the test ϕ_0 has maximum power among all tests whose sizes are no greater than the size of ϕ_0.*
(b) *(Existence). Given $\alpha \in (0, 1)$, there exist constants K and γ_0 such that the LRT defined by this K and $\gamma(x) = \gamma_0$ for all x has size exactly α.*
(c) *(Uniqueness). If the test ϕ has size α, and is of maximum power amongst all possible tests of size α, then ϕ is necessarily a likelihood ratio test, except possibly on a set of values of x which has probability 0 under both H_0 and H_1.*

Proof: assuming absolute continuity

(a) Let ϕ be any test for which $\mathbb{E}_{\theta_0}\phi(X) \leq \mathbb{E}_{\theta_0}\phi_0(X)$. Define $U(x) = \{\phi_0(x) - \phi(x)\}\{f_1(x) - K f_0(x)\}$. When $f_1(x) - K f_0(x) > 0$ we have $\phi_0(x) = 1$, so $U(x) \geq 0$. When $f_1(x) - K f_0(x) < 0$ we have $\phi_0(x) = 0$, so $U(x) \geq 0$. For $f_1(x) - K f_0(x) = 0$, of course $U(x) = 0$. Thus $U(x) \geq 0$ for all x. Hence

$$0 \leq \int \{\phi_0(x) - \phi(x)\}\{f_1(x) - K f_0(x)\}dx$$

$$= \int \phi_0(x) f_1(x)dx - \int \phi(x) f_1(x)dx + K \left\{ \int \phi(x) f_0(x)dx - \int \phi_0(x) f_0(x)dx \right\}$$

$$= \mathbb{E}_{\theta_1}\phi_0(X) - \mathbb{E}_{\theta_1}\phi(X) + K \left\{ \mathbb{E}_{\theta_0}\phi(X) - \mathbb{E}_{\theta_0}\phi_0(X) \right\}.$$

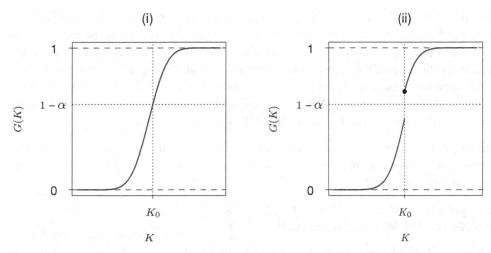

Figure 4.1 Diagram of the distribution function $G(K)$, showing the two possibilities

However, the expression in curly brackets is ≤ 0, because of the assumption that the size of ϕ is no greater than the size of ϕ_0. Thus

$$\int \phi_0(x) f_1(x) dx - \int \phi(x) f_1(x) dx \geq 0,$$

which establishes that the power of ϕ cannot be greater than the power of ϕ_0, as claimed.

(b) The probability distribution function $G(K) = \Pr_{\theta_0}\{\Lambda(X) \leq K\}$ is non-decreasing as K increases; it is also right-continuous (so that $G(K) = \lim_{y \searrow K} G(y)$ for each K). Try to find a value K_0 for which $G(K_0) = 1 - \alpha$. As can be seen from Figure 4.1, there are two possibilities: (i) such K_0 exists, or (ii) we cannot exactly solve the equation $G(K_0) = 1 - \alpha$ but we can find a K_0 for which $G_-(K_0) = \Pr_{\theta_0}\{\Lambda(X) < K_0\} \leq 1 - \alpha < G(K_0)$. In Case (i), we are done (set $\gamma_0 = 0$). In Case (ii), set

$$\gamma_0 = \frac{G(K_0) - (1 - \alpha)}{G(K_0) - G_-(K_0)}.$$

Then it is an easy exercise to demonstrate that this test has size exactly α, as required.

(c) Let ϕ_0 be the LRT defined by the constant K and function $\gamma(x)$, and suppose ϕ is another test of the same size α and the same power as ϕ_0. Define $U(x)$ as in (a). Then $U(x) \geq 0$ for all x, but, because ϕ and ϕ_0 have the same size and power, $\int U(x) dx = 0$. So the function $U(x)$ is non-negative and integrates to 0: hence $U(x) = 0$ for all x, except possibly on a set, S say, of values of x, which has probability zero under both H_0 and H_1. This in turn means that, except on the set S, $\phi(x) = \phi_0(x)$ or $f_1(x) = K f_0(x)$, so that $\phi(x)$ has the form of a LRT. This establishes the uniqueness result, and so completes the proof of the theorem. $\qquad\square$

4.3 Uniformly most powerful tests

A *uniformly most powerful* or UMP test of size α is a test $\phi_0(\cdot)$ for which

(i) $\mathbb{E}_\theta \phi_0(X) \leq \alpha$ for all $\theta \in \Theta_0$;

(ii) given any other test $\phi(\cdot)$ for which $\mathbb{E}_\theta \phi(X) \leq \alpha$ for all $\theta \in \Theta_0$, we have $\mathbb{E}_\theta \phi_0(X) \geq \mathbb{E}_\theta \phi(X)$ for all $\theta \in \Theta_1$.

In general, it is asking a very great deal to expect that UMP tests exist – in effect, it is asking that the Neyman–Pearson test for simple vs. simple hypotheses should be the same for *every* pair of simple hypotheses contained within H_0 and H_1. Nevertheless, for one-sided testing problems involving just a single parameter, for which $\Theta \subseteq \mathbb{R}$, there is a wide class of parametric families for which just such a property holds. Such families are said to have *monotone likelihood ratio* or MLR. The purpose of this section is to define such families and to explain exactly why they have the UMP property.

Example 4.2 Suppose $X_1, \ldots, X_n$ are independent, identically distributed from an exponential distribution with mean θ: $f(x; \theta) = \theta^{-1} e^{-x/\theta}$ for $0 < x < \infty$, where $0 < \theta < \infty$. Consider the following:

Problem 1: Test $H_0 : \theta = \theta_0$ against $H_1 : \theta > \theta_0$.
Problem 2: Test $H_0^* : \theta \le \theta_0$ against $H_1 : \theta > \theta_0$.

Here $\theta_0 > 0$ is a given number. We look at Problem 1 first. Consider the test of $\theta = \theta_0$ against $\theta = \theta_1$ for some $\theta_1 > \theta_0$. Since

$$f(x; \theta) = \frac{1}{\theta^n} \exp\left\{-\frac{1}{\theta} \sum x_i\right\},$$

we have

$$\frac{f(x; \theta_1)}{f(x; \theta_0)} = \exp\left\{\left(\frac{1}{\theta_0} - \frac{1}{\theta_1}\right) \sum x_i\right\}.$$

But $1/\theta_0 - 1/\theta_1 > 0$ and $\Lambda(x)$ is an increasing function of $\sum x_i$, so the Neyman–Pearson test will be: reject H_0 if $\sum X_i > k_\alpha$, where k_α is chosen so that $\Pr_{\theta_0}\{\sum X_i > k_\alpha\} = \alpha$.

Since $X_1 \sim \text{Gamma}(1, 1/\theta)$, we have by elementary properties of the Gamma distribution, $\sum_1^n X_i \sim \text{Gamma}(n, 1/\theta)$. Thus we require to find k_α such that

$$\frac{1}{\theta_0^n (n-1)!} \int_{k_\alpha}^{\infty} t^{n-1} e^{-t/\theta_0} dt = \alpha.$$

The integral may be evaluated from tables of the incomplete Gamma function, by repeated integration by parts or (for large n) approximately by the Central Limit Theorem.

However, the point of the example is not to worry too much about the computation of k_α, but to demonstrate that *the optimal test does not depend on which θ_1 is chosen*, so long as we do choose $\theta_1 > \theta_0$. In other words, the test is uniformly most powerful for all $\theta \in \Theta_1$.

To show that the same test is also UMP for problem 2, first note that

$$\Pr_\theta\left\{\sum X_i > k\right\} = \Pr_\theta\left\{\frac{\sum X_i}{\theta} > \frac{k}{\theta}\right\}$$

$$= \Pr_\theta\left\{Y > \frac{k}{\theta}\right\},$$

where $Y \sim \text{Gamma}(n, 1)$. This is a non-decreasing function of θ. Therefore, since this test ($\phi_0(X)$ say) has size α under the null hypothesis H_0, then it also has size α under the null hypothesis H_0^*.

Now let $\phi(X)$ be any other test of size α under H_0^*. Since H_0 is a smaller hypothesis than H_0^*, the test $\phi(X)$ must also have size $\le \alpha$ under H_0. But then, by the Neyman–Pearson Theorem, $\mathbb{E}_{\theta_1}\phi(X) \le \mathbb{E}_{\theta_1}\phi_0(X)$ for all $\theta_1 > \theta_0$. Thus the test $\phi_0(X)$ is UMP.

4.3.1 Monotone likelihood ratio

The reason that the above example works out so neatly is because the likelihood ratio for $\theta = \theta_0$ against $\theta = \theta_1$ is a monotone function of the statistic $t(x) = \sum x_i$. Thus the Neyman–Pearson test is always of the form

$$\phi_0(x) = \begin{cases} 1 & \text{if } t(x) > k_\alpha, \\ 0 & \text{if } t(x) \le k_\alpha, \end{cases}$$

regardless of the actual values of θ_0 and θ_1. We should therefore expect the same argument to work for any family of densities in which the likelihood ratio is a monotone function of some statistic $t(x)$. We now make this notion precise.

Definition *The family of densities $\{f(x; \theta), \ \theta \in \Theta \subseteq \mathbb{R}\}$ with real scalar parameter θ is said to be of monotone likelihood ratio (MLR for short) if there exists a function $t(x)$ such that the likelihood ratio*

$$\frac{f(x; \theta_2)}{f(x; \theta_1)}$$

is a non-decreasing function of $t(x)$ whenever $\theta_1 \le \theta_2$.

Note that any family for which the likelihood ratio turns out to be non-increasing (rather than non-decreasing) as a function of $t(x)$ is still MLR: simply replace $t(x)$ by $-t(x)$.

Example 4.3 Consider a simple one-parameter exponential family in which observations $X_1, \ldots, X_n$ are independent, identically distributed from the density

$$f(x; \theta) = c(\theta)h(x)e^{\theta \tau(x)}.$$

The exponential distribution mentioned above is an example of this class (on rewriting θ in place of $1/\theta$), but, as we shall see in Chapter 5, there are many examples of exponential families that go well beyond the original exponential distribution.

If we redefine X to be the vector $(X_1, \ldots, X_n)$ rather than just a single observation, then the density of X becomes

$$f(x; \theta) = c(\theta)^n \prod_i \{h(x_i)\} e^{\theta \sum \tau(x_i)}$$

$$= c(\theta)^n \prod_i \{h(x_i)\} e^{\theta t(x)},$$

where $t(x) = \sum \tau(x_i)$. Then for any $\theta_1 \le \theta_2$

$$\frac{f(x; \theta_2)}{f(x; \theta_1)} = \left\{ \frac{c(\theta_2)}{c(\theta_1)} \right\}^n \exp\{(\theta_2 - \theta_1)t(x)\}.$$

This is non-decreasing in $t(x)$, and so the family is MLR.

Example 4.4 Suppose $X_1, \ldots, X_n$ are independent, identically distributed from the uniform distribution on $(0, \theta)$, where $\theta > 0$ is unknown. The density is

$$f(x; \theta) = \begin{cases} \theta^{-n} & \text{if } \max(x_1, \ldots, x_n) \le \theta, \\ 0 & \text{if } \max(x_1, \ldots, x_n) > \theta. \end{cases}$$

Define $t(x) = \max(x_1, \ldots, x_n)$. Then, for $0 < \theta_1 \leq \theta_2$,

$$\frac{f(x; \theta_2)}{f(x; \theta_1)} = \begin{cases} \left(\frac{\theta_1}{\theta_2}\right)^n & \text{if } 0 \leq t(x) \leq \theta_1, \\ +\infty & \text{if } \theta_1 < t(x) \leq \theta_2. \end{cases}$$

This is again a non-decreasing function of $t(x)$. Note that we do not need to consider cases for which $t(x) > \theta_2$ because this is impossible under either model.

Example 4.5 As an example of a one-parameter family which is not MLR, suppose we have one observation from the Cauchy density

$$f(x; \theta) = \frac{1}{\pi\{1 + (x - \theta)^2\}}.$$

Then the likelihood ratio is

$$\frac{f(x; \theta_2)}{f(x; \theta_1)} = \frac{1 + (x - \theta_1)^2}{1 + (x - \theta_2)^2}.$$

It is easily seen that this likelihood ratio is not a monotonic function of x, and it may readily be deduced that infact it is not a monotonic function of $t(x)$ for *any* t.

We now state the main result of this section, that, for a one-sided test in a MLR family, an UMP test exists. For simplicity, we restrict ourselves to absolutely continuous distributions so as to avoid the complications of randomised tests.

Theorem 4.2 *Suppose X has a distribution from a family which is MLR with respect to a statistic $t(X)$, and that we wish to test $H_0 : \theta \leq \theta_0$ against $H_1 : \theta > \theta_0$. Suppose the distribution function of $t(X)$ is continuous.*

(a) The test

$$\phi_0(x) = \begin{cases} 1 & \text{if } t(x) > t_0, \\ 0 & \text{if } t(x) \leq t_0 \end{cases}$$

is UMP among all tests of size $\leq \mathbb{E}_{\theta_0}\{\phi_0(X)\}$.

(b) Given some α, where $0 < \alpha \leq 1$, there exists some t_0 such that the test in (a) has size exactly α.

Proof:

(i) For any $\theta_1 > \theta_0$, the Neyman–Pearson test of $H_0 : \theta = \theta_0$ against $H_1 : \theta = \theta_1$ is of the form $\phi_0(x)$ given in (a), for some t_0. The form of this test does not depend on θ_1, so we see at once that the test ϕ_0 is UMP for testing $H_0 : \theta = \theta_0$ against $H_1 : \theta > \theta_0$.

(ii) For any test ϕ_0 of the form given in (a), $\mathbb{E}_\theta\{\phi_0(X)\}$ is a non-decreasing function of θ. To see this, consider any $\theta_2 < \theta_1$ and suppose $\mathbb{E}_{\theta_2}\{\phi_0(X)\} = \beta$. Consider the following trivial test: $\phi(x) = \beta$ for all x. Thus ϕ chooses H_0 or H_1 with probabilities $1 - \beta$ and β regardless of the value of x. Obviously, $\mathbb{E}_{\theta_1}\{\phi(X)\} = \beta$. But the test ϕ_0 is the Neyman–Pearson test for testing θ_2 against θ_1, so it must be at least as good as ϕ. Therefore, $\mathbb{E}_{\theta_1}\{\phi_0(X)\} \geq \beta$, as required.

(iii) It then follows that, if the test constructed in (i) satisfies $\mathbb{E}_{\theta_0}\{\phi_0(X)\} = \alpha$, then ϕ_0 is also of size α under the larger null hypothesis $H_0 : \theta \leq \theta_0$. Suppose ϕ is any other test of size α under H_0. Then $\mathbb{E}_{\theta_0}\{\phi(X)\} \leq \alpha$. So, by the Neyman–Pearson Theorem, $\mathbb{E}_{\theta_1}\{\phi(X)\} \leq \mathbb{E}_{\theta_1}\{\phi_0(X)\}$ for any $\theta_1 > \theta_0$. This proves (a), that ϕ_0 is UMP among all tests of its size.

(iv) Finally if α is given, we need to show that there exists a t_0 such that $\mathrm{Pr}_{\theta_0}\{t(X) > t_0\} = \alpha$. But this follows at once from the assumption that the distribution function of $t(X)$ is continuous. Thus (b) is proved, and hence the whole theorem. $\qquad\square$

Remark

1 What if the MLR property is false? One remedy is to look at *locally best tests*, defined by taking θ_1 very close to θ_0.

Recalling that the Neyman–Pearson test rejects θ_0 in favour of θ_1 if

$$\frac{f(x;\theta_1)}{f(x;\theta_0)} > K$$

for some suitable K, or equivalently

$$\log f(x;\theta_1) - \log f(x;\theta_0) > \log K,$$

a first-order Taylor expansion (for details see Section 8.1.5) allows us to rewrite this approximately as

$$(\theta_1 - \theta_0)\frac{\partial \log f(x;\theta)}{\partial \theta}\bigg|_{\theta=\theta_0} > \log K,$$

which is equivalent to rejecting H_0 when the *score statistic*

$$S(x;\theta_0) = \frac{\partial \log f(x;\theta)}{\partial \theta}\bigg|_{\theta=\theta_0}$$

is too large. This assumes, of course, that $\log f(x;\theta)$ is indeed differentiable with respect to θ. As we shall see later in Chapter 8, this is one well-established universal method of constructing a hypothesis test, but by no means the only one.

2 None of this applies to two-sided tests, such as $H_0 : \theta = \theta_0$ against $H_1 : \theta \neq \theta_0$. In this case, if a distribution has the MLR property, it would still seem natural to base any test on the statistic $t(X)$, but there is no UMP test. For a detailed theory of two-sided tests from a classical point of view, the reader is referred to Chapter 7.

4.4 Bayes factors

In this section we briefly outline a Bayesian approach to hypothesis testing problems, which, as we shall quickly see, leads in quite different directions from the Neyman–Pearson theory.

4.4.1 Bayes factors for simple hypotheses

Consider first the case of simple H_0 vs. simple H_1. Suppose the prior probability that H_j is true is denoted by π_j for $j = 0, 1$, with $\pi_0 > 0$, $\pi_1 > 0$, $\pi_0 + \pi_1 = 1$, and let the respective densities of X under H_0 and H_1 be $f_0(x)$ and $f_1(x)$, as in Section 4.2. Bayes' law quickly leads to

$$\Pr\{H_0 \text{ true } | X = x\} = \frac{\pi_0 f_0(x)}{\pi_0 f_0(x) + \pi_1 f_1(x)}.$$

There is another way to write this, namely

$$\frac{\Pr\{H_0 \text{ true } | X = x\}}{\Pr\{H_1 \text{ true } | X = x\}} = \frac{\pi_0}{\pi_1} \frac{f_0(x)}{f_1(x)},$$

where π_0/π_1 is the *prior odds* in favour of H_0 over H_1, and $f_0(x)/f_1(x)$ is called the *Bayes factor*. In words,

$$\text{Posterior Odds} = \text{Prior Odds} \times \text{Bayes Factor}.$$

In a court of law, the prior and posterior odds might represent a juror's strength of conviction that the accused person is innocent, respectively before and after hearing the evidence (or a specific piece of evidence, such as the result of a DNA test). Then the Bayes factor represents 'the strength of the evidence'.

The first person to use Bayes factors extensively was Jeffreys, in his book *Theory of Probability* (first edition 1939). Indeed, the theory of Bayes factors may be regarded as Jeffreys' main contribution to the subject of statistics. Following Jeffreys, however, there were not many new methodological developments until the 1980s. Since then, however, there has been a great outpouring of research, and this is one of the most active areas of research in contemporary Bayesian statistics.

Jeffreys gave Table 4.1 to aid in the interpretation of a Bayes factor.

Table 4.1 *Interpretation of Bayes factors*

Bayes factor B	Interpretation
$B > 1$	Evidence supports H_0
$1 > B > 10^{-1/2}$	Slight evidence against H_0
$10^{-1/2} > B > 10^{-1}$	Substantial evidence against H_0
$10^{-1} > B > 10^{-3/2}$	Strong evidence against H_0
$10^{-3/2} > B > 10^{-2}$	Very strong evidence against H_0
$10^{-2} > B$	Decisive evidence against H_0

These interpretations might be considered disputable; for example, in a DNA fingerprinting case, not everyone would agree that a Bayes factor of 0.01 was 'decisive' evidence against the accused! But the point of this discussion is that we can think of the size of the Bayes factor as a meaningful measure in its own right, without resorting to significance levels and the like. It is this point of view that represents the real point of departure between the classical and Bayesian approaches to hypothesis testing.

From the point of view of decision theory, any Bayes rule will take the form: reject H_0 if $B < k$ for some constant k, otherwise accept H_0. Thus, if we ignore the possible need for randomised tests, the class of Bayes rules is exactly the same as the class of Neyman–Pearson rules. This is to be expected, in view of the close correspondence between Bayes rules and admissible rules which we saw in Chapter 2. However, the process by which we decide on the critical value k is fundamentally different.

4.4.2 Bayes factors for composite hypotheses

Suppose now the hypotheses H_0 and H_1 are composite. In order to calculate Bayes factors, we now need to know a complete prior distribution for θ. In other words, it is not sufficient to know the prior probabilities π_0 and π_1 that H_0 and H_1 are correct, but we must know the prior density for θ under each of these two hypotheses. Suppose θ has prior density $g_0(\theta)$, $\theta \in \Theta_0$ conditionally on H_0 being true, and $g_1(\theta)$, $\theta \in \Theta_1$ conditionally on H_1 being true. The Bayes factor in this case is defined as

$$B = \frac{\int_{\Theta_0} f(x;\theta)g_0(\theta)d\theta}{\int_{\Theta_1} f(x;\theta)g_1(\theta)d\theta}.$$

In the case of a simple $H_0 : \theta = \theta_0$ and composite H_1, such as $H_1 : \theta \neq \theta_0$, we may write

$$B = \frac{f(x;\theta_0)}{\int_{\Theta_1} f(x;\theta)g_1(\theta)d\theta}.$$

Note that there is nothing that requires the same parametrisation θ for the two hypotheses. More generally, suppose we have two candidate parametric models M_1 and M_2 for data X, and the two models have respective parameter vectors θ_1 and θ_2. Under prior densities $\pi_i(\theta_i)$, $i = 1, 2$ for the parameter vectors in the two models, the marginal distributions of X are found as

$$p(x \mid M_i) = \int f(x;\theta_i, M_i)\pi_i(\theta_i)d\theta_i, \ i = 1, 2,$$

and the Bayes factor is the ratio of these:

$$B = \frac{p(x \mid M_1)}{p(x \mid M_2)}.$$

The interpretation of B is the same as in Jeffreys' table above, but the actual calculation depends on the prior densities in a non-trivial way.

Example 4.6 Suppose $X_1, \ldots, X_n$ are independent, identically distributed from $N(\theta, \sigma^2)$, with σ^2 known. Consider $H_0 : \theta = 0$ against $H_1 : \theta \neq 0$. Also suppose the prior g_1 for θ under H_1 is $N(\mu, \tau^2)$. We have

$$B = \frac{P_1}{P_2},$$

where

$$P_1 = (2\pi\sigma^2)^{-n/2} \exp\left(-\frac{1}{2\sigma^2} \sum X_i^2\right),$$

$$P_2 = (2\pi\sigma^2)^{-n/2} \int_{-\infty}^{\infty} \exp\left\{-\frac{1}{2\sigma^2} \sum (X_i - \theta)^2\right\} \cdot (2\pi\tau^2)^{-1/2} \exp\left\{-\frac{(\theta - \mu)^2}{2\tau^2}\right\} d\theta.$$

Completing the square as in Example 3.3, we see that

$$\sum (X_i - \theta)^2 + k(\theta - \mu)^2 = (n + k)(\theta - \hat{\theta})^2 + \frac{nk}{n + k}(\bar{X} - \mu)^2 + \sum (X_i - \bar{X})^2,$$

for arbitrary k, where $\hat{\theta} = (n\bar{X} + k\mu)/(n + k)$. Thus

$$\frac{1}{\sigma^2} \sum (X_i - \theta)^2 + \frac{1}{\tau^2}(\theta - \mu)^2$$

$$= \frac{n\tau^2 + \sigma^2}{\sigma^2 \tau^2}(\theta - \hat{\theta})^2 + \frac{n}{n\tau^2 + \sigma^2}(\bar{X} - \mu)^2 + \frac{1}{\sigma^2}\sum(X_i - \bar{X})^2.$$

Also

$$\int_{-\infty}^{\infty} \exp\left\{ -\frac{n\tau^2 + \sigma^2}{2\sigma^2\tau^2}(\theta - \hat{\theta})^2 \right\} d\theta = \left(\frac{2\pi\sigma^2\tau^2}{n\tau^2 + \sigma^2} \right)^{1/2}.$$

Hence

$$P_2 = (2\pi\sigma^2)^{-n/2} \left(\frac{\sigma^2}{n\tau^2 + \sigma^2} \right)^{1/2} \exp\left[-\frac{1}{2}\left\{ \frac{n}{n\tau^2 + \sigma^2}(\bar{X} - \mu)^2 + \frac{1}{\sigma^2}\sum(X_i - \bar{X})^2 \right\} \right]$$

and so

$$B = \left(1 + \frac{n\tau^2}{\sigma^2} \right)^{1/2} \exp\left[-\frac{1}{2}\left\{ \frac{n\bar{X}^2}{\sigma^2} - \frac{n}{n\tau^2 + \sigma^2}(\bar{X} - \mu)^2 \right\} \right].$$

Defining $t = \sqrt{n}\bar{X}/\sigma$, $\eta = -\mu/\tau$, $\rho = \sigma/(\tau\sqrt{n})$, this may also be written

$$B = \left(1 + \frac{1}{\rho^2} \right)^{1/2} \exp\left[-\frac{1}{2}\left\{ \frac{(t - \rho\eta)^2}{1 + \rho^2} - \eta^2 \right\} \right].$$

However, the form of this solution illustrates a difficulty with the Bayes factor approach. As we saw in Chapter 3, many Bayesian solutions to point and interval estimation problems are approximately the same as classical solutions when the prior is diffuse. Unfortunately this is not the case here; if we let $\rho \to 0$ (corresponding to the limiting case of an uninformative prior, $\tau \to \infty$), then $B \to \infty$ – in other words, there is overwhelming evidence in support of H_0! One must therefore choose values of η and ρ that represent some reasonable judgement of where θ is likely to be when H_0 is false. There is no way of ducking the issue by recourse to vague priors.

Jeffreys himself proposed the use of a Cauchy prior for θ, rather than the normal prior studied here, arguing that this led to actions consistent with the way scientists intuitively interpret data from conflicting experiments. However, this does not remove the basic difficulty that the answer depends on the arbitrary parameters of the prior distribution. Modern authors have proposed a number of alternative approaches. For example, it might be reasonable to use very diffuse or improper priors but for a normalising constant – $(1 + 1/\rho^2)^{1/2}$ in the above example – which is very large or indeterminate. In one development, Smith and Spiegelhalter (1980) proposed replacing this constant by an arbitrary number c. To determine c, they proposed conducting a thought experiment, based on a very small number of observations, for which it would just be possible to discriminate between the two hypotheses. They then suggested setting $B = 1$ for this thought experiment, which determines c, and hence the Bayes factors for all other experiments of the same structure. Berger and

Sellke (1987) proposed an alternative approach in which bounds were obtained on the Bayes factor over a wide class of 'reasonable' choices of the prior density g_1. In recent years, there have been many alternative approaches under such names as partial Bayes factors, intrinsic Bayes factors and fractional Bayes factors. Two valuable recent articles are those by Kass and Raftery (1995) and Berger and Pericchi (1996).

One method which can provide a rough measure of the evidence in favour of one model over another without reference to any prior distribution is provided by the *Bayesian Information Criterion* (*BIC*). This is based on the result that, for large sample sizes n, an approximation to $-2 \log B$ is given by

$$\Delta BIC = W - (p_2 - p_1) \log n,$$

where p_i is the number of parameters in model M_i, $i = 1, 2$, and W is the *likelihood ratio statistic*

$$W = -2 \log \frac{\sup_{\theta_1} f(x; \theta_1, M_1)}{\sup_{\theta_2} f(x; \theta_2, M_2)}.$$

So a (crude) approximation to the Bayes factor B is given by

$$B \approx \exp(-\tfrac{1}{2} \Delta BIC), \tag{4.1}$$

which does not depend on the priors on the parameters in the two models.

Example 4.7 Here is another example which illustrates even more strikingly the difference between the classical significance testing and Bayes factor approaches. Suppose $X \sim \text{Bin}(n, \theta)$ and consider testing $H_0 : \theta = \theta_0$ against $H_1 : \theta \neq \theta_0$. Suppose under H_1, θ is uniformly distributed on $(0, 1)$. Then the Bayes factor on observing $X = x$ is $B = P_1/P_2$, where

$$P_1 = \binom{n}{x} \theta_0^x (1 - \theta_0)^{n-x},$$

$$P_2 = \binom{n}{x} \int_0^1 \theta^x (1 - \theta)^{n-x} d\theta$$

$$= \binom{n}{x} \frac{\Gamma(x + 1)\Gamma(n - x + 1)}{\Gamma(n + 2)}$$

$$= \binom{n}{x} \frac{x!(n - x)!}{(n + 1)!}.$$

Thus

$$B = \frac{(n + 1)!}{x!(n - x)!} \theta_0^x (1 - \theta_0)^{n-x}.$$

For large n and x, it is possible to use Stirling's approximation ($n! \sim \sqrt{2\pi} n^{n+1/2} e^{-n}$) to approximate this exact Bayes factor by

$$B \approx \left\{ \frac{n}{2\pi \theta_0 (1 - \theta_0)} \right\}^{1/2} \exp \left\{ -\frac{(x - n\theta_0)^2}{2n\theta_0 (1 - \theta_0)} \right\}. \tag{4.2}$$

Suppose for example $n = 100$, $\theta_0 = \frac{1}{2}$, $x = 60$. Then a classical procedure estimates $\hat{\theta} = 0.6$ with standard error 0.05 under H_0, leading to the approximately standard normal test statistic $z = (0.6 - 0.5)/0.05 = 2$, on the basis of which the classical procedure rejects H_0 at $\alpha = 0.05$. However the Bayes factor, as computed using (4.2), is $B = 1.09$, pointing to evidence in favour of H_0! The *BIC* approximation to the Bayes factor, as given by (4.1), is 1.33. The phenomenon of the Bayes factor contradicting the classical test is more pronounced at larger sample sizes: for example for $n = 400$, $x = 220$ we estimate $\hat{\theta} = 0.55$ with standard error 0.025, so again $z = 2$, but in this case $B = 2.16$ (with the *BIC* approximation giving 2.70), while with $n = 10\,000$, $x = 5100$, which again leads to $z = 2$, we find $B = 10.8$, with the *BIC* approximation 13.5, which Jeffreys would interpret as 'strong evidence' in support of H_0, even though a classical significance test would again lead to rejection!

This therefore points to an instance where Bayesian and classical approaches do lead to sharply conflicting conclusions. They are, however, based on quite different viewpoints of the problem. The classical significance test is relevant if it really is important to know whether θ has exactly the value θ_0, and no other. For example, if we tossed a supposedly fair coin 10 000 times, and observed 5100 heads, then we would have strong grounds for concluding that the coin is biased. It makes no difference whether the true θ is 0.499 or 0.51 or 0.67 or 0.11 ... all are evidence that there is something wrong with the coin. For a contrasting situation, suppose a manufacturer introduces a new drug, claiming that it is equally effective as, but much cheaper than, an existing drug for which it is known that the probability of no response within a specified time period T is 0.5. If $\theta = \Pr\{$New drug shows no response within time period $T\}$, then initially we have no information, other than the manufacturer's claim, about the true value of θ. However, if a very large number of trials resulted in an estimated θ of 0.51, it would seem reasonable to conclude that the manufacturer's claim is correct – the difference between 0.51 and 0.50 being unlikely, in this context, to be of much practical importance. Thus, in this situation, it seems more reasonable to use the Bayes factor as a measure of the strength of the evidence for or against the manufacturer's claim.

Finally we should point out that Bayes factors are not universally accepted, even by Bayesians. To those who believe that the only rational approach to statistics lies in the full specification of *subjective* prior probabilities and utilities, the Bayes factor looks like a way of avoiding part of the problem. A formal specification of a Bayes decision rule, along the lines laid out in Chapter 3, cannot depend solely on the Bayes factor.

4.5 Problems

4.1 A random variable X has one of two possible densities:

$$f(x; \theta) = \theta e^{-\theta x}, \quad x \in (0, \infty), \quad \theta \in \{1, 2\}.$$

Consider the family of decision rules

$$d_\mu(x) = \begin{cases} 1 & \text{if } x \geq \mu, \\ 2 & \text{if } x < \mu, \end{cases}$$

where $\mu \in [0, \infty]$. Calculate the risk function $R(\theta, d_\mu)$ for loss function $L(\theta, d) = |\theta - d|$, and sketch the parametrised curve $\mathcal{C} = \{(R(1, d_\mu), R(2, d_\mu)) : \mu \in [0, \infty]\}$ in $\mathbb{R}^2$.

Use the Neyman–Pearson Theorem to show that $\mathcal{C}$ corresponds precisely to the set of admissible decision rules.

For what prior mass function for θ does the minimax rule coincide with the Bayes rule?

4.2 Let $X_1, \ldots, X_n$ be independent $N(\mu, \sigma^2)$ random variables, where $\sigma^2 \in (0, \infty)$ is a known constant and $\mu \in \mathbb{R}$ is unknown. Show that $X = (X_1, \ldots, X_n)$ has monotone likelihood ratio. Given $\alpha \in (0, 1)$ and $\mu_0 \in \mathbb{R}$, construct a uniformly most powerful test of size α of $H_0 : \mu \leq \mu_0$ against $H_1 : \mu > \mu_0$, expressing the critical region in terms of the standard normal distribution function $\Phi(x)$. Verify directly that (as we already know from the theory) your test has monotone power function.

Now suppose it is μ that is known and σ^2 unknown. Let $\sigma_0^2 \in (0, \infty)$ be given. Construct a uniformly most powerful size α test of $H_0 : \sigma^2 \leq \sigma_0^2$ against $H_1 : \sigma^2 > \sigma_0^2$.

4.3 Let $X_1, \ldots, X_n$ be independent random variables with a common density function

$$f(x; \theta) = \theta e^{-\theta x}, x \geq 0,$$

where $\theta \in (0, \infty)$ is an unknown parameter. Consider testing the null hypothesis $H_0 : \theta \leq 1$ against the alternative $H_1 : \theta > 1$. Show how to obtain a uniformly most powerful test of size α.

4.4 Let $X_1, \ldots, X_n$ have the same joint distribution as in problem 4.3, but assume now we want to test $H_0 : \theta = 1$ against $H_1 : \theta \neq 1$. Define $S_n = X_1 + \cdots + X_n$.

  (i) Show that the test which rejects H_0 whenever $|S_n - n| > z_{\alpha/2}\sqrt{n}$, where $z_{\alpha/2}$ is the upper-$\alpha/2$ point of the standard normal distribution, has size approximately α for large n.

  (ii) Suppose that the prior distribution for θ, conditionally on H_1 being true, has a Gamma distribution with parameters a and b, so that the prior density is $b^a \theta^{a-1} e^{-b\theta} / \Gamma(a)$. Show that the Bayes factor for H_0 against H_1, conditional on $S_n = s_n$, is

$$\frac{\Gamma(a)}{\Gamma(a+n)} \frac{(b + s_n)^{a+n} e^{-s_n}}{b^a}.$$

  (iii) Suppose now $a = b = 1$ and write $s_n = n + z_n\sqrt{n}$, so that, if $z_n = \pm z_{\alpha/2}$, the test in (i) will be just on the borderline of rejecting H_0 at the two-sided significance level α. Show that, as $n \to \infty$, provided $z_n \to \infty$ sufficiently slowly,

$$B \sim \sqrt{\frac{n}{2\pi}} e^{1 - z_n^2/2}.$$

  (iv) Hence show that there exists a sequence $\{s_n, n \geq 1\}$ such that, for the sequence of problems with $S_n = s_n$ for all n, the null hypothesis H_0 is rejected at significance level α, for all sufficiently large n, *whatever* the value of $\alpha > 0$, but the Bayes factor for H_0 against H_1 tends to ∞ as $n \to \infty$.

4.5 Let $X_1, \ldots, X_n$ be an independent sample of size n from the uniform distribution on $(0, \theta)$. Show that there exists a uniformly most powerful size α test of $H_0 : \theta = \theta_0$ against $H_1 : \theta > \theta_0$, and find its form.

Let $T = \max(X_1, \ldots, X_n)$. Show that the test

$$\phi(x) = \begin{cases} 1, & \text{if } t > \theta_0 \text{ or } t \leq b, \\ 0, & \text{if } b < t \leq \theta_0, \end{cases}$$

where $b = \theta_0 \alpha^{1/n}$, is a uniformly most powerful test of size α for testing H_0 against $H_1' : \theta \neq \theta_0$.

(Note that in a 'more regular' situation, a UMP test of H_0 against H_1' does not exist.)

5

Special models

Two general classes of models particularly relevant in theory and practice are *exponential families* and *transformation families*. The purpose of this short and fairly technical chapter is to establish a suitable background on these two classes of model as a preliminary for the more statistical discussion to follow, in particular in Chapters 7 and 9.

5.1 Exponential families

The class of exponential families contains many well-known examples of parametric statistical distributions, including, for example, binomial, Poisson, normal, Gamma, Beta, negative binomial distributions and many more. The families have a number of properties which are extremely useful when they are used for more advanced testing and estimation procedures.

5.1.1 Definition and elementary examples

Suppose the random variable X depends on a parameter θ through a density $f(x;\theta)$, of the form

$$f(x;\theta) = c(\theta)h(x)\exp\left\{\sum_{i=1}^{k}\pi_i(\theta)\tau_i(x)\right\}, \quad x \in \mathcal{X}, \quad \theta \in \Theta, \tag{5.1}$$

with $\mathcal{X}$ not depending on θ. Then the model is said to be an *exponential family*. Here we suppose that $\Theta \subseteq \mathbb{R}^d$, so that $\theta = (\theta_1, \ldots, \theta_d)$ say, and note that the concept is defined for both scalar and vector X, and for discrete as well as continuous X, with f being interpreted as a probability mass function in the discrete case. The value of k may be reduced if either $\tau(x) = (\tau_1(x), \ldots, \tau_k(x))$ or $\pi(\theta) = (\pi_1(\theta), \ldots, \pi_k(\theta))$ satisfies a linear constraint, so we will assume that the representation (5.1) is minimal, in that k is as small as possible.

Example 5.1 The exponential density with mean θ and density function $f(x;\theta) = \theta^{-1}\exp(-x/\theta)$, $0 < x < \infty$, $0 < \theta < \infty$ is of exponential family form with $k = 1$.

Example 5.2 The Beta density

$$f(x;a,b) = \frac{\Gamma(a+b)}{\Gamma(a)\Gamma(b)}x^{a-1}(1-x)^{b-1}, \quad 0 < x < 1, \, a > 0, \, b > 0$$

is of exponential family form with $\theta = (a,b)$, if we define $\tau_1(x) = \log x$, $\tau_2(x) = \log(1-x)$.

Example 5.3 The $N(\mu, \sigma^2)$ density may be written in the form

$$\frac{1}{\sqrt{2\pi\sigma^2}} \exp\left\{-\frac{\mu^2}{2\sigma^2}\right\} \cdot \exp\left\{\frac{x\mu}{\sigma^2} - \frac{x^2}{2\sigma^2}\right\}.$$

Writing $\theta = (\mu, \sigma^2)$, let $\tau_1(x) = x$, $\tau_2(x) = x^2$, $\pi_1(\theta) = \mu/\sigma^2$, $\pi_2(\theta) = -1/(2\sigma^2)$.

Examples which are *not* of exponential family type include the uniform distribution on $[0, \theta]$, and the t distribution on θ degrees of freedom. The uniform distribution is excluded because it violates a key general property of our definition of exponential family distributions, worth highlighting, that the range of the distribution does not depend on unknown parameters.

5.1.2 *Means and variances*

First note that for *any* family $f(x; \theta)$ which is sufficiently well-behaved for the identity

$$\int_{\mathcal{X}} f(x; \theta) dx = 1$$

to be differentiated under the integral sign, we have (the proof is left as an exercise)

$$\mathbb{E}_\theta\left\{\frac{\partial \log f(X; \theta)}{\partial \theta_j}\right\} = 0, \ \forall j, \tag{5.2}$$

$$\mathbb{E}_\theta\left\{\frac{\partial^2 \log f(X; \theta)}{\partial \theta_j \partial \theta_l}\right\} = -\mathbb{E}_\theta\left\{\frac{\partial \log f(X; \theta)}{\partial \theta_j}\frac{\partial \log f(X; \theta)}{\partial \theta_l}\right\}, \ \forall j, l. \tag{5.3}$$

In the case of an exponential family with $d = k$, we have

$$\frac{\partial \log f(x; \theta)}{\partial \theta_j} = \frac{\partial}{\partial \theta_j} \log c(\theta) + \sum_{i=1}^{k} \frac{\partial \pi_i(\theta)}{\partial \theta_j} \tau_i(x), \ j = 1, \ldots, k.$$

Applying (5.2) we therefore have

$$0 = \frac{\partial}{\partial \theta_j} \log c(\theta) + \sum_{i=1}^{k} \frac{\partial \pi_i(\theta)}{\partial \theta_j} \mathbb{E}_\theta\{\tau_i(X)\}, \ j = 1, \ldots, k.$$

This results in k equations in the k unknowns, $\mathbb{E}\{\tau_i(X)\}$, $i = 1, \ldots, k$, which may in principle be solved to find the expected values of each of the $\tau_i(X)$.

This problem is much easier to solve if $\pi_i(\theta) = \theta_i$ for each i. In that case,

$$\mathbb{E}_\theta\{\tau_i(X)\} = -\frac{\partial}{\partial \theta_i} \log c(\theta).$$

Moreover, a direct extension from (5.3) also shows

$$\text{Cov}_\theta\{\tau_i(X), \tau_j(X)\} = -\frac{\partial^2}{\partial \theta_i \partial \theta_j} \log c(\theta).$$

A model with $\pi_i(\theta) = \theta_i$ for each i is said to be in its *natural parametrisation*. We define the *natural parameter space* Π by

$$\Pi = \{\pi \equiv (\pi_1(\theta), \ldots, \pi_k(\theta)), \theta \in \Theta \text{ such that } 0 < J(\pi) < \infty\},$$

where

$$J(\pi) = \int h(x) \exp\left\{ \sum_{i=1}^{k} \pi_i(\theta) \tau_i(x) \right\} dx.$$

In the case where θ is k-dimensional, we speak of a *full exponential family*, or a (k, k) exponential family. If $d < k$, we refer to a (k, d) exponential family or *curved exponential family*, noting that our definition requires that $\{\pi_1(\theta), \ldots, \pi_k(\theta)\}$ does not belong to a v-dimensional linear subspace of $\mathbb{R}^k$ with $v < k$. Think of the case $d = 1$, $k = 2$: $\{\pi_1(\theta), \pi_2(\theta)\}$ describes a *curve* in the plane as θ varies, rather than a straight line.

Interest in curved exponential families stems from two features, related to statistical concepts to be discussed later. In a curved exponential family, the maximum likelihood estimator is not a sufficient statistic (see Chapter 6), so that there is scope for conditioning, as a Fisherian stance suggests, on a so-called ancillary statistic: see Chapter 7. Also, it can be shown that any sufficiently smooth parametric family can be approximated, locally to the true parameter value, to some suitable order, by a curved exponential family.

5.1.3 The natural statistics and their exact and conditional distributions

Suppose now we have a random sample $X_1, \ldots, X_n$ from an exponential family with (not necessarily natural) parameter θ. We write X for the vector $(X_1, \ldots, X_n)$, with similar notation for other vectors appearing below.

The density of this X is therefore

$$c(\theta)^n \left\{ \prod_{j=1}^{n} h(x_j) \right\} \exp\left\{ \sum_{i=1}^{k} \pi_i(\theta) \sum_{j=1}^{n} \tau_i(x_j) \right\}.$$

Let us rewrite this in the form

$$c(\theta)^n \left\{ \prod_{j=1}^{n} h(x_j) \right\} \exp\left\{ \sum_{i=1}^{k} \pi_i(\theta) t_i(x) \right\},$$

where $t_i(x) = \sum_j \tau_i(x_j)$ for each i.

The random variables $t_i(X)$, $i = 1, \ldots, k$ are called the *natural statistics* formed from the random vector X.

Two important properties of the natural statistics are given by the following results. These properties make exponential families particularly attractive, as they allow inference about selected components of the natural parameter, in the absence of knowledge about the other components: we will develop this idea in Chapter 7.

Lemma 5.1 *The joint distribution of $t_1(X), \ldots, t_k(X)$ is of exponential family form with natural parameters $\pi_1(\theta), \ldots, \pi_k(\theta)$.*

Lemma 5.2 *For any $S \subseteq \{1, 2, \ldots, k\}$, the joint distribution of $\{t_i(X), i \in S\}$ conditionally on $\{t_i(X), i \notin S\}$, is of exponential family form, with a distribution depending only on $\{\pi_i(\theta), i \in S\}$.*

We shall prove these results only for the case when X is discrete. In principle, that ought to suffice for the continuous case as well, because any continuous random variable can be approximated arbitrarily closely (in an appropriate sense) by a discrete random variable. However, the technicalities needed to make that statement precise are somewhat awkward, so we shall not attempt it here. A more advanced text such as Lehmann (1986) can be consulted for the full details.

Proof of Lemma 5.1

For any $x = (x_1, \ldots, x_n)$,

$$\text{Pr}_\theta\{X = x\} = c(\theta)^n \left\{\prod_{j=1}^n h(x_j)\right\} \exp\left\{\sum_{i=1}^k \pi_i(\theta)t_i(x)\right\}.$$

Fix some vector $y = (y_1, \ldots, y_k)$ and let

$$\mathcal{T}_y = \{x : t_1(x) = y_1, \ldots, t_k(x) = y_k\}.$$

Then

$$\text{Pr}_\theta\{t_1(X) = y_1, \ldots, t_k(X) = y_k\} = \sum_{x \in \mathcal{T}_y} \mathbb{P}\{X = x\}$$

$$= c(\theta)^n \sum_{x \in \mathcal{T}_y} \left\{\prod_{j=1}^n h(x_j)\right\} \exp\left\{\sum_{i=1}^k \pi_i(\theta)t_i(x)\right\}$$

$$= c(\theta)^n h_0(y) \exp\left\{\sum_{i=1}^k \pi_i(\theta)y_i\right\},$$

where

$$h_0(y) = \sum_{x \in \mathcal{T}_y} \left\{\prod_{j=1}^n h(x_j)\right\}.$$

$\square$

Proof of Lemma 5.2

Write $T_1, \ldots, T_k$ for $t_1(X), \ldots, t_k(X)$. By Lemma 5.1,

$$\text{Pr}_\theta\{T_1 = y_1, \ldots, T_k = y_k\} = c(\theta)^n h_0(y) \exp\left\{\sum_{i=1}^k \pi_i(\theta)y_i\right\}.$$

Since the ordering of $T_1, \ldots, T_k$ is arbitrary, there is no loss of generality in assuming $S = \{1, \ldots, l\}$ for some l, $1 \le l < k$. Then

$$\text{Pr}_\theta\{T_1 = y_1, \ldots, T_l = y_l | T_{l+1} = y_{l+1}, \ldots, T_k = y_k\} = \frac{\text{Pr}_\theta\{T_1 = y_1, \ldots, T_k = y_k\}}{\sum_{y'} \text{Pr}_\theta\{T_1 = y_1', \ldots, T_k = y_k'\}},$$

where the sum in the denominator is taken over all vectors $y' = (y_1', \ldots, y_k')$ such that $y_i' = y_i$ for $i > l$.

This may be rewritten

$$\frac{h_0(y) \exp\left\{\sum_i \pi_i(\theta)y_i\right\}}{\sum_{y'} h_0(y') \exp\left\{\sum_i \pi_i(\theta)y_i'\right\}}.$$

However, terms in the exponent of the form $\pi_i(\theta)y_i$, $i > l$ are common to both the numerator and denominator, and therefore cancel. If we then define

$$\frac{1}{c'(\theta)} = \sum_{y'} h_0(y') \exp\left\{\sum_{i=1}^{l} \pi_i(\theta)y_i'\right\}$$

we find that

$$\mathrm{Pr}_\theta\left\{T_1 = y_1, \ldots, T_l = y_l | T_{l+1} = y_{l+1}, \ldots, T_k = y_k\right\} = c'(\theta)h_0(y) \exp\left\{\sum_{i=1}^{l} \pi_i(\theta)y_i\right\},$$

which is of the required form. Moreover, this depends on θ only through $\pi_1(\theta), \ldots, \pi_l(\theta)$, as claimed. $\qquad\square$

5.1.4 Some additional points

We note here two technical points for use later.

1 Let $T = (T_1, \ldots, T_k)$ denote the vector of natural statistics. Assume the natural parameter space Π contains an open rectangle in $\mathbb{R}^k$. Then, if $\phi(T)$ is any real function of T with

$$\mathbb{E}_\theta \phi(T) = 0 \quad \text{for all } \theta,$$

we must have $\phi(T) = 0$ with probability 1.

To see this, suppose the parameter space is the natural parameter space and write the condition $\mathbb{E}_\theta \phi(T) = 0$ in the form

$$\int_{\mathbb{R}^k} \phi(t) h_0(t) \exp\left\{\sum \theta_i t_i\right\} dt = 0 \quad \text{for all } \theta \in \Theta.$$

The left-hand side is simply the k-dimensional Laplace transform of $\phi(t)h_0(t)$, evaluated at $(\theta_1, \ldots, \theta_k)$. By the uniqueness of the inverse of a Laplace transform, which exists in an open rectangle, $\phi(t)h_0(t)$ must be 0 almost everywhere. However $h_0(t) > 0$ (at least with probability 1), so $\phi(t) = 0$ with probability 1.

2 Suppose X is taken from an exponential family in its natural parametrisation

$$f(x; \theta) = c(\theta)h(x) \exp\left\{\sum_{i=1}^{k} \theta_i \tau_i(x)\right\},$$

and let $\phi(x)$ be any bounded function of x. Consider

$$\int \phi(x)h(x) \exp\left\{\sum_{i=1}^{k} \theta_i \tau_i(x)\right\} dx.$$

It can be shown that we may differentiate under the integral sign with respect to $\theta_1, \ldots, \theta_k$: see, for example, Lehmann (1986). In fact, not only is the expression differentiable with respect to θ, it has derivatives of all orders. Now set $\phi(x) \equiv 1$ to see that this property must be true of $1/c(\theta)$, and hence $c(\theta)$ itself, provided $c(\theta)$ is bounded away from 0. Then apply the same reasoning to

$$\int \phi(x)c(\theta)h(x)\exp\left\{\sum_{i=1}^{k}\theta_i\tau_i(x)\right\}dx$$

to see that $\mathbb{E}_\theta\{\phi(X)\}$ is differentiable of all orders with respect to θ. Note that this does not require differentiability of ϕ – for example, it could apply to a test function.

5.2 Transformation families

The basic idea behind a transformation family is that of a group of transformations acting on the sample space, generating a family of distributions all of the same form, but with different values of the parameters.

Recall that a group G is a mathematical structure having a binary operation $\circ$ such that:

- if $g, g' \in G$, then $g \circ g' \in G$;
- if $g, g', g'' \in G$, then $(g \circ g') \circ g'' = g \circ (g' \circ g'')$;
- G contains an identity element e such that $e \circ g = g \circ e = g$, for each $g \in G$; and
- each $g \in G$ possesses an inverse $g^{-1} \in G$ such that $g \circ g^{-1} = g^{-1} \circ g = e$.

In the present context, we will be concerned with a group G of transformations acting on the sample space $\mathcal{X}$ of a random variable X, and the binary operation will simply be a composition of functions: we have $e(x) = x$, $(g_1 \circ g_2)(x) = g_1(g_2(x))$.

The group elements typically correspond to elements of a parameter space Θ, so that a transformation may be written as, say, g_θ. The family of densities of $g_\theta(X)$, for $g_\theta \in G$, is called a *(group) transformation family*.

Setting $x \approx x'$ if and only if there is a $g \in G$ such that $x = g(x')$ defines an equivalence relation, which partitions $\mathcal{X}$ into equivalence classes called *orbits*. These may be labelled by an index a, say. Two points x and x' on the same orbit have the same index, $a(x) = a(x')$. Each $x \in \mathcal{X}$ belongs to precisely one orbit, and might be represented by a (which identifies the orbit) and its position on the orbit.

5.2.1 Maximal invariant

We say that the statistic t is *invariant* to the action of the group G if its value does not depend on whether x or $g(x)$ was observed, for any $g \in G$: $t(x) = t(g(x))$. An example is the index a above.

The statistic t is *maximal invariant* if every other invariant statistic is a function of it, or equivalently, $t(x) = t(x')$ implies that $x' = g(x)$ for some $g \in G$. A maximal invariant can be thought of (Davison, 2003: Section 5.3) as a reduced version of the data that represent it as closely as possible, while remaining invariant to the action of G. In some sense, it is what remains of X once minimal information about the parameter values has been extracted.

5.2.2 Equivariant statistics and a maximal invariant

As described, typically there is a one-to-one correspondence between the elements of G and the parameter space Θ, and then the action of G on $\mathcal{X}$ requires that Θ itself constitutes a group, with binary operation $*$ say: we must have $g_\theta \circ g_\phi = g_{\theta * \phi}$. The group action on $\mathcal{X}$ induces a group action on Θ. If $\bar{G}$ denotes this induced group, then associated with each $g_\theta \in G$ there is a $\bar{g}_\theta \in \bar{G}$, satisfying $\bar{g}_\theta(\phi) = \theta * \phi$.

If t is an invariant statistic, the distribution of $t(X)$ is the same as that of $t(g(X))$, for all g. If, as we assume here, the elements of G are identified with parameter values, this means that the distribution of T does not depend on the parameter and is known in principle. T is said to be *distribution constant*.

A statistic $S = s(X)$ defined on $\mathcal{X}$ and taking values in the parameter space Θ is said to be *equivariant* if $s(g_\theta(x)) = \bar{g}_\theta(s(x))$ for all $g_\theta \in G$ and $x \in \mathcal{X}$. Often S is chosen to be an estimator of θ, and it is then called an *equivariant estimator*.

A key operational point is that an equivariant estimator can be used to construct a maximal invariant.

Consider $t(X) = g_{s(X)}^{-1}(X)$. This is invariant, since

$$t(g_\theta(x)) = g_{s(g_\theta(x))}^{-1}(g_\theta(x)) = g_{\bar{g}_\theta(s(x))}^{-1}(g_\theta(x)) = g_{\theta * s(x)}^{-1}(g_\theta(x))$$
$$= g_{s(x)}^{-1}\{g_\theta^{-1}(g_\theta(x))\} = g_{s(x)}^{-1}(x) = t(x).$$

If $t(x) = t(x')$, then $g_{s(x)}^{-1}(x) = g_{s(x')}^{-1}(x')$, and it follows that $x' = g_{s(x')} \circ g_{s(x)}^{-1}(x)$, which shows that $t(X)$ is a maximal invariant.

The statistical importance of a maximal invariant will be illuminated in Chapter 9. In a transformation family, a maximal invariant plays the role of the ancillary statistic in the conditional inference on the parameter of interest indicated by a Fisherian approach. The above direct construction of a maximal invariant from an equivariant estimator facilitates identification of an appropriate ancillary statistic in the transformation family context.

5.2.3 An example

An important example is the *location-scale model*. Let $X = \eta + \tau \epsilon$, where ϵ has a known density f, and the parameter $\theta = (\eta, \tau) \in \Theta = \mathbb{R} \times \mathbb{R}_+$. Define a group action by $g_\theta(x) = g_{(\eta, \tau)}(x) = \eta + \tau x$, so

$$g_{(\eta, \tau)} \circ g_{(\mu, \sigma)}(x) = \eta + \tau \mu + \tau \sigma x = g_{(\eta + \tau \mu, \tau \sigma)}(x).$$

The set of such transformations is closed with identity $g_{(0,1)}$. It is easy to check that $g_{(\eta, \tau)}$ has inverse $g_{(-\eta/\tau, \tau^{-1})}$. Hence, $G = \{g_{(\eta, \tau)} : (\eta, \tau) \in \mathbb{R} \times \mathbb{R}_+\}$ constitutes a group under the composition of functions operation $\circ$ defined above.

The action of $g_{(\eta, \tau)}$ on a random sample $X = (X_1, \ldots, X_n)$ is $g_{(\eta, \tau)}(X) = \eta + \tau X$, with $\eta \equiv \eta 1_n$, where 1_n denotes the $n \times 1$ vector of 1s, and X is written as an $n \times 1$ vector.

The induced group action on Θ is given by $\bar{g}_{(\eta, \tau)}((\mu, \sigma)) \equiv (\eta, \tau) * (\mu, \sigma) = (\eta + \tau \mu, \tau \sigma)$.

The sample mean and standard deviation are equivariant, because with $s(X) = (\bar{X}, V^{1/2})$, where $V = (n-1)^{-1}\sum(X_j - \bar{X})^2$, we have

$$s(g_{(\eta,\tau)}(X)) = \left(\overline{\eta + \tau X}, \left\{(n-1)^{-1}\sum(\eta + \tau X_j - \overline{(\eta + \tau X)})^2\right\}^{1/2}\right)$$

$$= \left(\eta + \tau \bar{X}, \left\{(n-1)^{-1}\sum(\eta + \tau X_j - \eta - \tau \bar{X})^2\right\}^{1/2}\right)$$

$$= \left(\eta + \tau \bar{X}, \tau V^{1/2}\right)$$

$$= \bar{g}_{(\eta,\tau)}(s(X)).$$

A maximal invariant is $A = g_{s(X)}^{-1}(X)$, and the parameter corresponding to $g_{s(X)}^{-1}$ is $(-\bar{X}/V^{1/2}, V^{-1/2})$. Hence a maximal invariant is the vector of residuals

$$A = (X - \bar{X})/V^{1/2} = \left(\frac{X_1 - \bar{X}}{V^{1/2}}, \dots, \frac{X_n - \bar{X}}{V^{1/2}}\right)^T,$$

called the *configuration*. It is easily checked directly that the distribution of A does not depend on θ. Any function of A is also invariant. The orbits are determined by different values a of the statistic A, and X has a unique representation as $X = g_{s(X)}(A) = \bar{X} + V^{1/2}A$.

5.3 Problems

5.1 Prove that random samples from the following distributions form (m, m) exponential families with either $m = 1$ or $m = 2$: Poisson, binomial, geometric, Gamma (index known), Gamma (index unknown). Identify the natural statistics and the natural parameters in each case. What are the distributions of the natural statistics?

The negative binomial distribution with both parameters unknown provides an example of a model that is not of exponential family form. Why?

(If Y has a Gamma distribution of known index k, its density function is of the form

$$f_Y(y; \lambda) = \frac{\lambda^k y^{k-1} e^{-\lambda y}}{\Gamma(k)}.$$

The gamma distribution with index unknown has both k and λ unknown.)

5.2 Let $Y_1, \dots, Y_n$ be independent, identically distributed $N(\mu, \mu^2)$. Show that this model is an example of a curved exponential family.

5.3 Find the general form of a conjugate prior density for θ in a Bayesian analysis of the one-parameter exponential family density

$$f(x; \theta) = c(\theta)h(x)\exp\{\theta t(x)\}, \quad x \in \mathbb{R}.$$

5.4 Verify that the family of gamma distributions of known index constitutes a transformation model under the action of the group of scale transformations. (This provides an example of a family of distributions which constitutes *both* an exponential family, *and* a transformation family. Are there any others?)

5.5 The *maximum likelihood estimator* $\widehat{\theta}(x)$ of a parameter θ maximises the *likelihood function* $L(\theta) = f(x; \theta)$ with respect to θ. Verify that maximum likelihood estimators are equivariant with respect to the group of one-to-one transformations. (Maximum likelihood estimation is the subject of Chapter 8.)

5.6 Verify directly that in the location-scale model the configuration has a distribution which does not depend on the parameters.

Sufficiency and completeness

This chapter is concerned primarily with point estimation of a parameter θ. For many parametric problems, including in particular problems with exponential families, it is possible to summarise all the information about θ contained in a random variable X by a function $T = T(X)$, which is called a *sufficient statistic*. The implication is that any reasonable estimator of θ will be a function of $T(X)$. However, there are many possible sufficient statistics – we would like to use the one which summarises the information as efficiently as possible. This is called the *minimal sufficient* statistic, which is essentially unique. *Completeness* is a technical property of a sufficient statistic. A sufficient statistic, which is also complete, must be minimal sufficient (the Lehmann–Scheffé Theorem). Another feature of a complete sufficient statistic T is that, if some function of T is an unbiased estimator of θ, then it must be the *unique* unbiased estimator which is a function of a sufficient statistic. The final section of the chapter demonstrates that, when the loss function is convex (including, in particular, the case of squared error loss function), there is a best unbiased estimator, which is a function of the sufficient statistic, and that, if the sufficient statistic is also complete, this estimator is unique. In the case of squared error loss this is equivalent to the celebrated *Rao–Blackwell Theorem* on the existence of minimum variance unbiased estimators.

6.1 Definitions and elementary properties

6.1.1 Likelihood

Suppose X is a random variable with density (or probability mass function) $f(x;\theta)$ depending on a finite-dimensional parameter $\theta \in \Theta$.

The function $L_x(\theta) = f(x;\theta)$, viewed as a function of θ for fixed x, is called the *likelihood function* of the parameter θ based on observed data $X = x$.

The function

$$\Lambda_x(\theta_1, \theta_2) = \frac{f(x;\theta_1)}{f(x;\theta_2)}$$

is the *likelihood ratio* for one parameter value $\theta = \theta_1$ relative to another $\theta = \theta_2$.

Lemma 6.1 *Let $t(x)$ denote some function of x. Then the following are equivalent:*

(i) There exist functions $h(x)$ and $g(t;\theta)$ such that

$$f(x;\theta) = h(x)g(t(x);\theta). \tag{6.1}$$

(ii) For any pair x, x' such that $t(x) = t(x')$,

$$\Lambda_x(\theta_1, \theta_2) = \Lambda_{x'}(\theta_1, \theta_2) \quad \text{for all } \theta_1, \theta_2. \tag{6.2}$$

Proof　That (i) implies (ii) is obvious.

Conversely, suppose (ii) holds. Fix some reference value θ_0. For any θ,

$$\frac{f(x; \theta)}{f(x; \theta_0)} = \Lambda_x(\theta, \theta_0) = g^*(t(x), \theta, \theta_0) \text{ say.}$$

Then

$$f(x; \theta) = f(x; \theta_0) g^*(t(x), \theta, \theta_0).$$

Then write $h(x) = f(x; \theta_0)$ and $g(t; \theta) = g^*(t, \theta, \theta_0)$ to get (i).　　□

6.1.2 Sufficiency

The statistic $T = T(X)$ is *sufficient* for θ if the distribution of X, conditional on $T(X) = t$, is independent of θ.

Note that $T(X)$ and θ may both be vectors, not necessarily of the same dimension.

Two criteria for sufficiency are:

(a) *Factorisation Theorem*　$T(X)$ is sufficient for θ if and only if (6.1) holds.
(b) *Likelihood Ratio Criterion*　$T(X)$ is sufficient for θ if and only if (6.2) holds.

Criterion (a) is straightforward to prove, at least in the discrete case. (In the continuous case, measure-theoretic conditions are required to define the conditional distribution of X given $T(X) = t$). If $T(X)$ is a sufficient statistic, the conditional distribution of X given T is free of θ:

$$f_{X|T}(x|t) \equiv f_{X,T}(x, t; \theta) / f_T(t; \theta) \tag{6.3}$$

is free of θ. But T is a function $t(X)$ of X, so the joint density of X and T in the numerator of (6.3) is zero except where $T = t(X)$, so the numerator is just $f_X(x; \theta)$. A factorisation of the form (6.1) then holds with g the density of T and h the conditional density of X given T. Conversely, if (6.1) holds, we can find the density of T at t by integrating (6.1) over x for which $t(x) = t$. In the discrete case, this involves summing over those x for which $t(x) = t$. This gives the density of T of the form $f_T(t; \theta) = g(t; \theta) \sum h(x)$, and therefore the conditional density of X given T is of the form

$$\frac{f_X(x; \theta)}{f_T(t; \theta)} = \frac{g(t(x); \theta) h(x)}{g(t; \theta) \sum h(x)} = \frac{h(x)}{\sum h(x)},$$

free of θ, so that T is sufficient.

Criterion (b) follows at once from Lemma 6.1.

6.1.3 Minimal sufficiency

As defined so far, sufficiency is too broad a concept, because a sufficient statistic may still contain superfluous information.

For example, if T is sufficient, so is (T, T^*), where T^* is any other function of X.

As another example, suppose X is a single observation from $N(0, \theta)$. Here X is sufficient, but so is $|X|$ – the sign of X is irrelevant to any inference about θ. This shows that it is not merely a matter of finding the sufficient statistic of the smallest dimension – here X and $|X|$ are both of dimension 1, but $|X|$ summarises the information more efficiently than X.

A sufficient statistic T is *minimal sufficient* if it is a function of every other sufficient statistic.

The immediate obvious questions are: does a minimal sufficient statistic exist, and is it unique?

In a strict sense the answer to the uniqueness question is no, because, if T is minimal sufficient, so is $2T$ or $42 - T$ or any other injective (one-to-one) function of T. However, if we allow that two statistics are equivalent if they are injective functions of each other, then the minimal sufficient statistic is unique.

Lemma 6.2 *If T and S are minimal sufficient statistics, then there exist injective functions g_1 and g_2 such that $T = g_1(S)$ and $S = g_2(T)$.*

Proof   The definition of minimal sufficiency implies that there must exist *some* functions g_1 and g_2 such that $T = g_1(S)$, $S = g_2(T)$. The task is to prove that g_1 and g_2 are injective on the ranges of S and T respectively.

Suppose x and x' are such that $g_1(S(x)) = g_1(S(x'))$. Then $T(x) = T(x')$ so $S(x) = g_2(T(x)) = g_2(T(x')) = S(x')$. This proves that g_1 is injective, and the same argument works for g_2. This completes the proof. $\square$

Now we give a condition for determining whether a sufficient statistic T is minimal sufficient.

Theorem 6.1 *A necessary and sufficient condition for a statistic $T(X)$ to be minimal sufficient is that*

$$T(x) = T(x') \text{ if and only if } \Lambda_x(\theta_1, \theta_2) = \Lambda_{x'}(\theta_1, \theta_2) \text{ for all } \theta_1, \theta_2. \tag{6.4}$$

Proof

(i) Suppose T satisfies (6.4) and S is sufficient. If T is not a function of S, then there must exist two values x, x' for which $S(x) = S(x')$ but $T(x) \neq T(x')$. By (6.2) applied to S, $\Lambda_x(\theta_1, \theta_2) = \Lambda_{x'}(\theta_1, \theta_2)$ for all θ_1 and θ_2. But (6.4) then implies that $T(x) = T(x')$, a contradiction. Therefore T is a function of S. But S was any sufficient statistic, that is T is a function of every sufficient statistic, so T is minimal sufficient.

(ii) Suppose T is minimal sufficient. We must show (6.4).

Suppose first that x, x' are such that $T(x) = T(x')$. Then $\Lambda_x(\theta_1, \theta_2) = \Lambda_{x'}(\theta_1, \theta_2)$ by (6.2) and the fact that T is sufficient. This proves one direction of the implication in (6.4).

Conversely suppose x' and x'' are such that $\Lambda_{x'}(\theta_1, \theta_2) = \Lambda_{x''}(\theta_1, \theta_2)$ for all θ_1 and θ_2, but $T(x') \neq T(x'')$. Define a new statistic S by $S(x) = T(x)$ except for those x where $T(x) = T(x'')$, where we set $S(x) = T(x')$. To see that S is sufficient, by (6.2), we must show that, whenever x and x^* are such that $S(x) = S(x^*)$, then $\Lambda_x = \Lambda_{x^*}$.

There is no problem if $T(x) = T(x^*)$, so we need only consider the case when $S(x) = S(x^*)$ but $T(x) \neq T(x^*)$. However this can only happen if one of $T(x)$ or $T(x^*)$ is equal to $T(x')$ and the other to $T(x'')$. Suppose $T(x) = T(x')$ and $T(x^*) = T(x'')$. In that case $\Lambda_x = \Lambda_{x'}$ (sufficiency of T), $\Lambda_{x^*} = \Lambda_{x''}$ (sufficiency of T), $\Lambda_{x'} = \Lambda_{x''}$ (by assumption). But then $\Lambda_x = \Lambda_{x^*}$, as required. Therefore S is sufficient. But S is a function of T, and T is minimal sufficient, therefore S must also be minimal sufficient and the relationship that takes T into S is injective. But, if the map is injective, then $S(x') = S(x'')$ implies $T(x') = T(x'')$. This contradiction establishes the reverse implication in (6.4), and completes the proof of the theorem. $\qquad\square$

Remark The representation (6.4) also establishes the following equivalence relation: $x \equiv x'$ if and only if x and x' define the same likelihood ratio. This partitions the sample space $\mathcal{X}$ into equivalence classes, and $T(x)$ may be any function which assigns a unique value to each equivalence class. In particular, this shows the *existence* of a minimal sufficient statistic. A completely rigorous proof of this statement requires some measure-theoretic technicalities, but these were established in a classical paper by Lehmann and Scheffé (1950) for the case of probability densities defined with respect to a single dominating measure on a finite-dimensional Euclidean space. Without such conditions, there are counterexamples to the existence of minimal sufficient statistics.

6.1.4 Examples

Example 6.1 Consider $X_1, \ldots, X_n$ independent, identically distributed from $N(\mu, \sigma^2)$. Then

$$f(x; \mu, \sigma^2) = (2\pi\sigma^2)^{-n/2} \exp\left\{ -\frac{\sum x_i^2}{2\sigma^2} + \frac{\mu \sum x_i}{\sigma^2} - \frac{n\mu^2}{2\sigma^2} \right\}.$$

Thus

$$\frac{f(x; \mu_1, \sigma_1^2)}{f(x; \mu_2, \sigma_2^2)} = b(\mu_1, \sigma_1^2, \mu_2, \sigma_2^2) \exp\left\{ \sum x_i^2 \left(\frac{1}{2\sigma_2^2} - \frac{1}{2\sigma_1^2} \right) + \sum x_i \left(\frac{\mu_1}{\sigma_1^2} - \frac{\mu_2}{\sigma_2^2} \right) \right\}.$$
$$(6.5)$$

We show that the minimal sufficient statistic is $T(x) = (\sum x_i, \sum x_i^2)$, or equivalently $(\bar{x}, \sum(x_i - \bar{x})^2)$, using the necessary and sufficient condition (6.4).

Suppose first that $x = (x_1, \ldots, x_n)$ and $x' = (x_1', \ldots, x_n')$ are two samples for which $T(x) = T(x')$. Then, writing $\theta_1 = (\mu_1, \sigma_1^2)$, $\theta_2 = (\mu_2, \sigma_2^2)$, we have, immediately from (6.5), that $\Lambda_x(\theta_1, \theta_2) = \Lambda_{x'}(\theta_1, \theta_2)$ for all θ_1, θ_2.

Verification of (6.4) is completed by confirming the reverse implication, which we do by showing that $T(x) \neq T(x')$ implies that there are *some* θ_1, θ_2 for which $\Lambda_x(\theta_1, \theta_2) \neq \Lambda_{x'}(\theta_1, \theta_2)$.

Suppose that $(\sum x_i, \sum x_i^2) \neq (\sum x_i', \sum (x_i')^2)$. Then again, we can certainly find a pair (α, β) for which

$$\alpha \sum x_i + \beta \sum x_i^2 \neq \alpha \sum x_i' + \beta \sum (x_i')^2.$$

Choose $\theta_1 = (\mu_1, \sigma_1^2)$ and $\theta_2 = (\mu_2, \sigma_2^2)$ such that

$$\left(\frac{\mu_1}{\sigma_1^2} - \frac{\mu_2}{\sigma_2^2}\right) = \alpha,$$

$$\left(\frac{1}{2\sigma_2^2} - \frac{1}{2\sigma_1^2}\right) = \beta.$$

Then, by (6.5), $\Lambda_x(\theta_1, \theta_2) \neq \Lambda_{x'}(\theta_1, \theta_2)$ for this (θ_1, θ_2). We have now verifed both impli-
cations in (6.4), and therefore that $T(x)$ is minimal sufficient.

If σ^2 is known, then $\bar{x}$ is minimal sufficient for μ.

If μ is known, then $\sum(x_i - \mu)^2$ is minimal sufficient for σ^2. In particular, in our earlier
example with $n = 1$, $\mu = 0$, we have that X_1^2 is minimal sufficient, which is equivalent to
$|X_1|$, that is this cannot be reduced any further.

Example 6.2 The reasoning in Example 6.1 is not specific to the normal distribution but applies whenever we have a full exponential family. Specifically, if we have an exponential family in its natural parametrisation with natural statistics $(T_1, \ldots, T_k) = (t_1(X), \ldots, t_k(X))$,

$$f(x; \theta) = c(\theta)h(x) \exp\left\{\sum_{i=1}^{k} \theta_i t_i(x)\right\},$$

then $(T_1, \ldots, T_k)$ is sufficient for θ by the factorisation theorem. If, in addition, Θ contains
an open rectangle in $\mathbb{R}^k$, then T is minimal sufficient – this follows by the same argument
as just given for the normal distribution.

A further property is that, if $S \subseteq \{1, \ldots, k\}$ and $\{\theta_i, \ i \in S\}$ are known, then $\{T_i, \ i \notin S\}$
are sufficient for $\{\theta_i, \ i \notin S\}$. We proved in Lemma 5.2 of Chapter 5 that the conditional
distribution of $\{T_i, \ i \in S\}$ given $\{T_i, \ i \notin S\}$ does not depend on $\{\theta_i, \ i \notin S\}$. A simple
adaptation of the proof shows that the conditional distribution of X given $\{T_i, \ i \notin S\}$ does
not depend on $\{\theta_i, \ i \notin S\}$.

6.2 Completeness

A sufficient statistic $T(X)$ is *complete* if for any real function g,

$$\mathbb{E}_\theta\{g(T)\} = 0 \text{ for all } \theta$$

implies

$$\Pr_\theta\{g(T) = 0\} = 1 \text{ for all } \theta.$$

This definition has a number of consequences. For instance, if there exists an unbiased
estimator of a scalar parameter θ, which is a function of a complete sufficient statistic T,
then it is the unique such estimator (except possibly on a set of measure 0). This follows because, if, for instance, $g_1(T)$ and $g_2(T)$ are two such estimators, then $\mathbb{E}_\theta\{g_1(T) - g_2(T)\} =
\theta - \theta = 0$, so $g_1(T) = g_2(T)$ with probability 1.

Lemma 6.3 *If $T = (T_1, \ldots, T_k)$ is the natural statistic for a full exponential family
in its natural parametrisation, and if Θ contains an open rectangle in $\mathbb{R}^k$, then T is
complete.*

Proof This follows from the first of the technical results given in Section 5.1.4. □

Example 6.3 Suppose the density of the statistic T satisfies

$$f(t; \theta) = \begin{cases} h(t)c(\theta) & \text{if } 0 \le t \le \theta, \\ 0 & \text{if } t > \theta, \end{cases} \tag{6.6}$$

with $h(t) \ne 0$. For example, the uniform density is of this form. If $X_1, \ldots, X_n$ are independent, identically distributed from the uniform density on $(0, \theta)$, then $T = \max\{X_1, \ldots, X_n\}$ is sufficient for θ (proof of this: use the factorisation theorem) and has density $nt^{n-1}\theta^{-n}$, $0 < t < \theta$.

If (6.6) holds and g is any function of T, we have

$$\mathbb{E}_\theta\{g(T)\} = c(\theta) \int_0^\theta h(t)g(t)dt.$$

If this is $\equiv 0$, then $\int_0^\theta h(t)g(t)dt = 0$ for all θ. Differentiating with respect to θ, $h(\theta)g(\theta) = 0$. But we assumed h was not 0, therefore g must be with probability 1.

This is an example of a non-exponential family for which a complete sufficient statistic exists. It shows, for instance, that $(n + 1)T/n$, which we can easily check to be an unbiased estimator of θ, is the unique unbiased estimator which is a function of T. Without completeness, we might speculate whether there was some complicated nonlinear function of T which gave a better unbiased estimator, but completeness shows that this is impossible.

6.3 The Lehmann–Scheffé Theorem

Theorem 6.2 *Suppose X has density $f(x; \theta)$ and $T(X)$ is sufficient and complete for θ. Then T is minimal sufficient.*

Proof We know from the Remark following Theorem 6.1 that there exists a minimal sufficient statistic. By Lemma 6.2 this is unique up to one-to-one transformations, so call this S. Then $S = g_1(T)$ for some function g_1.

Define $g_2(S) = \mathbb{E}\{T|S\}$. This does not depend on θ, because S is sufficient. Now consider

$$g(T) = T - g_2(S) = T - g_2(g_1(T)).$$

By the iterated expectation formula,

$$\mathbb{E}_\theta\{g(T)\} = \mathbb{E}_\theta\{T\} - \mathbb{E}_\theta\{\mathbb{E}(T|S)\} = \mathbb{E}_\theta\{T\} - \mathbb{E}_\theta\{T\} = 0.$$

So, by completeness of T, $g_2(S) = T$ with probability 1. But then S is a function of T, and T is a function of S – by the argument of Lemma 6.2 both functions must be injective, and so T and S are equivalent. In other words, T is also minimal sufficient. □

6.4 Estimation with convex loss functions

Jensen's inequality is a well-known result that is proved in elementary analysis texts. It states that, if $g : \mathbb{R} \to \mathbb{R}$ is a convex function (so that $g(\lambda x_1 + (1 - \lambda)x_2) \le \lambda g(x_1) + (1 - \lambda)g(x_2)$ for all x_1, x_2 and $0 < \lambda < 1$) and X is a real-valued random variable, then $\mathbb{E}\{g(X)\} \ge g\{\mathbb{E}(X)\}$.

Theorem 6.3 *Suppose we want to estimate a real-valued parameter θ with an estimator $d(X)$ say. Suppose the loss function $L(\theta, d)$ is a convex function of d for each θ. Let $d_1(X)$ be an unbiased estimator for θ and suppose T is a sufficient statistic. Then the estimator*

$$\chi(T) = \mathbb{E}\{d_1(X)|T\}$$

is also unbiased and is at least as good as d_1.

Note that the definition of $\chi(T)$ does not depend on θ, because T is sufficient.

Proof   That $\chi(T)$ is unbiased follows from the iterated expectation formula,

$$\mathbb{E}_\theta \chi(T) = \mathbb{E}_\theta\{\mathbb{E}(d_1(X)|T)\} = \mathbb{E}_\theta d_1(X) = \theta.$$

For the risk function, we have

$$
\begin{aligned}
R(\theta, d_1) &= \mathbb{E}_\theta\{L(\theta, d_1(X))\} \\
&= \mathbb{E}_\theta[\mathbb{E}\{L(\theta, d_1(X))|T\}] \\
&\geq \mathbb{E}_\theta\{L(\theta, \chi(T))\} \quad \text{(Jensen)} \\
&= R(\theta, \chi).
\end{aligned}
$$

This is true for all θ, hence χ is as good as d_1, as claimed. This completes the proof of the theorem.   □

Remark 1 The inequality above will be strict unless L is a linear function of d, or the conditional distribution of $d_1(X)$ given T is degenerate. In all other cases, $\chi(T)$ strictly dominates $d_1(X)$.

Remark 2 If T is also complete, then $\chi(T)$ is the *unique* unbiased estimator minimising the risk.

Remark 3 If $L(\theta, d) = (\theta - d)^2$, then this is the Rao–Blackwell Theorem. In this case the risk of an unbiased estimator is just its variance, so the theorem asserts that there is a unique minimum variance unbiased estimator which is a function of the complete sufficient statistic. However, it is still possible that there are biased estimators which achieve a smaller mean squared error: the example of a minimax estimator of the parameter of a binomial distribution given in Chapter 3 is one such, and Stein's paradox example is another.

6.5 Problems

6.1 Let $X_1, \ldots, X_n$ be independent, identically distributed $N(\mu, \mu^2)$ random variables. Find a minimal sufficient statistic for μ and show that it is not complete.

6.2 Find a minimal sufficient statistic for θ based on an independent sample of size n from each of the following distributions:

(i) the gamma distribution with density

$$f(x; \alpha, \beta) = \frac{\beta^\alpha x^{\alpha-1} e^{-\beta x}}{\Gamma(\alpha)}, \ x > 0,$$

with $\theta = (\alpha, \beta)$;

(ii) the uniform distribution on $(\theta - 1, \theta + 1)$;

(iii) the Cauchy distribution with density

$$f(x; a, b) = \frac{b}{\pi\{(x - a)^2 + b^2\}}, \quad x \in \mathbb{R},$$

with $\theta = (a, b)$.

6.3 Independent factory-produced items are packed in boxes each containing k items. The probability that an item is in working order is $\theta, 0 < \theta < 1$. A sample of n boxes are chosen for testing, and X_i, the number of working items in the ith box, is noted. Thus $X_1, \ldots, X_n$ are a sample from a binomial distribution, $\text{Bin}(k, \theta)$, with index k and parameter θ. It is required to estimate the probability, θ^k, that all items in a box are in working order. Find the minimum variance unbiased estimator, justifying your answer.

6.4 A married man who frequently talks on his mobile is well known to have conversations the lengths of which are independent, identically distributed random variables, distributed as exponential with mean $1/\lambda$. His wife has long been irritated by his behaviour and knows, from infinitely many observations, the exact value of λ. In an argument with her husband, the woman produces $t_1, \ldots, t_n$, the times of n telephone conversations, to prove how excessive her husband is. He suspects that she has randomly chosen the observations, conditional on their all being longer than the expected length of conversation. Assuming he is right in his suspicion, the husband wants to use the data he has been given to infer the value of λ. What is the minimal sufficient statistic he should use? Is it complete? Find the maximum likelihood estimator for λ.

Two-sided tests and conditional inference

This chapter is concerned with two separate but interrelated themes. The first has to do with extending the discussion of Chapter 4 to more complicated hypothesis testing problems, and the second is concerned with conditional inference.

We will consider first testing two-sided hypotheses of the form $H_0 : \theta \in [\theta_1, \theta_2]$ (with $\theta_1 < \theta_2$) or $H_0 : \theta = \theta_0$ where, in each case, the alternative H_1 includes all θ not part of H_0. For such problems we cannot expect to find a uniformly most powerful test in the sense of Chapter 4. However, by introducing an additional concept of *unbiasedness* (Section 7.1), we are able to define a family of *uniformly most powerful unbiased*, or UMPU, tests. In general, characterising UMPU tests for two-sided problems is a much harder task than characterising UMP tests for one-sided hypotheses, but for one specific but important example, that of a one-parameter exponential family, we are able to find UMPU tests. The details of this are the subject of Section 7.1.2.

The extension to multiparameter exponential families involves the notion of *conditional tests*, discussed in Section 7.2. In some situations, a statistical problem may be greatly simplified by working not with the unconditional distribution of a test statistic, but the conditional distribution given some other statistic. We discuss two situations where conditional tests naturally arise, one when there are *ancillary statistics*, and the other where conditional procedures are used to construct *similar tests*. The basic idea behind an ancillary statistic is that of a quantity with distribution not depending on the parameter of interest. The Fisherian paradigm then argues that *relevance* to the data at hand demands conditioning on the observed value of this statistic. The notion behind similarity is that of eliminating dependence on nuisance parameters. Having introduced the general concepts, we then specialise to the case of a multiparameter exponential family in which one particular parameter is of interest, while the remaining $k - 1$ are regarded as nuisance parameters. This method is also relevant in testing hypotheses about some linear combination of the natural parameters of an exponential family. In Section 7.3 we discuss *confidence sets*. A particular focus of the discussion is the notion of a duality between hypothesis tests and confidence sets. Based on data x, we may construct a confidence set of coverage level $1 - \alpha$ for a parameter θ of interest as the set of values θ_0 which would be *accepted* in an appropriate hypothesis test of $H_0 : \theta = \theta_0$ against $H_1 : \theta \neq \theta_0$ based on x. Good confidence sets may then be related directly to optimal hypothesis tests, as considered in the current chapter and Chapter 4.

7.1 Two-sided hypotheses and two-sided tests

We consider a general situation with a one-dimensional parameter $\theta \in \Theta \subseteq \mathbb{R}$. We are particularly interested in the case when the null hypothesis is $H_0 : \theta \in \Theta_0$, where Θ_0 is either the interval $[\theta_1, \theta_2]$ for some $\theta_1 < \theta_2$, or else the single point $\Theta_0 = \{\theta_0\}$, and $\Theta_1 = \mathbb{R} \setminus \Theta_0$.

In this situation, we cannot in general expect to find a UMP test, even for nice families, such as the monotone likelihood ratio (Section 4.3.1) or exponential family models (Section 5.1), etc. The reason is obvious: if we construct a Neyman–Pearson test of say $\theta = \theta_0$ against $\theta = \theta_1$ for some $\theta_1 \neq \theta_0$, the test takes quite a different form when $\theta_1 > \theta_0$ from when $\theta_1 < \theta_0$. We simply cannot expect one test to be most powerful in both cases simultaneously.

However, if we have an exponential family with natural statistic $T = t(X)$, or a family with MLR with respect to $t(X)$, we might still expect tests of the form

$$\phi(x) = \begin{cases} 1 & \text{if } t(x) > t_2 \text{ or } t(x) < t_1, \\ \gamma(x) & \text{if } t(x) = t_2 \text{ or } t(x) = t_1, \\ 0 & \text{if } t_1 < t(x) < t_2, \end{cases}$$

where $t_1 < t_2$ and $0 \leq \gamma(x) \leq 1$, to have good properties. Such tests are called *two-sided tests* and much of our purpose here is to investigate when two-sided tests are optimal in some sense.

7.1.1 Unbiased tests

Definition *A test ϕ of $H_0 : \theta \in \Theta_0$ against $H_1 : \theta \in \Theta_1$ is called* unbiased of size α if

$$\sup_{\theta \in \Theta_0} \mathbb{E}_\theta \{\phi(X)\} \leq \alpha$$

and

$$\mathbb{E}_\theta \{\phi(X)\} \geq \alpha \text{ for all } \theta \in \Theta_1.$$

An unbiased test captures the natural idea that the probability of rejecting H_0 should be higher when H_0 is false than when it is true.

Definition *A test which is uniformly most powerful amongst the class of all unbiased tests is called* uniformly most powerful unbiased, *abbreviated UMPU.*

The idea is illustrated by Figure 7.1, for the case $H_0 : \theta = \theta_0$ against $H_1 : \theta \neq \theta_0$. (In the figure, $\theta_0 = 0$.) The optimal UMP tests for the alternatives $H_1 : \theta > \theta_0$ and $H_1 : \theta < \theta_0$ each fails miserably to be unbiased, but there is a two-sided test whose power function is given by the dotted curve, and we may hope that such a test will be UMPU.

The requirement that a test be unbiased is one way of resolving the obvious conflict between the two sides of a two-sided alternative hypothesis. We shall use it in the remainder of this chapter as a criterion by which to assess two-sided tests. Nevertheless the objections to unbiasedness that we have noted in earlier chapters are still present – unbiasedness is not by itself an optimality criterion and, for any particular decision problem, there is no reason

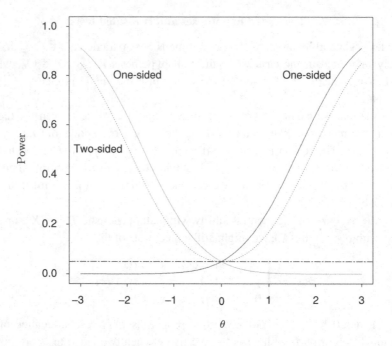

Figure 7.1 Power functions for one-sided and two-sided tests

why the optimal decision procedure should turn out to be unbiased. The principal role of unbiasedness is to restrict the class of possible decision procedures and hence to make the problem of determining an optimal procedure more manageable than would otherwise be the case.

7.1.2 UMPU tests for one-parameter exponential families

We shall not attempt to give general conditions for a two-sided test to be UMPU. However, in one particularly important special case, that of a one-parameter exponential family, we can do this, and the details of that are given here.

Consider an exponential family for a random variable X, which may as in previous examples be a vector of independent, identically distributed observations, with real-valued parameter $\theta \in \mathbb{R}$ and density of form

$$f(x; \theta) = c(\theta)h(x)e^{\theta t(x)},$$

where $T = t(X)$ is a real-valued natural statistic.

As we saw in Lemma 5.1 of Chapter 5, this implies that T also has an exponential family distribution, with density of form

$$f_T(t; \theta) = c(\theta)h_T(t)e^{\theta t}.$$

We shall assume that T is a continuous random variable with $h_T(t) > 0$ on the open set which defines the range of T. By restricting ourselves to families of this form we avoid the need for randomised tests and make it easy to prove the existence and uniqueness of two-sided tests, though in a more general version of the theory such assumptions are not required: see, for example Ferguson (1967).

We consider initially the case

$$\Theta_0 = [\theta_1, \theta_2], \quad \Theta_1 = (-\infty, \theta_1) \cup (\theta_2, \infty),$$

where $\theta_1 < \theta_2$.

Theorem 7.1 *Let ϕ be any test function. Then there exists a unique two-sided test ϕ' which is a function of T such that*

$$\mathbb{E}_{\theta_j} \phi'(X) = \mathbb{E}_{\theta_j} \phi(X), \quad j = 1, 2.$$

Moreover,

$$\mathbb{E}_\theta \phi'(X) - \mathbb{E}_\theta \phi(X) \begin{cases} \leq 0 & \text{for } \theta_1 < \theta < \theta_2, \\ \geq 0 & \text{for } \theta < \theta_1 \text{ or } \theta > \theta_2. \end{cases} \quad (7.1)$$

Corollary *For any $\alpha > 0$, there exists a UMPU test of size α, which is of two-sided form in T.*

Proof of Corollary (assuming Theorem 7.1) Consider the trivial test $\phi(x) = \alpha$ for all x. This has power α for all θ, so, by Theorem 7.1, there exists a unique two-sided test ϕ' for which

$$\mathbb{E}_\theta \{\phi'(X)\} \begin{cases} = \alpha & \text{if } \theta = \theta_1 \text{ or } \theta = \theta_2, \\ \leq \alpha & \text{if } \theta_1 < \theta < \theta_2, \\ \geq \alpha & \text{if } \theta < \theta_1 \text{ or } \theta > \theta_2. \end{cases}$$

Now suppose ϕ is *any* unbiased test of size α. By the second result of Section 5.1.4, the power function of ϕ is a continuous function of θ, so we must have $\mathbb{E}_\theta \{\phi(X)\} = \alpha$ when $\theta = \theta_1$ or θ_2. By Theorem 7.1, (7.1) holds for this ϕ and our two-sided test ϕ'. This establishes that ϕ' is UMP within the class of unbiased tests, which is what is required. $\square$

We now turn to the proof of Theorem 7.1 itself. First we note two preliminary results.

Lemma 7.1 (the generalised Neyman–Pearson Theorem) *Consider the test*

$$\phi'(x) = \begin{cases} 1 & \text{if } f_0(x) > k_1 f_1(x) + \cdots + k_m f_m(x), \\ \gamma(x) & \text{if } f_0(x) = k_1 f_1(x) + \cdots + k_m f_m(x), \\ 0 & \text{if } f_0(x) < k_1 f_1(x) + \cdots + k_m f_m(x), \end{cases}$$

where $f_0, \ldots, f_m$ are $m + 1$ possible densities for the random variable X, $k_1, \ldots, k_m$ are arbitrary constants (positive or negative) and $0 \leq \gamma(x) \leq 1$.

Then ϕ' maximises $\int \phi(x) f_0(x) dx$ among all tests ϕ for which $\int \phi(x) f_j(x) dx = \alpha_j$ for $j = 1, \ldots, m$, where $\alpha_1, \ldots, \alpha_m$ are prescribed constants.

The proof is omitted, being an elementary extension of the ordinary Neyman–Pearson Theorem from Chapter 4.

Lemma 7.2 *Let $\theta_a < \theta_b < \theta_c$ and consider the set*

$$S(K_1, K_2) = \{x \in \mathbb{R} : K_1 e^{\theta_a x} + K_2 e^{\theta_c x} > e^{\theta_b x}\}.$$

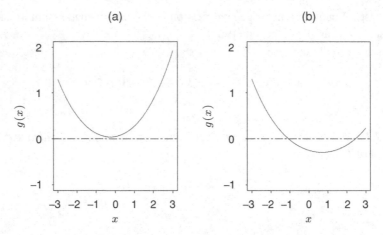

Figure 7.2 Plots of function $g(x)$, showing no zeros or two zeros

Then:

(i) *For any $K_1 > 0$, $K_2 > 0$, the set $S(K_1, K_2)$ is either the whole of $\mathbb{R}$ or else a union of two intervals of the form $(-\infty, x_1) \cup (x_2, \infty)$ for some x_1, x_2 such that $-\infty < x_1 < x_2 < \infty$.*

(ii) *Given x_1, x_2, there exists $K_1 > 0$, $K_2 > 0$ such that $S(K_1, K_2) = (-\infty, x_1) \cup (x_2, \infty)$.*

Proof

(i) Let $g(x) = K_1 \exp\{(\theta_a - \theta_b)x\} + K_2 \exp\{(\theta_c - \theta_b)x\} - 1$. Then g is a convex function, and $g \to +\infty$ as $x \to \pm\infty$. Thus g has no zeros or two (see Figure 7.2). Of course, in the latter case the two zeros may coincide (so that there is a multiple root of $g(x) = 0$), but then the set $\{x : g(x) \geq 0\}$ is the whole of $\mathbb{R}$.

Thus the set $\{x : g(x) > 0\}$ consists either of the whole of $\mathbb{R}$ or of two semi-infinite intervals, as claimed.

(ii) Fix x_1, x_2 and suppose $-\infty < x_1 < x_2 < \infty$. To make $S(K_1, K_2)$ have the required form, we need to solve the equations

$$K_1 e^{\theta_a x_j} + K_2 e^{\theta_c x_j} = e^{\theta_b x_j}, \quad j = 1, 2.$$

This is a pair of simultaneous linear equations for K_1, K_2, which we may solve directly to obtain

$$K_1 = \frac{\exp(\theta_b x_1 + \theta_c x_1)}{\exp(\theta_a x_1 + \theta_c x_1)} \cdot \frac{\exp\{\theta_c(x_2 - x_1)\} - \exp\{\theta_b(x_2 - x_1)\}}{\exp\{\theta_c(x_2 - x_1)\} - \exp\{\theta_a(x_2 - x_1)\}} > 0,$$

and similarly for K_2. This completes the proof.   □

Remark The cases $x_1 = -\infty$ or $x_2 = \infty$ are also covered by the result if we allow K_1 or K_2 to be 0; $x_1 = -\infty$ corresponds to $K_1 = 0$ and $x_2 = \infty$ corresponds to $K_2 = 0$.

Proof of Theorem 7.1* First we prove the existence of a two-sided test ϕ', depending on the natural statistic $T = t(X)$, for which $\mathbb{E}_\theta\{\phi'(X)\} = \mathbb{E}_\theta\{\phi(X)\}$ when $\theta = \theta_1$ or θ_2. We also write α_j for $\mathbb{E}_{\theta_j}\{\phi(X)\}$, $j = 1, 2$.

Let ϕ_w denote a one-sided test of the form

$$\phi_w(x) = \begin{cases} 1 & \text{if } t(x) \le t_w, \\ 0 & \text{if } t(x) > t_w, \end{cases}$$

where t_w is chosen so that $\mathbb{E}_{\theta_1}\{\phi_w(X)\} = w$. Because we have assumed that T has a positive density, it is easy to check that t_w is uniquely determined by w and that the map $w \to t_w$ is continuous.

Next consider the two-sided test

$$\phi_w'(x) = \phi_w(x) + 1 - \phi_{1-\alpha_1+w}(x),$$

where $0 \le w \le \alpha_1$. We see at once that

$$\mathbb{E}_{\theta_1}\{\phi_w'(X)\} = \alpha_1.$$

Consider $w = 0$. Then ϕ_w' is a Neyman–Pearson test for $H_0 : \theta = \theta_1$ against $H_1 : \theta = \theta_2$, and so maximises the power at θ_2 among all tests of fixed size. Therefore,

$$\mathbb{E}_{\theta_2}\{\phi_0'(X)\} \ge \alpha_2.$$

Now consider $w = \alpha_1$. In this case $1 - \phi_w'$ is a Neyman–Pearson test for θ_1 against θ_2, so ϕ_w' itself minimises the power at θ_2 for fixed size at θ_1. Thus

$$\mathbb{E}_{\theta_2}\{\phi_{\alpha_1}'(X)\} \le \alpha_2.$$

But $\mathbb{E}_{\theta_2}\{\phi_w'(X)\}$ is a continuous and strictly decreasing function of w, so there must be a unique w for which $\mathbb{E}_{\theta_2}\{\phi_w'(X)\} = \alpha_2$. This proves the existence of a (unique) two-sided test, which achieves the desired power at θ_1 and θ_2. Henceforth we drop the suffix w and denote the test as ϕ'.

We must now show (7.1). Suppose first $\theta < \theta_1$. Applying Lemma 7.2 with $(\theta_a, \theta_b, \theta_c) = (\theta, \theta_1, \theta_2)$, we find that there exist $K_1 > 0$, $K_2 > 0$ such that $\phi'(t) = 1$ corresponds to

$$K_1 e^{\theta t} + K_2 e^{\theta_2 t} > e^{\theta_1 t}.$$

Rewrite this inequality in the form

$$c(\theta)h(x)e^{\theta t(x)} > k_1 c(\theta_1)h(x)e^{\theta_1 t(x)} + k_2 c(\theta_2)h(x)e^{\theta_2 t(x)},$$

where

$$k_1 = \frac{c(\theta)}{K_1 c(\theta_1)}, \quad k_2 = -\frac{K_2 c(\theta)}{K_1 c(\theta_2)}.$$

Thus our test ϕ' rejects H_0 whenever

$$f(X; \theta) > k_1 f(X; \theta_1) + k_2 f(X; \theta_2).$$

However, we have seen from Lemma 7.1 that such a test maximises the power at θ among all tests of fixed size at θ_1 and θ_2, in other words,

$$\mathbb{E}_{\theta}\{\phi'(X)\} \ge \mathbb{E}_{\theta}\{\phi(X)\}.$$

A mirror-image argument applies when $\theta > \theta_2$.

For $\theta_1 < \theta < \theta_2$, Lemma 7.2 applied with $(\theta_a, \theta_b, \theta_c) = (\theta_1, \theta, \theta_2)$ shows that the test ϕ' is equivalent to rejecting H_0 when

$$f(X; \theta) < k_1 f(X; \theta_1) + k_2 f(X; \theta_2)$$

for suitable k_1 and k_2, that is with Lemma 7.1 applied to $1 - \phi'$, we see that ϕ' minimises the probability of rejecting H_0 at this θ, or in other words

$$\mathbb{E}_\theta\{\phi'(X)\} \le \mathbb{E}_\theta\{\phi(X)\}.$$

We have now proved (7.1), and hence the whole of Theorem 7.1. $\square$

7.1.3 Testing a point null hypothesis

Now consider the case $H_0 : \theta = \theta_0$ against $H_1 : \theta \ne \theta_0$ for a given value of θ_0. By analogy with the case just discussed, letting $\theta_2 - \theta_1 \to 0$, there exists a two-sided test ϕ' for which

$$\mathbb{E}_{\theta_0}\{\phi'(X)\} = \alpha, \quad \frac{d}{d\theta}\mathbb{E}_\theta\{\phi'(X)\}\Big|_{\theta=\theta_0} = 0. \tag{7.2}$$

Such a test is in fact UMPU, but we shall not prove this directly.

Example 7.1 Suppose $X_1, \ldots, X_n$ are independent, identically distributed $N(\theta, 1)$. Then the minimal sufficient statistic $T = \bar{X}$ has the $N(\theta, 1/n)$ distribution. Consider the test

$$\phi(x) = \begin{cases} 1 & \text{if } \bar{x} < t_1 \text{ or } \bar{x} > t_2, \\ 0 & \text{if } t_1 \le \bar{x} \le t_2. \end{cases}$$

Then $w(\theta) = \mathbb{E}_\theta\{\phi(X)\}$ is given by

$$w(\theta) = \Phi(\sqrt{n}(t_1 - \theta)) + 1 - \Phi(\sqrt{n}(t_2 - \theta)),$$

where Φ is the standard normal distribution function.

Suppose first $\Theta_0 = [\theta_1, \theta_2]$ for some $\theta_1 < \theta_2$. We must solve

$$\Phi(\sqrt{n}(t_1 - \theta_j)) + 1 - \Phi(\sqrt{n}(t_2 - \theta_j)) = \alpha, \quad j = 1, 2. \tag{7.3}$$

Suppose $\bar{\theta} = (\theta_1 + \theta_2)/2$, $d = (\theta_2 - \theta_1)/2$ and consider t_1, t_2 of the form $t_1 = \bar{\theta} - c, t_2 = \bar{\theta} + c$. It can quickly be checked that (7.3) is satisfied provided

$$\Phi(\sqrt{n}(d - c)) + 1 - \Phi(\sqrt{n}(d + c)) = \alpha. \tag{7.4}$$

However, with $c = 0$ the left-hand side of (7.4) is 1, and as $c \to \infty$ the left-hand side of (7.4) tends to 0. It is a continuous (indeed, strictly decreasing) function of c, so there exists a unique value of c for which (7.4) is satisfied, given any $\alpha \in (0, 1)$. The precise value of c must be determined numerically.

Now suppose $\Theta_0 = \{\theta_0\}$ for some given value of θ_0. In this case, from (7.2) we must satisfy

$$\Phi(\sqrt{n}(t_1 - \theta_0)) + 1 - \Phi(\sqrt{n}(t_2 - \theta_0)) = \alpha, \tag{7.5}$$

$$-\sqrt{n}\Phi'(\sqrt{n}(t_1 - \theta_0)) + \sqrt{n}\Phi'(\sqrt{n}(t_2 - \theta_0)) = 0. \tag{7.6}$$

The obvious try is $t_1 = \theta_0 - z_{\alpha/2}/\sqrt{n}$, $t_2 = \theta_0 + z_{\alpha/2}/\sqrt{n}$, where, as before, $z_{\alpha/2}$ is the upper-$\alpha/2$ point of the standard normal distribution, $\Phi(z_{\alpha/2}) = 1 - \alpha/2$. We leave it to the reader to verify that this choice of t_1 and t_2 does indeed satisfy (7.5) and (7.6), so that this is the UMPU test in this instance.

7.1.4 Some general remarks

The key to showing that two-sided tests are UMPU lies within Lemmas 7.1 and 7.2: Lemma 7.2 shows that two-sided tests satisfy the generalised Neyman–Pearson criterion, and Lemma 7.1 is then used to show that such a test is optimal in the sense of maximising the probability of rejecting H_0 for any $\theta \notin \Theta_0$, and minimising it for $\theta \in \Theta_0$. Clearly one might hope to use the same reasoning in other situations where one would like to show that a two-sided test is UMPU. Just as MLR families form a natural class of distributions for which one-sided tests are UMP, so there is a general class of distributions, the so-called Pólya type 3 distributions, for which two-sided tests are UMPU. We shall not attempt to provide any details of this, merely noting that the methodology of Lemma 7.2 is in principle applicable to other families besides the one-parameter exponential family considered here.

7.2 Conditional inference, ancillarity and similar tests

Consider the following hypothetical situation. An experiment is conducted to measure the carbon monoxide level in the exhaust of a car. A sample of exhaust gas is collected, and is taken along to the laboratory for analysis. Inside the laboratory are two machines, one of which is expensive and very accurate, the other an older model which is much less accurate. We will use the accurate machine if we can, but this may be out of service or already in use for another analysis. We do not have time to wait for this machine to become available, so if we cannot use the more accurate machine we use the other one instead (which is always available). Before arriving at the laboratory we have no idea whether the accurate machine will be available, but we do know that the probability that it is available is $\frac{1}{2}$ (independently from one visit to the next).

This situation may be formalised as follows: we observe (δ, X), where δ ($=1$ or 2) represents the machine used and X the subsequent observation. The distributions are $\Pr\{\delta = 1\} = \Pr\{\delta = 2\} = \frac{1}{2}$ and, given δ, $X \sim N(\theta, \sigma_\delta^2)$, where θ is unknown and σ_1, σ_2 are known, with $\sigma_1 < \sigma_2$. We want to test $H_0 : \theta \leq \theta_0$ against $H_1 : \theta > \theta_0$. Consider the following tests:

Procedure 1 Reject H_0 if $X > c$, where c is chosen so that the test has prescribed size α,

$$\Pr(X > c) = \Pr(X > c \mid \delta = 1)\Pr(\delta = 1) + \Pr(X > c \mid \delta = 2)\Pr(\delta = 2) = \alpha,$$

which requires

$$\frac{1}{2}\left\{1 - \Phi\left(\frac{c - \theta_0}{\sigma_1}\right)\right\} + \frac{1}{2}\left\{1 - \Phi\left(\frac{c - \theta_0}{\sigma_2}\right)\right\} = \alpha.$$

Procedure 2 Reject H_0 if $X > z_\alpha \sigma_\delta + \theta_0$.

Thus Procedure 1 sets a single critical level c, regardless of which machine is used, while Procedure 2 determines its critical level solely on the standard deviation for the machine

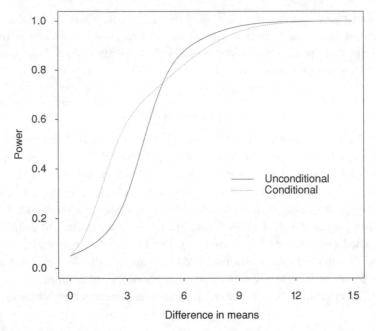

Figure 7.3 Power functions of tests for normal mixture problem

that was actually used, without taking the other machine into account at all. Procedure 2 is called a *conditional* test because it conditions on the observed value of δ. Note that the distribution of δ itself does not depend in any way on the unknown parameter θ, so we are not losing any information by doing this.

Intuitively, one might expect Procedure 2 to be more reasonable, because it makes sense to use all the information available and one part of that information is which machine was used. However, if we compare the two in terms of power, our main criterion for comparing tests up until now, it is not so clear-cut. Figure 7.3 shows the power curves of the two tests in the case $\sigma_1 = 1$, $\sigma_2 = 3$, $\alpha = 0.05$, for which $z_\alpha = 1.6449$ and it is determined numerically that $c = 3.8457 + \theta_0$. When the difference in means, $\theta_1 - \theta_0$, is small, Procedure 2 is much more powerful, but for larger values when $\theta_1 > \theta_0 + 4.9$, Procedure 1 is better.

At first sight this might seem counterintuitive, but a closer look shows what is going on. Let us compute $\alpha_j = \mathrm{Pr}_{\theta_0}\{X > c | \delta = j\}$ – we find $\alpha_1 = 0.00006$, $\alpha_2 = 0.09994$ (so that the overall size is $(\alpha_1 + \alpha_2)/2 = 0.05$). For large $\theta_1 - \theta_0$, this extra power when $\delta = 2$ is decisive in allowing Procedure 1 to perform better than Procedure 2. But is this really sensible? Consider the following scenario.

Smith and Jones are two statisticians. Smith works for the environmental health department of Cambridge City Council and Jones is retained as a consultant by a large haulage firm, which operates in the Cambridge area. Smith carries out a test of the exhaust fumes emitted by one of the lorries belonging to the haulage firm. On this particular day he has to use machine 2 and the observation is $X = \theta_0 + 4.0$, where θ_0 is the permitted standard. It has been agreed in advance that all statistical tests will be carried out at the 5% level and therefore, following Procedure 1 above, he reports that the company is in violation of the standard.

The company is naturally not satisfied with this conclusion and therefore sends the results to Jones for comment. The information available to Jones is that a test was conducted on a machine for which the standard deviation of all measurements is 3 units, that the observed measurement exceeded the standard by 4 units, and that therefore the null hypothesis (that the lorry is meeting the standard) is rejected at the 5% level. Jones calculates that the critical level should be $\theta_0 + 3z_{.05} = \theta_0 + 3 \times 1.645 = \theta_0 + 4.935$ and therefore queries why the null hypothesis was rejected.

The query is referred back to Smith who now describes the details of the test, including the existence of the other machine and Smith's preference for Procedure 1 over Procedure 2 on the grounds that Procedure 1 is of higher power when $|\theta_1 - \theta_0|$ is large. This however is all news to Jones, who was not previously aware that the other machine even existed.

The question facing Jones now is: should she revise her opinion on the basis of the new information provided by Smith? She does not see why she should. There is no new information about either the sample that was collected or the way that it was analysed. All that is new is that there was another machine which might have been used for the test, but which in the event was unavailable. Jones cannot see why this is relevant. Indeed, given the knowledge that there are two machines and that the probability of a false positive test (when the company is complying with the standard) is much higher using machine 2 than machine 1, she might be inclined to query the circumstances under which machine 2 was chosen to test her company's sample. She therefore advises the company to challenge the test in court.

The conclusion we can draw from this discussion is that, while maximising power is a well-established principle for choosing among statistical tests, there are occasions when it can lead to conclusions that appear to contradict common sense. We now develop some general principles underlying this discussion.

Suppose the minimal sufficient statistic T can be partitioned as $T = (S, C)$, where the distribution of C does not depend on the unknown parameter θ. Then C is called an *ancillary statistic*, and S is sometimes called *conditionally sufficient* (given C). In the above example, $S = X$ and $C = \delta$.

We have argued that inference about θ should be based on the conditional distribution of S given C, and this is an instance of a general principle known as *the conditionality principle*.

A more general definition of ancillarity arises in the case of nuisance parameters. We consider situations in which $\theta = (\psi, \lambda)$, where ψ is a parameter (scalar or vector) of interest to us, and λ is a nuisance parameter. For example, if we have a random sample from $N(\mu, \sigma^2)$, with μ and σ both unknown, and we want to test some hypothesis about μ, then we would identify μ with ψ and σ with λ. We assume that the parameter space Θ consists of all possible combinations of (ψ, λ) in their respective spaces, so that $\Theta = \Psi \times \Lambda$, where $\psi \in \Psi$ and $\lambda \in \Lambda$.

Definition *Suppose the minimal sufficient statistic T is partitioned as $T = (S, C)$ where:*

(a) the distribution of C depends on λ but not on ψ,
(b) the conditional distribution of S given $C = c$ depends on ψ but not on λ, for each c.

Then C is an ancillary statistic for ψ, and S is conditionally sufficient for ψ given C.

We can now formulate a principle appropriate to this case.

Conditionality principle (nuisance parameter case) *Inference about ψ should be based on the conditional distribution of S given C.*

There are other definitions of ancillarity. For example, some authors do not insist that an ancillary statistic should be part of a minimal sufficient statistic, but the above definition is the one most widely agreed upon historically. In Chapter 9, we will follow the modern convention of taking the term ancillary to mean a distribution constant statistic, which, together with the maximum likelihood estimator, discussed in Chapter 8, constitutes a minimal sufficient statistic.

In Chapter 6 we discussed the concept of sufficient statistics at some length. For any given problem the sufficient statistic is not unique, but we saw that it is possible to define a minimal sufficient statistic, which is unique up to one-to-one transformations. It turns out that ancillary statistics are not unique either. It would be appealing to define an analogous concept of a *maximal ancillary* statistic. We would like sufficient statistics to be as small as possible so as to eliminate all the irrelevant information. In contrast, we would like ancillary statistics to be as large as possible so that we are taking into account all the relevant conditioning variables. There are some examples where a maximal ancillary statistic exists.

Example 7.2 Consider a location-scale family: $X_1, \ldots, X_n$ are independent, identically distributed with common density $\sigma^{-1} f_0((x - \mu)/\sigma))$, where μ and σ are unknown location and scale constants and f_0 is a known density. Let $X_{(1)} \leq X_{(2)} \leq \ldots \leq X_{(n)}$ denote the order statistics of the sample. Suppose there is no further reduction of the problem by sufficiency, so that $(X_{(1)}, \ldots, X_{(n)})$ is the minimal sufficient statistic. The $(n - 2)$-dimensional vector $C = \{(X_{(3)} - X_{(1)})/(X_{(2)} - X_{(1)}), (X_{(4)} - X_{(1)})/(X_{(2)} - X_{(1)}), \ldots, (X_{(n)} - X_{(1)})/(X_{(2)} - X_{(1)})\}$ has a distribution independent of μ and σ and so is an ancillary statistic. Moreover, it can be shown that this ancillary statistic cannot be expanded in any way and that any other ancillary statistic can be mapped into this, that is it is a maximal ancillary statistic. In a remarkable paper, Fisher (1934) showed how to perform *exact* conditional inference for μ and σ given C, for any n and any distribution f_0. Note that the normal distribution is excluded from this discussion, or is a trivial special case of it, because in that case there is a two-dimensional minimal sufficient statistic and hence no possibility of simplifying the problem by conditioning. However, the Cauchy distribution, for instance, is an example of a problem for which the minimal sufficient statistic is the set of order statistics.

In other cases, however, maximal ancillary statistics do not exist. It is quite possible that two ancillary statistics C_1 and C_2 exist, but the combined statistic (C_1, C_2) is not ancillary. There is nothing the least bit pathological about this: it simply reflects the fact that the marginal distributions of two random variables do not determine their joint distribution. In this case, however, there is no uniquely specified course of action. Adoption of the conditionality principle implies that conditioning on either one of C_1 or C_2 would be superior to conditioning on neither, but it still leaves us with at least two plausible procedures and no indication of which is preferable. Some further ideas on this point are presented in Problem 9.16. In general, identifying an appropriate ancillary statistic in the first place may be more challenging than the issue of choosing between competing ancillaries. As we shall see in Chapter 9, much of modern statistical theory is based on notions of inference which respect the conditionality principle without requiring explicit specification of the conditioning ancillary.

7.2.1 Discussion

The need for a conditionality principle highlights a weakness in the emphasis on the power of tests, which is characteristic of the Neyman–Pearson theory, and, more generally, in the emphasis on the risk function, which is central to non-Bayesian decision theory. If one uses power as the sole criterion for deciding between two tests, then in our example concerning laboratory testing there are at least some circumstances where one would prefer to use Procedure 1, but this may not be sensible for other reasons. The historical disagreement between Fisher and Neyman (recall our brief discussion of this in Chapter 3) centred on Fisher's opinion that the Neyman–Pearson theory did not take adequate account of the need for conditional tests in this kind of situation. Another point of view might be to adopt a Bayesian approach. As we saw in Chapter 3, Bayesian procedures always try to minimise the expected loss based on the observed data and do not take account of other experiments that might have been conducted but were not. Thus in the situation with two machines discussed above, a Bayesian procedure would always act conditionally on which machine was actually used, so the kind of conflict that we saw between the two statisticians would not arise. However, Fisher did not accept Bayesian methods, because of the seeming arbitrariness of choosing the prior distribution, and so this would not have resolved the difficulty for him.

Fisher's own solution to this dilemma was ultimately to create a further theory of statistical inference, which he called *fiducial theory*. In certain cases – one of them being the location-scale problem mentioned above, in which it is possible to find an $(n-2)$-dimensional maximal ancillary statistic – this theory leads to a very precise and clear-cut solution. However, in other cases the justification for fiducial inference is much less clear-cut, and as a result there are few modern scholars who adhere to this point of view. Although many modern statisticians support the principles of the Fisherian approach (including, in particular, the principle of conditionality), there is no universal agreement about how best to implement it.

7.2.2 A more complicated example

The following example, although more artificial than the preceding discussion, illustrates a situation in which failure to observe the Conditionality Principle may lead to a seemingly absurd conclusion.

Let $X_1, \ldots, X_n$ be independent, identically distributed with a uniform distribution on $(\theta - \frac{1}{2}, \theta + \frac{1}{2})$, where θ is an unknown real parameter. Let $W_1 = \min(X_1, \ldots, X_n)$, $W_2 = \max(X_1, \ldots, X_n)$. The joint density of $(X_1, \ldots, X_n)$ is 1 when $\theta - \frac{1}{2} \leq W_1 \leq W_2 \leq \theta + \frac{1}{2}$ and 0 otherwise. This is a function of (W_1, W_2), so (W_1, W_2) is a sufficient statistic. It is in fact the minimal sufficient statistic. It is not complete because if, for example, we let $c = \mathbb{E}\{W_2 - W_1\}$ (in fact $c = \frac{n-1}{n+1}$, but the exact value of c is not important, merely the fact that it does not depend on θ), then $T = W_2 - W_1 - c$ is an example of a statistic which is a function of the minimal sufficient statistic, which satisfies $\mathbb{E}\{T\} = 0$, but for which it is obviously not the case that $T = 0$ with probability 1.

It is readily verified that the joint density of (W_1, W_2), evaluated at (w_1, w_2), is

$$n(n-1)(w_2-w_1)^{n-2}, \quad \theta - \frac{1}{2} \leq w_1 \leq w_2 \leq \theta + \frac{1}{2}.$$

Define $Y_1 = \frac{1}{2}(W_1 + W_2) - \theta$, $Y_2 = (W_2 - W_1)$. Note that, if the observed value of Y_2 is close to 1, θ is very precisely determined, under the assumed model, as necessarily being close to $\frac{1}{2}(W_1 + W_2)$. However, if Y_2 is close to 0, θ could reasonably be any value between $\frac{1}{2}(W_1 + W_2) - 1$ and $\frac{1}{2}(W_1 + W_2) + 1$. Intuitively, therefore, it seems appropriate to condition the inference on the observed value of Y_2, to ensure relevance of any probability calculation to the data at hand. This simple discussion therefore carries force as illustrating in a quantitative way the argument for adoption of the Conditionality Principle: whether a large or small value of Y_2 is observed is an outcome not in itself informative about θ, but affecting the *precision* which we can achieve in the inference. But, in fact, from the perspective of hypothesis testing, failure to condition can lead to illogical conclusions, as we now demonstrate.

The transformation from (W_1, W_2) to (Y_1, Y_2) has Jacobian 1, so the joint density of (Y_1, Y_2) is

$$f_{Y_1,Y_2}(y_1, y_2) = n(n-1)y_2^{n-2}$$

over the region for which $f_{Y_1,Y_2}(y_1, y_2) > 0$, but we have to be careful to define that region accurately: it is $\{(y_1, y_2) : 0 < y_2 < 1, |y_1| < \frac{1}{2}(1 - y_2)\}$ or equivalently $\{(y_1, y_2) : |y_1| < \frac{1}{2}, 0 < y_2 < 1 - 2|y_1|\}$. Integrating out to obtain the marginal densities of Y_1 and Y_2, we find

$$f_{Y_1}(y_1) = n(1 - 2|y_1|)^{n-1}, \quad |y_1| < \frac{1}{2},$$

$$f_{Y_2}(y_2) = n(n-1)y_2^{n-2}(1 - y_2), \quad 0 < y_2 < 1.$$

Dividing the last expression into the joint density of (Y_1, Y_2), we also obtain the conditional density of Y_1 given Y_2:

$$f_{Y_1|Y_2=y_2}(y_1) = \frac{1}{1 - y_2}, \quad |y_1| < \frac{1}{2}(1 - y_2).$$

In other words, given $Y_2 = y_2$, the conditional distribution of Y_1 is uniform on $(-\frac{1}{2}(1 - y_2), \frac{1}{2}(1 - y_2))$.

Suppose we want a size α test of the hypothesis $H_0 : \theta = \theta_0$ against $H_1 : \theta \neq \theta_0$ for some given value θ_0. The statistic Y_1 is a function of the minimal sufficient statistic, and its distribution does not depend on the parameter, so it looks as though a reasonable test would be of the form

$$\text{Reject } H_0 \text{ if } |Y_1| > k_{n,\alpha},$$

where $k_{n,\alpha}$ is chosen so that the size of the test is α. Based on the marginal density of Y_1, it is easily seen that this is

$$k_{n,\alpha} = \frac{1}{2}(1 - \alpha^{1/n}).$$

However, if we follow the Conditionality Principle with $T = (W_1, W_2)$, $S = W_1 + W_2$, $C = W_2 - W_1$, we see that the test should be based on the conditional distribution of Y_1 given Y_2. This leads us to the test criterion

$$\text{Reject } H_0 \text{ if } |Y_1| > \frac{1 - \alpha}{2}(1 - Y_2).$$

Thus we have two plausible test procedures, one based on the marginal distribution of Y_1, the other on the conditional distribution of Y_1 given Y_2. Which one is better?

The distinction between the two procedures is most clearly seen in the situation where the observed values of Y_1 and Y_2 satisfy

$$\frac{1}{2}(1 - Y_2) < |Y_1| < k_{n,\alpha}.$$

In this case the second inequality implies that the unconditional test will accept H_0, but the first inequality is incompatible with the known constraints on (Y_1, Y_2) when H_0 is true. In other words, in this situation we can be certain that H_0 is false, yet the unconditional test is still telling us to accept H_0! This seemingly absurd conclusion is easily avoided if we adopt the conditional test.

7.2.3 Similar tests

Sometimes part (b) of the general definition of an ancillary statistic holds, but not part (a). In such cases, there may still be reason to construct a test based on the conditional distribution of S given C. The reason is that such a test will then be *similar*.

Definition *Suppose $\theta = (\psi, \lambda)$ and the parameter space is of the form $\Theta = \Psi \times \Lambda$, as in the definition of an ancillary statistic. Suppose we wish to test the null hypothesis $H_0 : \psi = \psi_0$ against the alternative $H_1 : \psi \neq \psi_0$, with λ treated as a nuisance parameter. Suppose $\phi(x)$, $x \in \mathcal{X}$ is a test of size α for which*

$$\mathbb{E}_{\psi_0, \lambda}\{\phi(X)\} = \alpha \text{ for all } \lambda \in \Lambda.$$

Then ϕ is called a similar *test of size α.*

More generally, if the parameter space is $\theta \in \Theta$ and the null hypothesis is of the form $\theta \in \Theta_0$, where Θ_0 is a subset of Θ, then a similar test is one for which $\mathbb{E}_\theta\{\phi(X)\} = \alpha$ on the boundary of Θ_0.

By analogy with UMPU tests, if a test is uniformly most powerful among the class of all similar tests, we call it *UMP similar.*

The concept of similar tests has something in common with that of unbiased tests. In particular, if the power function is continuous in θ (a property which, as we have seen, holds automatically for exponential families), then any unbiased test of size α must have power exactly α on the boundary between Θ_0 and Θ_1, that is such a test must be similar. In such cases, if we can find a UMP similar test, and if this test turns out also to be unbiased, then it is necessarily UMPU.

Moreover, in many cases where part (b) of the definition of an ancillary statistic holds, but not part (a), we can demonstrate that a test, which is UMP among all tests based on the conditional distribution of S given C, is UMP amongst all similar tests. In particular, this statement will be valid when C is a complete sufficient statistic for λ.

The upshot of this discussion is that there are many cases when a test which is UMP (one-sided) or UMPU (two-sided), based on the conditional distribution of S given C, is in fact UMP similar or UMPU among the class of all tests.

Thus we have seen two quite distinct arguments for conditioning. In the first, when the conditioning statistic is ancillary, we have seen that the failure to condition may lead to paradoxical situations in which two analysts may form completely different viewpoints of the same data, though we also saw that the application of this principle may run counter to the strict Neyman–Pearson viewpoint of maximising power. The second point of view is based on power, and shows that under certain circumstances a conditional test may satisfy the conditions needed to be UMP similar or UMPU.

7.2.4 Multiparameter exponential families

We shall now see that the general ideas we have discussed have particular application in the case of multiparameter exponential families.

Consider a full exponential family model in its natural parametrisation,

$$f(x;\theta) = c(\theta)h(x)\exp\left(\sum_{i=1}^{k} t_i(x)\theta_i\right),$$

where x represents the value of a data vector X and $t_i(X)$, $i = 1, \ldots, k$ are the natural statistics. We also write T_i in place of $t_i(X)$.

Suppose our main interest is in one particular parameter, which we may without loss of generality take to be θ_1. Consider the test $H_0 : \theta_1 \leq \theta_1^*$ against $H_1 : \theta_1 > \theta_1^*$, where θ_1^* is prescribed. Take $S = T_1$ and $C = (T_2, \ldots, T_k)$. Then by Lemma 5.2 of Chapter 5, the conditional distribution of S given C is also of exponential family form and does not depend on $\theta_2, \ldots, \theta_k$. Therefore, C is sufficient for $\lambda = (\theta_2, \ldots, \theta_k)$ and since it is also complete (from the general property that the natural statistics are complete sufficient statistics for exponential families) the arguments concerning similar tests in Section 7.2.3 suggest that we ought to construct tests for θ_1 based on the conditional distribution of S given C.

In fact such tests do turn out to be UMPU, though we shall not attempt to fill in the details of this: the somewhat intricate argument is given by Ferguson (1967). Finally, it sometimes (though not always) turns out that C is an ancillary statistic for θ_1, and, when this happens, there is a far stronger argument based on the conditionality principle that says we ought to condition on C.

In cases where the distribution of T_1 is continuous, the optimal one-sided test will then be of the following form. Suppose we observe $T_1 = t_1, \ldots, T_k = t_k$. Then we reject H_0 if and only if $t_1 > t_1^*$, where t_1^* is calculated from

$$\text{Pr}_{\theta_1^*}\{T_1 > t_1^* | T_2 = t_2, \ldots, T_k = t_k\} = \alpha.$$

It can be shown that this test is UMPU of size α.

In similar fashion, if we want to construct a two-sided test of $H_0 : \theta_1^* \leq \theta_1 \leq \theta_1^{**}$ against the alternative, $H_1 : \theta_1 < \theta_1^*$ or $\theta_1 > \theta_1^{**}$, where $\theta_1^* < \theta_1^{**}$ are given, we can proceed by defining the conditional power function of a test ϕ based on T_1 as

$$w_{\theta_1}(\phi; t_2, \ldots, t_k) = \mathbb{E}_{\theta_1}\{\phi(T_1) | T_2 = t_2, \ldots, T_k = t_k\}.$$

Note that it is a consequence of our previous discussion that this quantity depends only on θ_1 and not on $\theta_2, \ldots, \theta_k$.

We can then consider a two-sided conditional test of the form

$$\phi'(t_1) = \begin{cases} 1 & \text{if } t_1 < t_1^* \text{ or } t_1 > t_1^{**}, \\ 0 & \text{if } t_1^* \leq t_1 \leq t_1^{**}, \end{cases}$$

where t_1^* and t_1^{**} are chosen such that

$$w_{\theta_1}(\phi'; t_2, \ldots, t_k) = \alpha \quad \text{when } \theta_1 = \theta_1^* \text{ or } \theta_1 = \theta_1^{**}. \tag{7.7}$$

If the hypotheses are of the form $H_0 : \theta_1 = \theta_1^*$ against $H_1 : \theta_1 \neq \theta_1^*$, then the test is of the same form but with (7.7) replaced by

$$w_{\theta_1^*}(\phi'; t_2, \ldots, t_k) = \alpha, \quad \frac{d}{d\theta_1}\left\{ w_{\theta_1}(\phi'; t_2, \ldots, t_k) \right\}\bigg|_{\theta_1 = \theta_1^*} = 0.$$

It can be shown that these tests are also UMPU of size α.

7.2.5 Combinations of parameters

Consider the same set-up as in Section 7.2.4, but suppose now we are interested in testing a hypothesis about $\sum_i c_i \theta_i$, where $c_1, \ldots, c_k$ are given constants and without loss of generality $c_1 \neq 0$.

Let $\psi_1 = \sum_i c_i \theta_i$, $\psi_i = \theta_i$ for $i = 2, \ldots, k$. Then $\theta_1 = (\psi_1 - \sum_2^k c_i \psi_i)/c_1$. We may write

$$\sum_i \theta_i t_i = \left(\psi_1 - \sum_{i=2}^k c_i \psi_i \right) \frac{t_1}{c_1} + \sum_{i=2}^k \psi_i t_i$$

$$= \psi_1 \frac{t_1}{c_1} + \sum_{i=2}^k \psi_i \left(t_i - \frac{c_i t_1}{c_1} \right),$$

which shows that, under a reparametrisation, the model is also of exponential family form in $\psi_1, \ldots, \psi_k$ with natural statistics

$$\frac{T_1}{c_1}, \left\{ T_i - \frac{c_i T_1}{c_1}, \ i = 2, \ldots, k \right\}.$$

Thus we can apply the same tests as in Section 7.2.4 under this reparametrisation of the model.

Example 7.3 Suppose X and Y are independent Poisson random variables with means λ and μ respectively, and we want to test $H_0 : \lambda \leq \mu$ against $H_1 : \lambda > \mu$.

The joint probability mass function is

$$f(x, y) = e^{-(\lambda + \mu)} \frac{e^{x \log \lambda + y \log \mu}}{x! y!},$$

which is of exponential family form with natural parameters $(\log \lambda, \log \mu)$. We identify $(T_1, T_2, \theta_1, \theta_2)$ with $(X, Y, \log \lambda, \log \mu)$ and consider $c_1 = 1, c_2 = -1, \psi_1 = c_1 \theta_1 + c_2 \theta_2 = \log(\lambda/\mu)$ so that the desired test becomes $H_0 : \psi_1 \leq 0$ against $H_1 : \psi_1 > 0$. Thus in this case inference about ψ_1 should be based on the conditional distribution of T_1 given $T_2 - c_2 T_1 / c_1 = T_1 + T_2$. According to theory, such a test will be UMP similar and indeed UMPU.

It is now an easy matter to check that

$$\Pr\{T_1 = t_1, T_2 = t_2\} = \frac{\lambda^{t_1} \mu^{t_2} e^{-\lambda - \mu}}{t_1! t_2!},$$

$$\Pr\{T_1 + T_2 = t_1 + t_2\} = \frac{(\lambda + \mu)^{t_1 + t_2} e^{-\lambda - \mu}}{(t_1 + t_2)!},$$

and so

$$\Pr\{T_1 = t_1 | T_1 + T_2 = t_1 + t_2\} = \binom{t_1 + t_2}{t_1} \left(\frac{\lambda}{\lambda + \mu}\right)^{t_1} \left(\frac{\mu}{\lambda + \mu}\right)^{t_2},$$

which is $\mathrm{Bin}(t_1 + t_2, p)$, where $p = e^{\psi_1}/(1 + e^{\psi_1})$. This confirms (as we already knew from the general theory) that the conditional distribution of T_1 depends only on ψ_1, and allows us to reformulate the testing problem in terms of $H_0 : p \leq \frac{1}{2}$ against $H_1 : p > \frac{1}{2}$.

Note that, in contrast to what has been assumed throughout the rest of the chapter, in this case the distribution is discrete and therefore it may be necessary to consider randomised tests, but this point has no effect on the general principles concerning the desirability of a conditional test.

Note that if we adopt the slightly different parametrisation $(\psi_1, \lambda + \mu)$ in this example, then the conditionality principle also implies that the appropriate test for inference about ψ_1 should condition on $T_1 + T_2$. We have that $T_1 + T_2$ has a Poisson distribution with mean $\lambda + \mu$, which does not depend on ψ_1, so that $T_1 + T_2$ is ancillary for ψ_1.

7.3 Confidence sets

7.3.1 Construction of confidence intervals via pivotal quantities

As in previous discussion, we assume a finite parameter model in which the distribution of a random variable X depends on a parameter θ. At least for the beginning of our discussion, it will be convenient to assume that θ is a scalar.

Definition *A random function $T(X, \theta)$ is said to be* pivotal, *or a* pivotal quantity, *if it is a function of both X and θ whose distribution does not depend on θ.*

More generally, the term *pivot* is often used to denote a general random function of X and θ, with distribution which may depend on θ. We will use this terminology extensively in Chapter 11.

Suppose we are given $\alpha \in (0, 1)$ and a pivotal quantity $T(X, \theta)$. Then we can find constants c_1 and c_2 such that

$$\Pr_\theta\{c_1 \leq T(X, \theta) \leq c_2\} = 1 - \alpha \quad \text{for all } \theta. \tag{7.8}$$

Provided $T(X, \theta)$ is of a reasonably manageable functional form, we can rewrite this relationship in the form

$$\Pr_\theta\{L(X) \le \theta \le U(X)\} = 1 - \alpha \quad \text{for all } \theta, \tag{7.9}$$

where $L(X)$ and $U(X)$ are functions of X alone. We then say that $[L(X), U(X)]$ is a $(1 - \alpha)$-level or $100(1 - \alpha)\%$ *confidence interval* for θ.

In more complicated cases, and especially when we extend the discussion to allow θ to be multidimensional, (7.8) may not invert to an interval, but if for fixed X we define

$$S(X) = \{\theta : c_1 \le T(X, \theta) \le c_2\}$$

then (7.9) is replaced by

$$\Pr_\theta\{S(X) \ni \theta\} = 1 - \alpha \quad \text{for all } \theta,$$

and in this case $S(X)$ is a $(1 - \alpha)$-level *confidence set* for θ.

Example 7.4 Let $X_1, \ldots, X_n$ be independent, identically distributed $N(\theta, \sigma^2)$, where θ and σ^2 are both unknown but our interest is in θ (so already we are making a slight extension of the above framework, by allowing the nuisance parameter σ^2).

Defining $\bar{X} = \sum X_i/n$, $s_X^2 = \sum(X_i - \bar{X})^2/(n - 1)$ to be the usual sample mean and variance, we have

$$\frac{\sqrt{n}(\bar{X} - \theta)}{s_X} \sim t_{n-1}.$$

So $T = \sqrt{n}(\bar{X} - \theta)/s_X$ is pivotal. Usually we define $c_1 = -c_2$, where $\Pr\{T > c_2\} = \alpha/2$, calculated from tables of the t_{n-1} distribution, and hence

$$\Pr_\theta\left\{-c_2 \le \frac{\sqrt{n}(\bar{X} - \theta)}{s_X} \le c_2\right\} = 1 - \alpha. \tag{7.10}$$

However we can invert (7.10) to write

$$\Pr_\theta\left\{\bar{X} - \frac{c_2 s_X}{\sqrt{n}} \le \theta \le \bar{X} + \frac{c_2 s_X}{\sqrt{n}}\right\} = 1 - \alpha,$$

so that $[\bar{X} - c_2 s_X/\sqrt{n}, \bar{X} + c_2 s_X/\sqrt{n}]$ is the desired confidence interval.

Example 7.5 Suppose $X_1, \ldots, X_n$ are independent, identically distributed from the uniform distribution on $(0, \theta)$ for unknown $\theta > 0$.

A complete sufficient statistic is $T = \max\{X_1, \ldots, X_n\}$ and we calculate

$$\Pr_\theta\left\{\frac{T}{\theta} \le y\right\} = \Pr_\theta\{X_1 \le y\theta, \ldots, X_n \le y\theta\} = y^n,$$

for any $y \in [0, 1]$. Thus T/θ is pivotal.

Given α, let $c_1 < c_2$ be in $[0, 1]$ such that $c_2^n - c_1^n = 1 - \alpha$. Then the interval

$$\left[\frac{T}{c_2}, \frac{T}{c_1}\right]$$

is a $(1 - \alpha)$-level confidence interval for θ.

For example, one choice is to take $c_2 = 1$, $c_1 = \alpha^{1/n}$. We then have that $[T, T\alpha^{-1/n}]$ is a suitable confidence interval.

The main difficulty with building an entire theory around pivotal quantities is that for many problems no pivotal quantity exists. This prompts us to seek more general constructions.

7.3.2 A general construction

For each X, we require to find a set $S(X)$ such that

$$\mathrm{Pr}_\theta\{\theta \in S(X)\} = 1 - \alpha \quad \text{for all } \theta \in \Theta,$$

where $\alpha \in (0, 1)$ is given.

Suppose, for each $\theta_0 \in \Theta$, we can find a non-randomised test $\phi_{\theta_0}(X)$ of exact size α for testing $H_0 : \theta = \theta_0$ against $H_1 : \theta \neq \theta_0$. Let $A(\theta_0)$ denote the corresponding acceptance region, so that

$$A(\theta_0) = \{X : \phi_{\theta_0}(X) = 0\}.$$

Thus

$$\mathrm{Pr}_\theta\{X \in A(\theta)\} = 1 - \alpha \quad \text{for all } \theta \in \Theta.$$

Now define, for each $X \in \mathcal{X}$,

$$S(X) = \{\theta : X \in A(\theta)\}.$$

Then it evidently follows that

$$\mathrm{Pr}_\theta\{\theta \in S(X)\} = 1 - \alpha \quad \text{for all } \theta \in \Theta.$$

Thus $S(X)$ fulfils the criteria to be a valid $(1 - \alpha)$-level confidence set for θ.

Conversely, if we are given a family of $(1 - \alpha)$-level confidence sets $\{S(X), X \in \mathcal{X}\}$, we may construct a family of tests of size α by defining acceptance regions by

$$A(\theta) = \{X : \theta \in S(X)\}.$$

Thus there is a one-to-one correspondence between confidence sets of a given level and hypothesis tests of the corresponding size.

In cases where there are nuisance parameters, there is a good argument for using similar tests, because in this case the coverage level of the confidence set for the parameter of interest will not depend on the true values of the nuisance parameters.

7.3.3 Criteria for good confidence sets

Just as we have a number of criteria for deciding what are good tests of hypotheses, so we can construct corresponding criteria for deciding between different families of confidence sets of a given level.

(a) Nesting
One desirable property is that if $S_1(X)$ and $S_2(X)$ are two confidence sets corresponding to levels $1 - \alpha_1$ and $1 - \alpha_2$ respectively, and $\alpha_1 < \alpha_2$, then we would expect $S_2 \subseteq S_1$.

For instance, we would expect a 95% confidence interval to be contained within a 99% confidence interval for the same problem.

The corresponding criterion on test regions is that, if $A_1(\theta_0)$ and $A_2(\theta_0)$ are acceptance regions of $H_0 : \theta = \theta_0$ of sizes α_1 and α_2 respectively, then $A_2(\theta_0) \subseteq A_1(\theta_0)$; that is, if we reject the null hypothesis at one size, then we will reject it at any larger size.

(b) Minimisation of error

Among all $(1 - \alpha)$-level confidence sets $\{S(X), \ X \in \mathcal{X}\}$, we should like to choose one for which $\Pr_{\theta_0}\{\theta \in S(X)\}$ is small when $\theta \neq \theta_0$. This, of course, corresponds to choosing a test which has high power when $\theta \neq \theta_0$.

One motivation for doing this in the context of confidence sets is as follows. Suppose the confidence set is a confidence interval for a one-dimensional parameter θ. Then the length of the interval is

$$L = \int_{-\infty}^{\infty} I\{\theta \in S(X)\}d\theta,$$

where $I\{\cdot\}$ is the indicator function. Then

$$\mathbb{E}_{\theta_0}\{L\} = \int_{-\infty}^{\infty} \Pr_{\theta_0}\{\theta \in S(X)\}d\theta.$$

Thus, if we can minimise $\Pr_{\theta_0}\{\theta \in S(X)\}$ uniformly for $\theta \neq \theta_0$, then this is equivalent to minimising the expected length of the confidence interval. In the more general case, where a confidence interval is replaced by a confidence set, the expected length of the interval is replaced by the expected measure of the set.

One thing we can try to do is to find UMP tests – for example, when a family has the MLR property (Chapter 4), it is possible to find UMP one-sided tests. However, one-sided tests lead to one-sided confidence intervals (that is, in (7.9), either $L(X) = -\infty$ or $U(X) = +\infty$), which is sometimes what is required, but usually not.

The alternative is to construct two-sided intervals from two-sided tests. In this case, we saw that the concept of *unbiasedness* was useful in restricting the class of tests considered, and in that case we can often find UMPU tests.

In the confidence set context, we make the following definition.

Definition *A family of confidence sets* $\{S(X), \ X \in \mathcal{X}\}$ *is* unbiased *if* $\Pr_{\theta_0}\{\theta \in S(X)\}$ *is maximised with respect to* θ *when* $\theta = \theta_0$, *for each* θ_0.

If we can find a UMPU test, and invert this to obtain an unbiased confidence set, then this test will minimise the expected measure of the confidence set among all unbiased confidence sets of a given level.

7.4 Problems

7.1 Let $X_1, \ldots, X_n$ be an independent sample from a normal distribution with mean 0 and variance σ^2. Explain in as much detail as you can how to construct a UMPU test of $H_0 : \sigma = \sigma_0$ against $H_1 : \sigma \neq \sigma_0$.

7.2 Let $X_1, \ldots, X_n$ be an independent sample from $N(\mu, \mu^2)$. Let $T_1 = \bar{X}$ and $T_2 = \sqrt{(1/n)\sum X_i^2}$. Show that $Z = T_1/T_2$ is ancillary. Explain why the Conditionality

Principle would lead to inference about μ being drawn from the conditional distribution of T_2 given Z. Find the form of this conditional distribution.

7.3 Let Y_1, Y_2 be independent Poisson random variables, with means $(1 - \psi)\lambda$ and $\psi\lambda$ respectively, with λ *known*.

Explain why the Conditionality Principle leads to inference about ψ being drawn from the conditional distribution of Y_2, given $Y_1 + Y_2$. What is this conditional distribution?

Suppose now that λ is *unknown*. How would you test $H_0 : \psi = \psi_0$ against $H_1 : \psi \neq \psi_0$?

7.4 Suppose X is normally distributed as $N(\theta, 1)$ or $N(\theta, 4)$, depending on whether the outcome, Y, of tossing a fair coin is heads ($y = 1$) or tails ($y = 0$). It is desired to test $H_0 : \theta = -1$ against $H_1 : \theta = 1$. Show that the most powerful (unconditional) size $\alpha = 0.05$ test is the test with rejection region given by $x \geq 0.598$ if $y = 1$ and $x \geq 2.392$ if $y = 0$.

Suppose instead that we condition on the outcome of the coin toss in construction of the tests. Verify that, given $y = 1$, the resulting most powerful size $\alpha = 0.05$ test would reject if $x \geq 0.645$, while, given $y = 0$, the rejection region would be $x \geq 2.290$.

7.5 A local councillor suspects that traffic conditions have become more hazardous in Ambridge than in Borchester, so she records the numbers A and B of accidents occurring in each place in the course of a month. Assuming that A and B are independent Poisson random variables with parameters λ and μ, it is desired to construct an unbiased test of size $\frac{1}{16}$ of $H_0 : \lambda \geq \mu$ against $H_1 : \lambda < \mu$.

Show that $A + B$ is distributed as Poisson with parameter $\lambda + \mu$, and that conditional on $A + B = n$, $A \sim \text{Bin}(n, p)$, where $p = \lambda/(\lambda + \mu)$.

Show that, if $X \sim \text{Bin}(n, p)$, then the UMPU size α test of $H_0 : p \geq 1/2$ against $H_1 : p < 1/2$ has test function of the form

$$\phi(k, n) = \begin{cases} 1 & \text{if } 0 \leq k < \kappa_n(\alpha), \\ \gamma_n(\alpha) & \text{if } k = \kappa_n(\alpha), \\ 0 & \text{if } \kappa_n(\alpha) < k \leq n, \end{cases}$$

where $\kappa_n(\alpha)$ and $\gamma_n(\alpha)$ are chosen so that $\mathbb{E}_{1/2}\phi(X, n) = \alpha$.

Show that the test of the original hypotheses defined by choosing H_1 with probability $\phi(A, A + B)$ is unbiased with size α. Can you verify from first principles that the test is in fact UMPU?

Carry out the test when $A = 5$ and $B = 2$, and also when $A = 3$ and $B = 0$.

7.6 A Bayesian friend of the local councillor of Problem 7.5 has independent exponential priors on λ and μ, each of mean 1. Compute the posterior probability that $\lambda \geq \mu$ in the two cases, $A = 2, B = 5$ and $A = 0, B = 3$.

7.7 Let $X \sim \text{Bin}(m, p)$ and $Y \sim \text{Bin}(n, q)$, with X and Y independent. Show that, as p and q range over $[0, 1]$, the joint distributions of X and Y form an exponential family. Show further that, if $p = q$, then

$$\Pr(X = x \mid X + Y = x + y) = \binom{m}{x}\binom{n}{y} \bigg/ \binom{m + n}{x + y}.$$

Hence find the form of a UMPU test of the null hypothesis $H_0 : p \leq q$ against $H_1 :$
$p > q$.

In an experiment to test the efficacy of a new drug for treatment of stomach ulcers, five patients are given the new drug and six patients are given a control drug. Of the patients given the new drug, four report an improvement in their condition, while only one of the patients given the control drug reports improvement. Do these data suggest, at level $\alpha = 0.1$, that patients receiving the new drug are more likely to report improvement than patients receiving the control drug? (This is the *hypergeometric distribution* and the test presented here is conventionally referred to as *Fisher's exact test* for a 2×2 table.)

7.8 Let $X_1, \ldots, X_n$ be independent, exponential random variables of mean θ. Show that $Y = 2 \sum_{i=1}^{n} X_i / \theta$ is a pivotal quantity, and construct a $(1 - \alpha)$-level confidence interval for θ.

7.9 Let $X \sim N(\mu, 1)$ and $Y \sim N(\nu, 1)$ be independent, and suppose we wish to construct a confidence interval for $\theta = \mu/\nu$. Show that $(X - \theta Y)/(1 + \theta^2)^{1/2}$ is a pivotal quantity, and that the confidence set obtained from it could consist of the whole real line, a single interval or two disjoint intervals. Is the confidence set ever empty?

7.10 It is generally believed that long-life light bulbs last no more than twice as long on average as standard light bulbs. In an experiment, the lifetimes X of a standard bulb and Y of a long-life bulb are recorded, as $X = 1$, $Y = 5$. Modelling the lifetimes as exponential random variables, of means λ and μ respectively, construct: (i) a test of size $\alpha = 0.05$ of the hypothesis $\lambda > \frac{1}{2}\mu$, (ii) a 95% confidence interval for λ/μ.

8

Likelihood theory

This chapter is concerned with one of the central principles of modern statistics – that of maximum likelihood.

Some definitions and basic properties of maximum likelihood estimators are given in Section 8.1. The basic idea is that the maximum likelihood estimator $\widehat{\theta}$ is the value of θ, which maximises the likelihood function $L(\theta)$. In *regular* cases this is given by solving the *likelihood equation(s)* formed by setting the first-order derivative(s) of $\log L$ to 0.

We establish in Section 8.2 the *Cramér–Rao Lower Bound* (CRLB), which gives a lower bound on the variance of an unbiased estimator under very general conditions. Under certain circumstances the CRLB is attained exactly – there is an interesting connection with the theory of exponential families here – though in general the CRLB is only an approximate guide to what is attainable in practice.

After some preliminaries about convergence of random variables in Section 8.3, we establish in Section 8.4 the *consistency, asymptotic normality and asymptotic efficiency* of a maximum likelihood estimator under quite general (regular family) conditions. Detailed proofs are given only for certain one-dimensional cases, though the basic results apply in any dimension. These results are at the centre of much modern theory based on maximum likelihood estimation.

Section 8.5 describes a methodology for conducting tests of hypotheses using maximum likelihood estimates, including *Wilks' Theorem* concerning the asymptotic χ^2 distribution of the log-likelihood ratio test statistic. Section 8.6 considers in greater depth multiparameter problems and introduces the notion of a *profile likelihood* for a parameter of interest, in the presence of nuisance parameters.

8.1 Definitions and basic properties

8.1.1 Maximum likelihood estimator

Suppose data x are the observed value of a random variable X from a parametric family of densities or mass functions, $X \sim f(x; \theta)$, where in general θ is multidimensional, $\theta = (\theta_1, \ldots, \theta_d) \in \Theta \subseteq \mathbb{R}^d$ for some $d \geq 1$. We allow X to be multidimensional as well, and in particular we will consider in detail the case when $X = (X_1, \ldots, X_n)$, an independent, identically distributed sample from some parametric model. After observing x, the *likelihood function* is defined by

$$L(\theta) \equiv L(\theta; x) = f(x; \theta),$$

viewed as a function of θ for the fixed x. The *maximum likelihood estimate* (MLE) $\widehat{\theta}(x)$ is defined to be the value of θ which maximises $L(\theta)$.

Usually we work with the *log-likelihood*

$$l(\theta) \equiv l(\theta; x) = \log L(\theta).$$

To stress that it is constructed from a sample of size n we may write the log-likelihood as $l_n(\theta)$, and write the MLE correspondingly as $\widehat{\theta}_n$. A key component of the current chapter is the study of the properties of the log-likelihood considered as a random variable

$$l(\theta; X) = \log f(X; \theta),$$

and also of the MLE as a random variable, $\widehat{\theta}(X)$, which is usually referred to as the *maximum likelihood estimator*.

In most cases l is differentiable and $\widehat{\theta}$ is obtained by solving the *likelihood equation*

$$l'(\theta) = 0, \tag{8.1}$$

or in the multiparameter case when $\theta = (\theta_1, \ldots, \theta_d) \in \mathbb{R}^d$,

$$\nabla_\theta l(\theta; x) = 0, \tag{8.2}$$

where $\nabla_\theta = (\partial/\partial\theta_1, \ldots, \partial/\partial\theta_d)^T$.

A number of questions immediately arise:

(i) Do the likelihood equations (8.1) or (8.2) have a solution?
(ii) If so, is the solution unique?
(iii) Is it a local maximum? To answer this we need to check second derivatives.
(iv) Is it a global maximum?

Each of the questions (i)–(iv) may have a negative answer, though we shall be concentrating on so-called *regular problems* for which the maximum likelihood estimator is indeed given by a local maximum of the log-likelihood function.

8.1.2 Examples

Example 8.1 Consider $X_1, \ldots, X_n$ independent, identically distributed from $N(\mu, \tau)$, with μ and τ both unknown. Then

$$L(\mu, \tau) = (2\pi\tau)^{-n/2} \exp\left\{-\frac{1}{2}\sum_i \frac{(X_i - \mu)^2}{\tau}\right\}.$$

It is an easy exercise to check that

$$\widehat{\mu} = \bar{X}, \quad \left(= \frac{1}{n}\sum X_i\right),$$

$$\widehat{\tau} = \frac{1}{n}\sum(X_i - \bar{X})^2.$$

The answers to questions (i)–(iv) above are all yes in this instance.

Note that $\widehat{\tau}$ is not the usual sample variance, for which the denominator is $n - 1$ rather than n. The MLE is biased, because $\mathbb{E}\{\widehat{\tau}\} = (n - 1)\tau/n$. For many problems it turns out that the MLE is biased though it is asymptotically unbiased and efficient (see Section 8.4 below).

Example 8.2 Consider $X_1, \ldots, X_n$, independent, identically distributed from the uniform distribution on $(0, \theta]$, where θ is the unknown parameter. The likelihood function is

$$L(\theta) = \frac{1}{\theta^n} I\{X_1 \le \theta, \ldots, X_n \le \theta\}, \tag{8.3}$$

where $I(\cdot)$ is the indicator function. Equation (8.3) is strictly decreasing in θ over the range for which it is non-zero, which is for $\theta \ge \max\{X_1, \ldots, X_n\}$. Therefore the MLE in this case is $\widehat{\theta} = \max\{X_1, \ldots, X_n\}$, which is also the complete sufficient statistic for this problem. So in this case the MLE is *not* given by solving the likelihood equation.

Example 8.3 Suppose X_1, X_2 are independent, identically distributed from the Cauchy density

$$f(x; \theta) = \frac{1}{\pi\{1 + (x - \theta)^2\}}, \quad -\infty < x < \infty.$$

The likelihood equation is

$$\frac{(X_1 - \theta)}{1 + (X_1 - \theta)^2} + \frac{(X_2 - \theta)}{1 + (X_2 - \theta)^2} = 0. \tag{8.4}$$

The following statements are left as an exercise for the reader:

(a) If $|X_1 - X_2| \le 2$ then there is a unique solution to (8.4) given by $\widehat{\theta} = (X_1 + X_2)/2$ and this maximises the likelihood function.

(b) If $|X_1 - X_2| > 2$ then there are three solutions to (8.4), where $(X_1 + X_2)/2$ is a local minimum of the likelihood function and there are two local maxima given as the (real, distinct) solutions of the equation

$$1 + (X_1 - \theta)(X_2 - \theta) = 0.$$

The difficulties created by this example remain present in larger sample sizes: the probability that there exist multiple maxima is positive for all sample sizes n, and does not tend to 0 as $n \to \infty$.

8.1.3 Score function and information

Recall from Chapter 5 that we define the *score function* by

$$u(\theta; x) = \nabla_\theta l(\theta; x).$$

In terms of the score function, the likelihood equation may be written

$$u(\widehat{\theta}; x) = 0.$$

To study the score function as a random variable we will write

$$U(\theta) = u(\theta; X),$$

and write the components of $U(\theta)$ as $U(\theta) = (U_1(\theta), \ldots, U_d(\theta))^T$.

We have already seen, in Chapter 5, that for regular problems for which the order of differentiation with respect to θ and integration over the sample space can be reversed, we have

$$\mathbb{E}_\theta\{U(\theta)\} = 0. \tag{8.5}$$

Also, the argument used in Chapter 5 further allows us to calculate the covariances between the components of the score function. We have

$$\text{cov}_\theta\{U_r(\theta), U_s(\theta)\}$$
$$= \mathbb{E}_\theta\left\{\frac{\partial l(\theta; X)}{\partial \theta_r}\frac{\partial l(\theta; X)}{\partial \theta_s}\right\}$$
$$= \mathbb{E}_\theta\left\{-\frac{\partial^2 l(\theta; X)}{\partial \theta_r \partial \theta_s}\right\}.$$

More compactly, the covariance matrix of U is

$$\text{cov}_\theta\{U(\theta)\} = \mathbb{E}_\theta\{-\nabla_\theta \nabla_\theta^T l\}.$$

This matrix is called the expected information matrix for θ, or, more usually, the *Fisher information matrix*, and will be denoted by $i(\theta)$. The Hessian matrix $-\nabla_\theta \nabla_\theta^T l$ of second-order partial derivatives of l is called the *observed information matrix*, and is denoted by $j(\theta)$. Note that $i(\theta) = \mathbb{E}_\theta\{j(\theta)\}$.

8.1.4 Some discussion

1 The definitions above are expressed in terms of arbitrary random variables X. Often the components X_j are mutually independent, in which case both the log-likelihood and the score function are sums of contributions

$$l(\theta; x) = \sum_{j=1}^n l(\theta; x_j),$$

$$u(\theta; x) = \sum_{j=1}^n \nabla_\theta l(\theta; x_j) = \sum_{j=1}^n u(\theta; x_j),$$

say, and where $l(\theta; x_j)$ is found from the density of X_j.

Quite generally, even for dependent random variables, if $X_{(j)} = (X_1, \ldots, X_j)$, we may write

$$l(\theta; x) = \sum_{j=1}^n l_{X_j | X_{(j-1)}}(\theta; x_j \mid x_{(j-1)}),$$

the jth term being computed from the conditional density of X_j given $X_{(j-1)}$.

2 Although maximum likelihood estimators are widely used because of their good sampling properties, they are also consistent with a broader principle known as the *likelihood principle*.

In statistical inference, the objective is to draw conclusions about the underlying distribution of a random variable X, on the basis of its observed value x. In a parametric model, this means drawing inference about the unknown value of the parameter θ. The likelihood function $L(\theta)$ measures how likely different values of θ are to be the true value. It may then be argued that the general problem of inference for θ is solved by simply examining the likelihood function. The *likelihood principle* is a formal expression of this idea.

Suppose we have an observation x from a model $\{f(\cdot; \theta), \theta \in \Theta\}$, and an observation y from another model $\{g(\cdot; \theta), \theta \in \Theta\}$, with the parameter θ having the same meaning in both models. Let $L(\theta)$ and $\tilde{L}(\theta)$ be the likelihood functions for θ for the two models. Then, if given values x and y, $L(\theta) = C\tilde{L}(\theta)$ for all θ, where C is a constant, then according to the likelihood principle identical conclusions regarding θ should be drawn from x and y. Under this principle, quantities that depend on the sampling distribution of a statistic, which is in general not a function of the likelihood function alone, are irrelevant for statistical inference. Thus, the likelihood principle can be regarded as giving further justification for the use of maximum likelihood estimators, though, to the extent that L and $\tilde{L}$ may be based on different sampling models and therefore have different sampling properties, it also stands somewhat in conflict with the repeated sampling principle as a justification for using one estimator over another.

8.1.5 Some mathematical reminders

We provide here some reminders of definitions and results used later in this chapter and in Chapter 9.

The *Taylor expansion* for a function $f(x)$ of a single real variable about $x = a$ is given by

$$f(x) = f(a) + f^{(1)}(a)(x - a) + \frac{1}{2!}f^{(2)}(a)(x - a)^2 + \cdots + \frac{1}{n!}f^{(n)}(a)(x - a)^n + R_n,$$

where

$$f^{(l)}(a) = \frac{d^l f(x)}{dx^l}\Big|_{x=a},$$

and the remainder R_n is of the form

$$\frac{1}{(n + 1)!}f^{(n+1)}(c)(x - a)^{n+1},$$

for some $c \in [a, x]$.

The Taylor expansion is generalised to a function of several variables in a straightforward manner. For example, the expansion of $f(x, y)$ about $x = a$ and $y = b$ is given by

$$f(x, y) = f(a, b) + f_x(a, b)(x - a) + f_y(a, b)(y - b)$$
$$+ \frac{1}{2!}\{f_{xx}(a, b)(x - a)^2 + 2f_{xy}(a, b)(x - a)(y - b) + f_{yy}(a, b)(y - b)^2\} + \cdots,$$

where

$$f_x(a, b) = \frac{\partial f}{\partial x}\Big|_{x=a, y=b},$$

$$f_{xy}(a, b) = \frac{\partial^2 f}{\partial x \partial y}\Big|_{x=a, y=b},$$

and similarly for the other terms.

The *sign function* sgn is defined by

$$\text{sgn}(x) = \begin{cases} 1, & \text{if } x > 0, \\ 0, & \text{if } x = 0, \\ -1, & \text{if } x < 0. \end{cases}$$

Suppose we partition a matrix A so that $A = \begin{bmatrix} A_{11} & A_{12} \\ A_{21} & A_{22} \end{bmatrix}$, with A^{-1} correspondingly

written $A^{-1} = \begin{bmatrix} A^{11} & A^{12} \\ A^{21} & A^{22} \end{bmatrix}$. If A_{11} and A_{22} are non-singular, let

$$A_{11.2} = A_{11} - A_{12}A_{22}^{-1}A_{21},$$

and

$$A_{22.1} = A_{22} - A_{21}A_{11}^{-1}A_{12}.$$

Then

$$A^{11} = A_{11.2}^{-1}, \quad A^{22} = A_{22.1}^{-1}, \quad A^{12} = -A_{11}^{-1}A_{12}A^{22},$$
$$A^{21} = -A_{22}^{-1}A_{21}A^{11}.$$

8.2 The Cramér–Rao Lower Bound

Let $W(X)$ be *any* estimator of θ and let $m(\theta) = \mathbb{E}_\theta\{W(X)\}$. For the purpose of this section we shall restrict ourselves to scalar θ, though analogous results are available for vector θ. Define

$$Y = W(X), \quad Z = \frac{\partial}{\partial\theta}\log f(X;\theta).$$

The elementary inequality that the correlation between any two random variables lies between -1 and 1, $-1 \le \text{corr}(Y, Z) \le 1$, leads to

$$\{\text{cov}(Y, Z)\}^2 \le \text{var}(Y)\,\text{var}(Z). \tag{8.6}$$

Now we have

$$\text{cov}(Y, Z) = \int w(x)\left\{\frac{\partial}{\partial\theta}\log f(x;\theta)\right\}f(x;\theta)dx$$
$$= \frac{\partial}{\partial\theta}\left\{\int w(x)f(x;\theta)dx\right\} \tag{8.7}$$
$$= m'(\theta).$$

Note that the second line here requires that we interchange the order of integration with respect to x and differentiation with respect to θ, and the validity of this step is an assumption of the calculation. It is valid for most regular problems, but there are counterexamples!

We also have by (8.5), again under assumptions about the validity of interchanging integration and differentiation,

$$\mathrm{var}(Z) = \mathbb{E}\left\{\left(\frac{\partial \log f(X;\theta)}{\partial \theta}\right)^2\right\} = \mathbb{E}\left\{-\frac{\partial^2 \log f(X;\theta)}{\partial \theta^2}\right\} = i(\theta). \qquad (8.8)$$

Putting (8.6)–(8.8) together,

$$\mathrm{var}\{W(X)\} \geq \frac{\{m'(\theta)\}^2}{i(\theta)}. \qquad (8.9)$$

Equation (8.9) is known as the *Cramér–Rao Lower Bound*, or CRLB for short. In particular, if $W(X)$ is an unbiased estimator of θ, so that $m(\theta) = \theta$, it reduces to the simpler form

$$\mathrm{var}\{W(X)\} \geq \frac{1}{i(\theta)}. \qquad (8.10)$$

Thus any unbiased estimator which achieves the lower bound in (8.10) is immediately seen to be a MVUE, a minimum variance unbiased estimator. However, there is no guarantee that any estimator exists which achieves this lower bound exactly, a point to which we return momentarily.

In cases where $X = (X_1, \ldots, X_n)$, where $X_1, \ldots, X_n$ are independent, identically distributed from some density $f(\cdot;\theta)$, we write f_n for the density of X, i_n for the Fisher information and note that because

$$\log f_n(X;\theta) = \sum_{i=1}^n \log f(X_i;\theta),$$

(8.8) leads at once to

$$i_n(\theta) = n i_1(\theta).$$

Example 8.4 Suppose $X_1, \ldots, X_n$ are independent, identically distributed from an exponential distribution with mean θ, so that the common density is $f(x;\theta) = \theta^{-1} \exp(-x/\theta)$, $0 < x < \infty$, $0 < \theta < \infty$. Let $S_n = X_1 + \cdots + X_n$ denote the complete sufficient statistic. Then the log-likelihood based on sample size n is

$$l_n(\theta) = -n \log \theta - \frac{S_n}{\theta},$$

$$l_n'(\theta) = -\frac{n}{\theta} + \frac{S_n}{\theta^2},$$

$$l_n''(\theta) = \frac{n}{\theta^2} - 2\frac{S_n}{\theta^3}.$$

It then follows that the likelihood equation has a unique solution when

$$\widehat{\theta} = \frac{S_n}{n}.$$

We also check that

$$\mathbb{E}_\theta\left\{\left(\frac{\partial \log f_n(X;\theta)}{\partial \theta}\right)^2\right\} = \frac{1}{\theta^4}\mathrm{var}(S_n) = \frac{n}{\theta^2},$$

and

$$\mathbb{E}_\theta \left\{ -\frac{\partial^2 \log f_n(X;\theta)}{\partial\theta^2} \right\} = -\frac{n}{\theta^2} + \frac{2}{\theta^3}\mathbb{E}(S_n) = \frac{n}{\theta^2},$$

which verifies directly in this instance that the two definitions of $i_n(\theta)$ lead to the same thing, while

$$\mathrm{var}(\widehat{\theta}) = \mathrm{var}\left(\frac{S_n}{n}\right) = \frac{\theta^2}{n}$$

so that the CRLB is attained in this example. Note, however, that, if we had parametrised the problem in terms of $1/\theta$ instead of θ, writing the density as $f(x;\theta) = \theta e^{-\theta x}$, $x > 0$, then the CRLB would not be attained by the MLE – this statement is left as an exercise for the reader.

Example 8.4 naturally raises the question of when it is possible for the CRLB to be an equality rather than an inequality. Let $W(X)$ be an unbiased estimator of θ. We can argue as follows. The inequality in (8.6) is an equality if and only if $\mathrm{corr}(Y, Z) = \pm 1$, and this can only occur if Y and Z are proportional to one another (as functions of X for each θ). Thus

$$\frac{\partial}{\partial\theta} \log f(X;\theta) = a(\theta)(W(X) - \theta)$$

for some function $a(\theta)$, and on performing an indefinite integration,

$$\log f(X;\theta) = A(\theta)W(X) + B(\theta) + C(X)$$

for some functions A, B and C. Thus the distribution must be of exponential family form with natural statistic $W(X)$, parametrised so that $\mathbb{E}\{W(X)\} = \theta$. Note that this is different from the natural parametrisation referred to throughout Chapters 5–7.

While there have been many occasions when we have used exponential family models as convenient examples, this is the first time we have indicated that some property of a problem *requires* that the model be of exponential family form.

8.3 Convergence of sequences of random variables

In this section we review some basic properties of convergence of random variables, including a technical result (Slutsky's Lemma), which is particularly useful in proving theorems about convergence of MLEs.

Recall the three modes of convergence for a sequence of random variables $\{Y_n,\ n \geq 1\}$:

A sequence of random variables $\{Y_1, Y_2, \ldots\}$ is said to *converge in probability* to $a \in \mathbb{R}$ if, given $\epsilon > 0$ and $\delta > 0$, there exists an $n_0 \equiv n_0(\delta, \epsilon)$ such that, for all $n > n_0$,

$$\mathrm{Pr}(|Y_n - a| > \epsilon) < \delta.$$

We write $Y_n \xrightarrow{p} a$.

A sequence of random variables $\{Y_1, Y_2, \ldots\}$ is said to *converge almost surely* to $a \in \mathbb{R}$ if, given $\epsilon > 0$ and $\delta > 0$, there exists an $n_0 \equiv n_0(\delta, \epsilon)$ such that

$$\mathrm{Pr}(|Y_n - a| > \epsilon \text{ for some } n > n_0) < \delta.$$

We write $Y_n \xrightarrow{a.s} a$.

A sequence of random variables *converges in distribution* if there exists a distribution function F such that

$$\lim_{n \to \infty} \Pr(Y_n \le y) = F(y),$$

for all y that are continuity points of the limiting distribution F. If F is the distribution function of the random variable Y, we write $Y_n \xrightarrow{d} Y$.

Let $X_1, X_2, \ldots, X_n$ be independent, identically distributed random variables with finite mean μ.

The *strong law of large numbers* (SLLN) says that the sequence of random variables $Y_n = n^{-1}(X_1 + \cdots + X_n)$ converges almost surely to μ, if and only if $\mathbb{E}|X_i|$ is finite.

We shall also make use of the *weak law of large numbers* (WLLN), which says that, if the X_i have finite variance, $Y_n \xrightarrow{p} \mu$.

The *Central Limit Theorem* (CLT) says that, under the condition that the X_i are of finite variance σ^2, then a suitably standardised version of Y_n, $Z_n = \sqrt{n}(Y_n - \mu)/\sigma$, converges in distribution to a random variable Z having the standard normal distribution $N(0, 1)$.

A very useful result is *Slutsky's Theorem*, which states that, if $Y_n \xrightarrow{d} Y$ and $Z_n \xrightarrow{p} c$, where c is a finite constant, then, if g is a continuous function, $g(Y_n, Z_n) \xrightarrow{d} g(Y, c)$. In particular: (i) $Y_n + Z_n \xrightarrow{d} Y + c$, (ii) $Y_n Z_n \xrightarrow{d} cY$, (iii) $Y_n / Z_n \xrightarrow{d} Y/c$, if $c \ne 0$.

Remark A trivial extension (which we shall use later) is that, if $Y_n \xrightarrow{d} Y$, $Z_n \xrightarrow{p} c$, $W_n \xrightarrow{p} d$, then $Y_n Z_n W_n \xrightarrow{d} cdY$. Finally there are multivariate extensions: if $\{Y_n\}$ is a sequence of random row vectors converging in distribution to a random vector Y and if Z_n is a sequence of random matrices converging to a fixed matrix C (in the sense that each entry of Z_n converges in probability to the corresponding entry of C), then $Y_n Z_n$ converges in distribution to YC.

8.4 Asymptotic properties of maximum likelihood estimators

Although, as we have seen, maximum likelihood estimators have some nice properties in small samples, it is really for their large sample properties that they are popular, so we turn now to this topic.

For all the discussion which follows we require some regularity conditions on the family, though the exact conditions required differ from one result to the next – for example, consistency (Section 8.4.1) holds under weaker conditions than asymptotic normality and efficiency (Section 8.4.2). At a minimum the latter properties require that the MLE is given by the solution of the likelihood equation and that the conditions required for the CRLB are satisfied. These requirements exclude, for instance, Example 8.2 of Section 8.1 – for that example the asymptotic distribution of $\widehat{\theta}_n$ can be calculated exactly but is not normal: see Problem 8.5.

Also, throughout our discussion we shall assume θ is one-dimensional but this is purely for simplicity of presentation – virtually all the results extend to the case of multidimensional θ and at various points we shall indicate the form of the extension.

8.4.1 Consistency

Let $\widehat{\theta}_n$ denote an estimator of a parameter θ based on a sample of size n. We say that $\widehat{\theta}_n$ is *weakly consistent* if $\widehat{\theta}_n \xrightarrow{p} \theta$ and *strongly consistent* if $\widehat{\theta}_n \xrightarrow{a.s} \theta$. Where the word 'consistent' is used without qualification, it is usually understood to mean weakly consistent.

Suppose $f(x;\theta)$ is a family of probability densities or probability mass functions and let θ_0 denote the true value of the parameter θ. For any $\theta \neq \theta_0$ we have by Jensen's inequality (recall Section 6.4)

$$\mathbb{E}_{\theta_0} \left\{ \log \frac{f(X;\theta)}{f(X;\theta_0)} \right\} \leq \log \mathbb{E}_{\theta_0} \left\{ \frac{f(X;\theta)}{f(X;\theta_0)} \right\} = 0, \tag{8.11}$$

since, for example, in the case of continuous X,

$$\mathbb{E}_{\theta_0} \left\{ \frac{f(X;\theta)}{f(X;\theta_0)} \right\} = \int_{\mathcal{X}} \left\{ \frac{f(x;\theta)}{f(x;\theta_0)} \right\} f(x;\theta_0)dx = \int_{\mathcal{X}} f(x;\theta)dx = 1$$

with an analogous argument in the case of discrete X. Moreover, the inequality in (8.11) is strict unless $f(X;\theta)/f(X;\theta_0) = 1$ (almost everywhere), as a function of X.

Fix $\delta > 0$ and let

$$\mu_1 = \mathbb{E}_{\theta_0} \left\{ \log \frac{f(X;\theta_0 - \delta)}{f(X;\theta_0)} \right\} < 0, \quad \mu_2 = \mathbb{E}_{\theta_0} \left\{ \log \frac{f(X;\theta_0 + \delta)}{f(X;\theta_0)} \right\} < 0.$$

By the SLLN,

$$\frac{l_n(\theta_0 - \delta) - l_n(\theta_0)}{n} \xrightarrow{a.s} \mu_1$$

and so, with probability 1, $l_n(\theta_0 - \delta) < l_n(\theta_0)$, for all n sufficiently large. Similarly, with probability 1, $l_n(\theta_0 + \delta) < l_n(\theta_0)$, for all n sufficiently large. Hence, for all n sufficiently large, there exists an estimator $\widehat{\theta}_n$ which maximises the log-likelihood on $(\theta_0 - \delta, \theta_0 + \delta)$ for any $\delta > 0$ – in this sense, the MLE is a strongly consistent estimator.

Note that this is a very general argument which does not require differentiability of the log-likelihood. For example, in the case of Example 8.2 of Section 8.1, the MLE is strongly consistent even though it is not found by solving the likelihood equation. If one assumes, however, that $l_n(\theta)$ is differentiable for θ in some neighbourhood of θ_0, then the above argument shows that there is a *local* maximum of the likelihood function, which is given by solving the likelihood equations and which is then a consistent estimator. This argument still says nothing about the uniqueness of such an estimator but an obvious consequence is that, if it can be shown by other means that with probability 1 the solution of the likelihood equations is unique for all sufficiently large n, then the solution defines a strongly consistent MLE.

8.4.2 The asymptotic distribution of the maximum likelihood estimator

Now we shall assume that the log-likelihood function $l_n(\theta)$ is twice continuously differentiable on a neighbourhood of θ_0. By the arguments just given in Section 8.4.1, there exists a sequence of local maxima $\widehat{\theta}_n$ such that $l'_n(\widehat{\theta}_n) = 0$ and such that $\widehat{\theta}_n \xrightarrow{a.s} \theta_0$ – we do not make any assumption at this point about the uniqueness of $\widehat{\theta}_n$ though it does turn out that the MLE is unique on any sufficiently small neighbourhood of θ_0.

A Taylor expansion tells us that

$$-l'_n(\theta_0) = l'_n(\widehat{\theta}_n) - l'_n(\theta_0) = (\widehat{\theta}_n - \theta_0)l''_n(\theta_n^*),$$

where θ_n^* lies between θ_0 and $\widehat{\theta}_n$. Thus

$$\widehat{\theta}_n - \theta_0 = -\frac{l'_n(\theta_0)}{l''_n(\theta_n^*)}.$$

Let us now write

$$\sqrt{ni_1(\theta_0)}(\widehat{\theta}_n - \theta_0) = \frac{l'_n(\theta_0)}{\sqrt{ni_1(\theta_0)}} \cdot \frac{l''_n(\theta_0)}{l''_n(\theta_n^*)} \cdot \left\{ -\frac{l''_n(\theta_0)}{ni_1(\theta_0)} \right\}^{-1}.$$

If we can show that

$$\frac{l'_n(\theta_0)}{\sqrt{ni_1(\theta_0)}} \xrightarrow{d} N(0,1), \tag{8.12}$$

$$\frac{l''_n(\theta_0)}{l''_n(\theta_n^*)} \xrightarrow{p} 1, \tag{8.13}$$

$$-\frac{l''_n(\theta_0)}{ni_1(\theta_0)} \xrightarrow{p} 1, \tag{8.14}$$

then, by the slightly extended form of the Slutsky Lemma given at the end of Section 8.3, it will follow that

$$\sqrt{ni_1(\theta_0)}(\widehat{\theta}_n - \theta_0) \xrightarrow{d} N(0,1). \tag{8.15}$$

However, (8.12) is a consequence of the Central Limit Theorem applied to $\sum(\partial/\partial\theta)\log f(X;\theta)$ (each component of which has mean 0 and variance $i_1(\theta)$, by (5.1) and (8.8)) and (8.14) follows from the law of large numbers (in this case the WLLN suffices) also using (8.8). To show (8.13), one condition (sufficient, but not necessary) is to assume that

$$\left| \frac{\partial^3 \log f(x;\theta)}{\partial\theta^3} \right| \leq g(x)$$

uniformly for θ in some neighbourhood of θ_0, where $\mathbb{E}_{\theta_0}\{g(X)\} < \infty$. In that case

$$\left| \frac{l''_n(\theta_n^*) - l''_n(\theta_0)}{n} \right| \leq |\theta_n^* - \theta_0| \cdot \frac{\sum_{i=1}^n g(X_i)}{n}. \tag{8.16}$$

The second factor tends to a constant by the WLLN applied to $\{g(X_i),\ i \geq 1\}$, while the first tends to 0 in probability by the consistency of $\widehat{\theta}_n$ and the obvious fact that $|\theta_n^* - \theta_0| < |\widehat{\theta}_n - \theta_0|$.

Thus the left-hand side of (8.16) tends in probability to 0 and it follows that

$$\frac{l''_n(\theta_n^*)}{l''_n(\theta_0)} - 1 = \frac{l''_n(\theta_n^*) - l''_n(\theta_0)}{n} \cdot \left\{ \frac{l''_n(\theta_0)}{n} \right\}^{-1}, \tag{8.17}$$

where the first factor tends in probability to 0 and the second to $-1/i_1(\theta_0)$. Hence the expression in (8.17) tends in probability to 0, which establishes (8.13), and hence the result (8.15).

8.4.3 Discussion

1 In more informal language, (8.15) says that $\widehat{\theta}_n$ is approximately normally distributed with mean θ_0 and variance $1/(ni_1(\theta_0))$. The latter is of course the CRLB – thus, even though we cannot guarantee that the CRLB is achieved by any estimator for finite n, we have shown that it is asymptotically achieved by the maximum likelihood estimator as $n \to \infty$. In this sense, the MLE is *asymptotically efficient*. We have therefore established the three fundamental properties which are primarily responsible for the great popularity of the maximum likelihood method in practical statistics: consistency, asymptotic normality and asymptotic efficiency. These properties are valid under quite weak conditions, but the conditions (on differentiability of the log-likelihood, validity of the CRLB etc.) are not trivial and there exist interesting and complicated *non-regular* problems, where some or all of these properties do not hold, of which the uniform distribution with unknown endpoint (Example 8.2 of Section 8.1) is one of the best-known examples. Dependent data problems, random processes etc. provide other examples of non-regular problems.

2 Most of what we have said is valid for multiparameter problems as well. For each $\theta = (\theta_1, \ldots, \theta_d) \in \mathbb{R}^d$ define $-i_{jk}(\theta)$ for $1 \le j, k \le d$ by either of the equivalent expressions in (5.3) and let $i_1(\theta)$ be the *Fisher information matrix* with (j, k) entry $i_{jk}(\theta)$. Then the multidimensional analogue of (8.15) states that, when θ_0 is the true value of θ, a consistent sequence of local maximum likelihood estimators $\{\widehat{\theta}_n, \ n \ge 1\}$ exists such that $\sqrt{n}(\widehat{\theta}_n - \theta_0)$ converges in distribution to a multivariate normal vector with mean 0 and covariance matrix $i_1(\theta_0)^{-1}$. It is part of the assumption that $i_1(\theta_0)$ is invertible.

Note that, as before, $i_1(\theta)$ denotes the Fisher information matrix based on a single observation. The Fisher information matrix based on an independent sample of size n is $i_n(\theta) \equiv ni_1(\theta)$, and will be abbreviated to $i(\theta)$ where convenient, as in Section 8.1.3.

3 In practice, in all but very simple problems the maximum likelihood estimators must be found numerically and there exist many computer packages to do this. Most of them use some form of Newton or quasi-Newton optimisation. In order to calculate confidence intervals and tests of hypotheses, we need to know or estimate $i_1(\theta_0)$ as well, and there are two approaches to this (both of which assume that $i_1(\theta)$ is continuous at θ_0).

(a) Sometimes, it will be analytically possible to calculate the theoretical Fisher information matrix $i_1(\theta)$: to estimate $i_1(\theta_0)$, this would usually be evaluated at $\theta = \widehat{\theta}_n$.
(b) A second approach makes use of the observed information matrix defined in Section 8.1.3, the matrix of second-order derivatives of $-l_n(\theta)$, again evaluated at $\theta = \widehat{\theta}_n$, if θ_0 is unspecified.

At first sight it may seem that it would be preferable to use the theoretical Fisher information matrix, if it were easy to calculate, but in fact an extensive body of theory and practice suggests that in most cases the inverse of the observed information matrix gives a better approximation to the true covariance matrix of the estimators, and this is therefore the preferred method in most applications. An especially important reference here is Efron and Hinkley (1978).

8.5 Likelihood ratio tests and Wilks' Theorem

In this section we consider hypothesis testing within the framework of likelihood methods. It is possible to use the results of Section 8.4.2, on the asymptotic distribution of maximum likelihood estimators, directly to construct tests of hypotheses about the parameters, but this is often not the most convenient or the most accurate procedure to adopt. Instead, for certain types of hypotheses, there is a very direct and straightforward method based on *likelihood ratio tests*. The distribution of a likelihood ratio test statistic, when the null hypothesis is true, can in rare cases be computed exactly, but in most cases one has to resort to asymptotic theory. *Wilks' Theorem* is a general result giving this asymptotic distribution. The result involves arguments similar to those in Section 8.4.2, however, and in particular the regularity conditions needed are essentially the same.

Consider a multiparameter problem in which $\theta = (\theta_1, \ldots, \theta_d) \in \Theta$ forming an open subset of $\mathbb{R}^d$ and suppose we want to test a hypothesis of the form

$$H_0 : \theta_1 = \theta_1^0, \ldots, \theta_m = \theta_m^0 \tag{8.18}$$

against the alternative H_1 in which $\theta_1, \ldots, \theta_d$ are unrestricted. Here $1 \le m \le d$ and $\theta_1^0, \ldots, \theta_m^0$ are known prescribed values. With suitable reparametrisation and reordering of the components, many hypotheses can be expressed in this way. In particular, if Θ represents the full model and we are interested in knowing whether the model can be reduced to some $(d - m)$-dimensional subspace of Θ, then it is quite generally possible to express this by means of a null hypothesis of the form of (8.18).

Letting $L(\theta)$ denote the likelihood function and Θ_0 the subset of Θ satisfying the constraints (8.18), we write

$$L_0 = \sup\{L(\theta) : \theta \in \Theta_0\}, \quad L_1 = \sup\{L(\theta) : \theta \in \Theta\}, \tag{8.19}$$

and define the *likelihood ratio statistic*

$$T_n = 2 \log \left(\frac{L_1}{L_0} \right),$$

where the notation indicates dependence on the sample size n. Note that, if we have the wherewithal to calculate maximum likelihood estimates, then there is hardly any additional difficulty in calculating likelihood ratio statistics as well, since the maximisations in (8.19) involve simply calculating the maximum likelihood estimates within the respective models.

The key result is now as follows:

Wilks' Theorem *Suppose the model satisfies the same regularity conditions as are needed for the asymptotic properties of maximum likelihood estimators in Section 8.4.2 – in particular, we require that $L(\theta)$ be at least twice continuously differentiable in all its components in some neighbourhood of the true value of θ, and that the Fisher information matrix be well-defined and invertible. Suppose H_0 is true. Then, as $n \to \infty$,*

$$T_n \xrightarrow{d} \chi_m^2.$$

Proof in the case $d = m = 1$.   So θ is a scalar and we may write $H_0 : \theta = \theta_0$ against $H_1 : \theta$ unrestricted, for some prescribed θ_0.

In this case we have, by taking a Taylor expansion about the maximum likelihood estimate $\widehat{\theta}_n$,

$$
\begin{aligned}
T_n &= 2\left\{l_n(\widehat{\theta}_n) - l_n(\theta_0)\right\} \\
&= 2(\widehat{\theta}_n - \theta_0)l_n'(\widehat{\theta}_n) - (\widehat{\theta}_n - \theta_0)^2 l_n''(\theta_n^\dagger),
\end{aligned}
$$

where $\theta_n^\dagger$ is some other value of θ which lies between θ_0 and $\widehat{\theta}_n$.

However $l_n'(\widehat{\theta}_n) = 0$ by definition, so we quickly see that

$$
\begin{aligned}
T_n &= ni_1(\theta_0)(\widehat{\theta}_n - \theta_0)^2 \cdot \frac{l_n''(\theta_n^\dagger)}{l_n''(\theta_0)} \cdot \frac{l_n''(\theta_0)}{\{-ni_1(\theta_0)\}} \\
&= T_n^{(1)} \cdot T_n^{(2)} \cdot T_n^{(3)} \text{ say.}
\end{aligned}
$$

However $T_n^{(1)}$ is asymptotically the square of a standard normal random variable, and hence distributed as χ_1^2, while $T_n^{(2)}$ and $T_n^{(3)}$ both tend to 1 in probability, by the same arguments as used in Section 8.4.2. Slutsky's Lemma then establishes the required result. □

Discussion

1 The proof for general d and m uses similar methodology, involving Taylor expansion about the true value of θ followed by arguments based on the (multidimensional) Central Limit Theorem and laws of large numbers.

2 An alternative to this approach is to use the asymptotic distribution of the maximum likelihood estimator directly.

Consider just the one-parameter case for simplicity. Assume that (8.15) holds and suppose $\widehat{i}_1$ is a consistent estimator of $i_1(\theta_0)$. Usually we use either the Fisher information evaluated at $\widehat{\theta}_n$, or else the observed information, as in the discussion of Section 8.4.3. Provided $\widehat{i}_1/i_1(\theta_0) \xrightarrow{P} 1$, a further application of Slutsky's Lemma gives

$$
\sqrt{n\widehat{i}_1}(\widehat{\theta}_n - \theta_0) \xrightarrow{d} N(0, 1).
$$

An approximate size α test of $H_0 : \theta = \theta_0$ against $H_1 : \theta \neq \theta_0$ is then: reject H_0 if $\sqrt{n\widehat{i}_1}|\widehat{\theta}_n - \theta_0| > z_{\alpha/2}$, where, as previously, z_β denotes the upper-β point of the standard normal distribution.

Once again, the extension to the multiparameter case is essentially straightforward: some details are given in Section 8.6 below. This procedure of using the asymptotic distribution of the MLE to construct a test is sometimes called *Wald's test*.

There is yet a third procedure, which is essentially the locally most powerful test mentioned briefly in Chapter 4. This test is based on the value of $U(\theta_0)$, the score function whose asymptotic distribution follows directly from the Central Limit Theorem. We have, under H_0, that $U(\theta_0)$ is asymptotically normally distributed, with mean 0 and variance $ni_1(\theta_0)$. The *score test* is based on referring $U(\theta_0)/\sqrt{n\widehat{i}_1}$ to the standard normal distribution, with, as before, $\widehat{i}_1$ a consistent estimator of $i_1(\theta_0)$. Again, the extension to the multiparameter case is straightforward: see Section 8.6.

3 For a scalar θ, a test of $H_0 : \theta = \theta_0$ may be based on

$$
r(\theta_0) = \text{sgn}(\widehat{\theta} - \theta_0)\sqrt{T_n},
$$

the *signed root likelihood ratio statistic*.

It may be shown that $r \xrightarrow{d} N(0, 1)$, so again a test may be carried out by referring $r(\theta_0)$ to the standard normal distribution.

Also, $i(\widehat{\theta})^{1/2}(\widehat{\theta} - \theta)$ is asymptotically $N(0, 1)$, so that an approximate $100(1 - \alpha)\%$ confidence interval for θ is

$$\widehat{\theta} \mp i(\widehat{\theta})^{-1/2}\Phi^{-1}(1 - \alpha/2),$$

in terms of the $N(0, 1)$ distribution function Φ.

4 Finally it should be mentioned that in recent years a number of improvements to the likelihood ratio procedure have been suggested, generally designed to ensure improved agreement between asymptotic and exact distributions. Chapter 9 gives a review of such procedures. Many of these tests implicitly involve conditioning on exactly or approximately ancillary statistics, so indirectly calling on arguments of the kind given in Section 7.2.

8.6 More on multiparameter problems

Consider again the multiparameter problem in which $\theta = (\theta_1, \ldots, \theta_d) \in \Theta$, an open subset of $\mathbb{R}^d$.

8.6.1 No nuisance parameter case

Suppose first that there are no nuisance parameters and that we are interested in testing $H_0 : \theta = \theta_0$. There are many ways of testing H_0. We describe three tests, as outlined above for the case of scalar θ:

1 the likelihood ratio test based on the statistic

$$w(\theta_0) \equiv T_n = 2\{l_n(\widehat{\theta}) - l_n(\theta_0)\},$$

2 the score test based on the statistic

$$w_s(\theta_0) = U(\theta_0)^T i^{-1}(\theta_0) U(\theta_0), \tag{8.20}$$

3 the Wald test based on the statistic

$$w_w(\theta_0) = (\widehat{\theta} - \theta_0)^T i(\theta_0)(\widehat{\theta} - \theta_0). \tag{8.21}$$

In each case the asymptotic null distribution of the statistic is χ_d^2. Further, in $w_s(\theta_0)$ and $w_w(\theta_0)$, $i(\theta_0)$ can be replaced by a consistent estimator: if $i(\theta)$ is continuous at θ_0, any of $i(\widehat{\theta})$, $j(\widehat{\theta})$, $j(\theta_0)$, $i(\theta^*)$ or $j(\theta^*)$, where θ^* is a consistent estimator of θ and $j(\theta)$ is the observed information matrix, may be used.

Confidence regions at level $1 - \alpha$ may be formed approximately as, for example,

$$\{\theta : w(\theta) \leq \chi_{d,\alpha}^2\},$$

where $\chi_{d,\alpha}^2$ is the upper α point of the relevant chi-squared distribution χ_d^2.

8.6.2 Nuisance parameter case: profile likelihood

Typically, interest lies in inference for a subparameter or parameter function $\psi = \psi(\theta)$. The *profile likelihood* $L_p(\psi)$ for ψ is defined by

$$L_p(\psi) = \sup_{\{\theta:\psi(\theta)=\psi\}} L(\theta),$$

so that the supremum of $L(\theta)$ is taken over all θ that are consistent with the given value of ψ.

The profile log-likelihood is $l_p = \log L_p$. Again, it may be written as l_{np} if it is to be stressed that it is based on a sample of size n.

Often ψ is a component of a given partition $\theta = (\psi, \chi)$ of θ into sub-vectors ψ and χ of dimension m and $d - m$ respectively, and we may then write

$$L_p(\psi) = L(\psi, \widehat{\chi}_\psi),$$

where $\widehat{\chi}_\psi$ denotes the maximum likelihood estimate of χ for a given value of ψ. We assume this is the case from now on, and denote the maximum likelihood estimator of θ by $\widehat{\theta} = (\widehat{\psi}, \widehat{\chi})$.

The profile likelihood $L_p(\psi)$ can, to a considerable extent, be thought of and used as if it were a genuine likelihood. In particular, the maximum profile likelihood estimate of ψ equals $\widehat{\psi}$. Further, the profile log-likelihood ratio statistic $2\{l_p(\widehat{\psi}) - l_p(\psi_0)\}$ equals the log-likelihood ratio statistic for $H_0 : \psi = \psi_0$,

$$2\{l_p(\widehat{\psi}) - l_p(\psi_0)\} \equiv 2\{l(\widehat{\psi}, \widehat{\chi}) - l(\psi_0, \widehat{\chi}_0)\},$$

where $l \equiv l_n$ is the log-likelihood and we have written $\widehat{\chi}_0$ for $\widehat{\chi}_{\psi_0}$. The asymptotic null distribution of the profile log-likelihood ratio statistic is therefore given by Wilks' Theorem of Section 8.5.

Patefield (1977) gave the important result that the inverse of the observed profile information equals the ψ component of the full observed inverse information evaluated at $(\psi, \widehat{\chi}_\psi)$,

$$j_p^{-1}(\psi) = j^{\psi\psi}(\psi, \widehat{\chi}_\psi).$$

Here j_p denotes observed profile information, that is minus the matrix of second-order derivatives of l_p, and $j^{\psi\psi}$ is the $\psi\psi$-block of the inverse of the full observed information j, obtained by partitioning $j(\theta)$ and its inverse as

$$j(\theta) = \begin{bmatrix} j_{\psi\psi}(\psi, \chi) & j_{\psi\chi}(\psi, \chi) \\ j_{\chi\psi}(\psi, \chi) & j_{\chi\chi}(\psi, \chi) \end{bmatrix},$$

$$j^{-1}(\theta) = \begin{bmatrix} j^{\psi\psi}(\psi, \chi) & j^{\psi\chi}(\psi, \chi) \\ j^{\chi\psi}(\psi, \chi) & j^{\chi\chi}(\psi, \chi) \end{bmatrix},$$

corresponding to the partitioning of θ.

For scalar ψ, this result follows on differentiating $l_p(\psi) = l(\psi, \widehat{\chi}_\psi)$ twice with respect to ψ. Let l_ψ and l_χ denote the partial derivatives of $l(\psi, \chi)$ with respect to ψ, χ respectively. The profile score is $l_\psi(\psi, \widehat{\chi}_\psi)$, on using the chain rule to differentiate $l_p(\psi)$ with respect to ψ, noting that $l_\chi(\psi, \widehat{\chi}_\psi) = 0$. The second derivative is, following the notation,

$l_{\psi\psi}(\psi, \widehat{\chi}_\psi) + l_{\psi\chi}(\psi, \widehat{\chi}_\psi)\frac{\partial}{\partial\psi}\widehat{\chi}_\psi$. Now use the result that

$$\partial\widehat{\chi}_\psi/\partial\psi = -j_{\psi\chi}(\psi, \widehat{\chi}_\psi)j_{\chi\chi}^{-1}(\psi, \widehat{\chi}_\psi).$$

This latter formula follows by differentiating the likelihood equation $l_\chi(\psi, \widehat{\chi}_\psi) = 0$ with respect to ψ. This gives

$$l_{\chi\psi}(\psi, \widehat{\chi}_\psi) + l_{\chi\chi}(\psi, \widehat{\chi}_\psi)\frac{\partial}{\partial\psi}\widehat{\chi}_\psi = 0,$$

from which

$$\frac{\partial}{\partial\psi}\widehat{\chi}_\psi = -(l_{\chi\chi}(\psi, \widehat{\chi}_\psi))^{-1}l_{\chi\psi}(\psi, \widehat{\chi}_\psi).$$

It follows that

$$j_{\mathrm{p}}(\psi) = -(l_{\psi\psi} - l_{\psi\chi}(l_{\chi\chi})^{-1}l_{\chi\psi}),$$

where all the derivatives are evaluated at $(\psi, \widehat{\chi}_\psi)$. Then, using the formulae for the inverse of a partitioned matrix, as given in Section 8.1.5, the result is proved. The vector case follows similarly.

When ψ is scalar, this implies that the curvature of the profile log-likelihood is directly related to the precision of $\widehat{\psi}$. We have seen that a key property of the log-likelihood $l(\theta)$ when there are no nuisance parameters is that the observed information $j(\widehat{\theta})$ can be used as an estimate of the inverse asymptotic covariance matrix of $\widehat{\theta}$, which is actually $i(\theta)$. The above result shows that the corresponding function computed from the profile log-likelihood,

$$j_{\mathrm{p}}(\widehat{\psi}) = -[\nabla_\psi\nabla_\psi^T l_{\mathrm{p}}(\psi)]_{\psi=\widehat{\psi}}$$

determines an estimate of the inverse asymptotic covariance matrix for $\widehat{\psi}$.

8.6.3 Further test statistics

For testing $H_0 : \psi = \psi_0$, in the presence of a nuisance parameter χ, the forms of the score statistic (8.20) and the Wald statistic (8.21) corresponding to the profile log-likelihood ratio statistic are obtained by partitioning the maximum likelihood estimate, the score vector $U(\theta) \equiv l_n'(\theta)$, the information matrix $i(\theta)$ and its inverse in the way considered in the previous section:

$$U(\theta) = \begin{pmatrix} U_\psi(\psi, \chi) \\ U_\chi(\psi, \chi) \end{pmatrix},$$

$$i(\theta) = \begin{bmatrix} i_{\psi\psi}(\psi, \chi) & i_{\psi\chi}(\psi, \chi) \\ i_{\chi\psi}(\psi, \chi) & i_{\chi\chi}(\psi, \chi) \end{bmatrix},$$

$$i^{-1}(\theta) = \begin{bmatrix} i^{\psi\psi}(\psi, \chi) & i^{\psi\chi}(\psi, \chi) \\ i^{\chi\psi}(\psi, \chi) & i^{\chi\chi}(\psi, \chi) \end{bmatrix}.$$

Because of the asymptotic normality of $U(\theta)$, we have that $U_\psi(\psi_0, \chi_0)$ is asymptotically normal with zero mean and inverse covariance matrix $i^{\psi\psi}(\psi_0, \chi_0)$, under the true $\theta_0 = (\psi_0, \chi_0)$. Replacing χ_0 by $\widehat{\chi}_0$ we obtain a version of the score statistic (8.20) for testing

$H_0 : \psi = \psi_0$:

$$w_{sp}(\psi_0) = U_\psi(\psi_0, \widehat{\chi}_0)^T i^{\psi\psi}(\psi_0, \widehat{\chi}_0) U_\psi(\psi_0, \widehat{\chi}_0). \qquad (8.22)$$

This test has the advantage, over the likelihood ratio statistic for example, that χ has to be estimated only under H_0.

Similarly, $\widehat{\psi}$ is asymptotically normally distributed with mean ψ_0 and covariance matrix $i^{\psi\psi}(\psi_0, \chi_0)$, which can be replaced by $i^{\psi\psi}(\psi_0, \widehat{\chi}_0)$, yielding a version of the Wald test statistic (8.21) for this nuisance parameter case:

$$w_{wp}(\psi_0) = (\widehat{\psi} - \psi_0)^T [i^{\psi\psi}(\psi_0, \widehat{\chi}_0)]^{-1} (\widehat{\psi} - \psi_0). \qquad (8.23)$$

Both $w_{sp}(\psi_0)$ and $w_{wp}(\psi_0)$ have asymptotically a chi-squared distribution with m degrees of freedom.

The three test procedures – the likelihood ratio test, Wald's test and the score test – are all asymptotically equivalent, as can easily be checked (at least in the one-parameter case) by Taylor expansion of the log-likelihood. In practice, when the three tests are evaluated numerically, they often lead to substantially different answers. There is no clear-cut theory to establish which procedure is best, but there is a substantial body of literature pointing towards the conclusion that, of the three, the likelihood ratio procedure has the best agreement between the true and asymptotic distributions. Cox and Hinkley (1974) surveyed the literature on this field up to the time that book was published; a much more up-to-date but also much more advanced treatment is Barndorff-Nielsen and Cox (1994).

8.7 Problems

8.1 Let $X = (X_1, \ldots, X_n)$ be a random sample of size $n \geq 3$ from the exponential distribution of mean $1/\theta$. Find a sufficient statistic $T(X)$ for θ and write down its density. Obtain the maximum likelihood estimator $\widehat{\theta}_n$ based on the sample of size n for θ and show that it is biased, but that a multiple of it is not.

Calculate the Cramér–Rao Lower Bound for the variance of an unbiased estimator, and explain why you would not expect the bound to be attained in this example. Confirm this by calculating the variance of your unbiased estimator and comment on its behaviour as $n \to \infty$.

8.2 You are given a coin, which you are going to test for fairness. Let the probability of a head be p, and consider testing $H_0 : p = 1/2$ against $H_1 : p > 1/2$.

(i) You toss the coin 12 times and observe nine heads, three tails. Do you reject H_0 in a test of size $\alpha = 0.05$?

(ii) You toss the coin until you observe the third head, and note that this occurs on the 12th toss. Do you reject H_0 in a test of size $\alpha = 0.05$?

In (i), the number of heads has a binomial distribution, while in (ii) the number of tosses performed has a negative binomial distribution. What is the likelihood function for p in the two cases? The likelihood principle would demand that identical inference about p would be drawn in the two cases. Comment.

8.3 Consider the (exponential) family of distributions on $(0, \infty)$ with densities

$$f(x; \theta) \propto \exp(-\theta x^a),$$

$\theta > 0$, where $a > 1$ is fixed. Find the normalising constant for the density, and calculate the maximum likelihood estimator $\widehat{\theta}_n$ based on an independent sample of size n from this distribution. What is the asymptotic distribution of the maximum likelihood estimator?

8.4 The family of densities

$$f(x; \theta) = \frac{2}{\pi} \frac{e^{\theta x} \cos(\theta \pi / 2)}{\cosh(x)}, \quad x \in \mathbb{R},$$

constitutes an exponential family. Find the maximum likelihood estimator of θ based on an independent sample of size n, and compute the Fisher information.

8.5 Let $X_1, \ldots, X_n$ be independent and uniformly distributed on $(0, \theta)$, $\theta > 0$. Find the maximum likelihood estimator $\widehat{\theta}_n$ of θ, its expectation, and its variance. Is it consistent? What is the asymptotic distribution of $\{\mathbb{E}(\widehat{\theta}_n) - \widehat{\theta}_n\} / \sqrt{\text{var}\{\widehat{\theta}_n\}}$?

8.6 Let $X_1, \ldots, X_n$ be independent and identically distributed with density of the form $\rho e^{-\rho x}$, $x \geq 0$. Find the forms of the Wald, score and likelihood ratio statistics for testing the hypothesis $H_0 : \rho = \rho_0$ against the unrestricted alternative $H_1 : \rho > 0$, and verify their asymptotic equivalence.

8.7 Suppose that $x_1, x_2, \ldots, x_n$ are fixed real numbers such that

$$\Sigma x_k = 0, \quad \Sigma x_k^2 = n.$$

Let $Y_1, Y_2, \ldots, Y_n$ be independent random variables such that Y_k is normally distributed with mean $\alpha + \beta x_k$ and variance V.

(a) Suppose first that the variance V is known to be 1, but that α and β are unknown parameters. It is desired to test the null hypothesis $H_0 : (\alpha, \beta) = (0, 0)$ against the unrestricted alternative. Let $\widehat{L}$ be the likelihood of the observations $Y_1, Y_2, \ldots, Y_n$ when (α, β) takes the value $(\widehat{\alpha}, \widehat{\beta})$ of the maximum likelihood estimator; and let L_0 be the likelihood under the null hypothesis H_0. Prove that, under the null hypothesis, the *exact* distribution of $2 \log(\widehat{L}/L_0)$ is the χ^2 distribution with 2 degrees of freedom.

(b) Suppose now that V is not known. Explain how Wilks' asymptotic likelihood ratio test would be used to test $(\alpha, \beta, V) = (0, 0, 1)$ against an unrestricted alternative. How would you test $(\alpha, \beta) = (0, 0)$, with V as nuisance parameter?

8.8 Let $Y_1, \ldots, Y_n$ be independent and identically distributed $N(\mu, \sigma^2)$, and let the parameter of interest be μ, with σ^2 unknown. Obtain the form of the profile log-likelihood for μ, and show how to construct a confidence interval of approximate coverage $1 - \alpha$ from the profile log-likelihood. How would you construct a confidence interval of *exact* coverage $1 - \alpha$?

8.9 Let $Y_1, \ldots, Y_n$ be independent, identically distributed with joint density $f(y; \theta)$, with θ a scalar parameter.

Suppose a one-to-one differentiable transformation is made from θ to $\phi = \phi(\theta)$. What is the relationship between the Fisher information $i(\phi)$ about ϕ contained in the sample and $i(\theta)$, the Fisher information about θ?

In Bayesian inference for θ, it was suggested by Jeffreys that a uniform prior should be assumed for that function ϕ of θ for which the Fisher information $i(\phi)$ is constant. By considering the relationship between the prior density of θ and that of ϕ, show that the suggestion leads to a prior density for θ itself which is proportional to $i(\theta)^{1/2}$.

Let X be a single observation from a binomial(k, θ) distribution. What is the *Jeffreys' prior* for θ?

8.10 A random sample of size n is taken from the normal distribution $N(\mu, 1)$. Show that the maximum likelihood estimator $\widehat{\mu}_n$ of μ is the minimum variance unbiased estimate, and find its distribution and variance.

Now define a second estimator T_n by

$$T_n = \begin{cases} \widehat{\mu}_n & \text{when } |\widehat{\mu}_n| \geq n^{-1/4}, \\ \frac{1}{2}\widehat{\mu}_n & \text{when } |\widehat{\mu}_n| < n^{-1/4}. \end{cases}$$

Compare the asymptotic distributions of T_n and $\widehat{\mu}_n$, (a) when $\mu = 0$, (b) when $\mu > 0$. Compare the mean squared errors of $\widehat{\mu}_n$ and T_n for finite n, numerically if appropriate. Is T_n a sensible estimator in practice?

(This is an example of *asymptotic superefficiency*. Asymptotically, T_n outperforms the minimum variance unbiased estimator at the point of superefficiency $\mu = 0$, but for fixed n does much worse than $\widehat{\mu}_n$ for values close to $\mu = 0$. This is a general feature of an asymptotically superefficient estimator of a one-dimensional parameter.)

8.11 Let $X_1, \ldots, X_n$ be independent and distributed as $N(\mu, 1)$; let $Y_i = |X_i|$, $i = 1, \ldots, n$, and $\theta = |\mu|$. Show that the log-likelihood function for θ, given the data $Y_i = y_i$, $i = 1, \ldots, n$, may be written, ignoring an additive constant, as

$$l(\theta) = -\frac{1}{2}\Sigma_i(y_i^2 + \theta^2) + \Sigma_i \log(e^{\theta y_i} + e^{-\theta y_i}),$$

and that in the neighbourhood of $\theta = 0$

$$l(\theta) = l(0) + \frac{1}{2}\theta^2(\Sigma_i y_i^2 - n) - \frac{1}{12}\theta^4\Sigma_i y_i^4 + O(\theta^6).$$

Deduce that the equation for maximising $l(\theta)$ has a solution $\widehat{\theta}_n$, where:

(a) if $\Sigma_i y_i^2 \leq n, \widehat{\theta}_n = 0$;
(b) if $\Sigma_i y_i^2 > n, \widehat{\theta}_n^2 = 3(\Sigma_i y_i^2 - n)/\Sigma_i y_i^4$.

Hence show that, if $\theta = 0$, the asymptotic distribution of $\widehat{\theta}_n^2\sqrt{n}$ is the positive half of $N(0, 2)$, together with an atom of probability $(p = \frac{1}{2})$ at 0. What is the asymptotic distribution of $\widehat{\theta}_n$ in the case $\theta \neq 0$?

(In the case $\theta = 0$, the true parameter value lies on the boundary of the parameter space, so the conventional asymptotic normality asymptotics do not hold.)

9

Higher-order theory

In Chapter 8, we sketched the asymptotic theory of likelihood inference. It is our primary purpose in this chapter to describe refinements to that asymptotic theory, our discussion having two main origins. One motivation is to improve on the first-order limit results of Chapter 8, so as to obtain approximations whose asymptotic accuracy is higher by one or two orders. The other is the Fisherian proposition that inferences on the parameter of interest should be obtained by conditioning on an ancillary statistic, rather than from the original model. We introduce also in this chapter some rather more advanced ideas of statistical theory, which provide important underpinning of statistical methods applied in many contexts.

Some mathematical preliminaries are described in Section 9.1, most notably the notion of an *asymptotic expansion*. The concept of *parameter orthogonality*, and its consequences for inference, are discussed in Section 9.2. Section 9.3 describes ways of dealing with a nuisance parameter, through the notions of *marginal likelihood* and *conditional likelihood*. *Parametrisation invariance* (Section 9.4) provides an important means of discrimination between different inferential procedures. Two particularly important forms of asymptotic expansion, *Edgeworth expansion* and *saddlepoint expansion*, are described in Section 9.5 and Section 9.6 respectively. The *Laplace approximation* method for approximation of integrals is described briefly in Section 9.7. The remainder of the chapter is concerned more with inferential procedures. Section 9.8 presents a highlight of modern likelihood-based inference: the so-called p^* approximation to the *conditional density of a maximum likelihood estimator*, given an ancillary statistic. This formula leads to adjusted forms of the signed root likelihood ratio statistic, relevant to inference on a scalar parameter of interest and distributed as $N(0, 1)$ to a high degree of accuracy. Conditional inference to eliminate nuisance parameters in exponential families, as discussed already in Section 7.2.4, is considered further in Section 9.9. A striking means of improving the χ^2 approximation to the distribution of the likelihood ratio statistic, *Bartlett correction*, is considered in Section 9.10, while Section 9.11 defines a *modified profile likelihood*, constructed to behave more like a genuine likelihood function than the profile likelihood introduced in Section 8.6. The chapter concludes in Section 9.12 with a brief review of the asymptotic theory of Bayesian inference.

9.1 Preliminaries

9.1.1 Mann–Wald notation

In asymptotic theory, the so-called Mann–Wald or '*o* and *O*' notation is useful, to describe the order of magnitude of specified quantities. For two sequences of positive constants $(a_n), (b_n)$, we write $a_n = o(b_n)$ when $\lim_{n \to \infty}(a_n/b_n) = 0$, and $a_n = O(b_n)$ when $\limsup_{n \to \infty}(a_n/b_n) = K < \infty$. For sequences of random variables $\{Y_n\}$, we write $Y_n = o_p(a_n)$ if $Y_n/a_n \xrightarrow{p} 0$ as $n \to \infty$ and $Y_n = O_p(a_n)$ when Y_n/a_n is bounded in probability as $n \to \infty$, that is given $\epsilon > 0$ there exist $k > 0$ and n_0 such that, for all $n > n_0$,

$$\Pr(|Y_n/a_n| < k) > 1 - \epsilon.$$

In particular, $Y_n = c + o_p(1)$ means that $Y_n \xrightarrow{p} c$.

To illustrate the use of the notation, key results from Chapter 8 may be described in terms of this notation as follows. We suppose for simplicity the case of inference for a scalar parameter θ, with no nuisance parameter: extensions to the multivariate case and to cases involving nuisance parameters involve only notational complication. We have $\sqrt{i(\theta)}(\widehat{\theta} - \theta) = Z + O_p(n^{-1/2})$, where Z is $N(0, 1)$. Similarly, the likelihood ratio statistic $w(\theta) = 2\{l(\widehat{\theta}) - l(\theta)\} = W + O_p(n^{-1/2})$, where W is χ_1^2. For testing the hypothesis $H_0 : \theta = \theta_0$, the three test statistics, likelihood ratio, Wald and score, differ by $O_p(n^{-1/2})$. When considering the distribution functions, if we let $r(\theta) = \text{sgn}(\widehat{\theta} - \theta)w(\theta)^{1/2}$ be the signed root likelihood ratio statistic we have $\Pr(r \leq r_0) = \Phi(r_0) + O(n^{-1/2})$, while $\Pr(w \leq w_0) = G(w_0) + O(n^{-1})$, where $G(\cdot)$ is the distribution function of χ_1^2. These results are easily verified by careful accounting, in the analysis given in Chapter 8 of the orders of magnitude of various terms in expansions. The key to this is to note that the score function $U(\theta)$ and the Fisher information $i(\theta)$ refer to the *whole* vector X of dimension n, and that as $n \to \infty$, subject to suitable regularity,

$$U(\theta) = O_p(n^{1/2}),$$
$$i(\theta) = O(n),$$
$$\widehat{\theta} - \theta = O_p(n^{-1/2}).$$

Actually, it is necessary that these quantities are *precisely* of these stated orders for the results of Chapter 8 to hold.

9.1.2 Moments and cumulants

The moment generating function of a scalar random variable X is defined by $M_X(t) = \mathbb{E}\{\exp(tX)\}$ whenever this expectation exists. Note that $M_X(0) = 1$, and that the moment generating function is defined in some interval containing 0. If $M_X(t)$ exists for t in an open interval around 0, then all the moments $\mu_r' = \mathbb{E}X^r$ exist, and we have the Taylor expansion

$$M_X(t) = 1 + \mu_1't + \mu_2'\frac{t^2}{2!} + \cdots + \mu_r'\frac{t^r}{r!} + O(t^{r+1}),$$

as $t \to 0$.

The cumulant generating function $K_X(t)$ is defined by $K_X(t) = \log\{M_X(t)\}$, defined on the same interval as $M_X(t)$. Provided $M_X(t)$ exists in an open interval around 0, the Taylor series expansion

$$K_X(t) = \kappa_1 t + \kappa_2 \frac{t^2}{2!} + \cdots + \kappa_r \frac{t^r}{r!} + O(t^{r+1}),$$

as $t \to 0$, defines the rth cumulant κ_r.

The rth cumulant κ_r can be expressed in terms of the rth and lower-order moments by equating coefficients in the expansions of $\exp\{K_X(t)\}$ and $M_X(t)$. We have, in particular, $\kappa_1 = \mathbb{E}(X) = \mu_1'$ and $\kappa_2 = \text{var}(X) = \mu_2' - \mu_1'^2$. The third and fourth cumulants are called the skewness and kurtosis respectively. For the normal distribution, all cumulants of third and higher order are 0.

Note that, for $a, b \in \mathbb{R}$, $K_{aX+b}(t) = bt + K_X(at)$, so that, if $\tilde{\kappa}_r$ is the rth cumulant of $aX + b$, then $\tilde{\kappa}_1 = a\kappa_1 + b, \tilde{\kappa}_r = a^r \kappa_r, r \geq 2$. Also, if $X_1, \ldots, X_n$ are independent and identically distributed random variables with cumulant generating function $K_X(t)$, and $S_n = X_1 + \ldots + X_n$, then $K_{S_n}(t) = n K_X(t)$.

Extension of these notions to multivariate X involves no conceptual complication: see Pace and Salvan (1997: Chapter 3).

9.1.3 Asymptotic expansions

Various technical tools are of importance in the development of statistical theory. Key methods, which we describe in subsequent sections, used to obtain higher-order approximations to densities and distribution functions are Edgeworth expansion, saddlepoint approximation and Laplace's method. Here we consider first two important general ideas, those of asymptotic expansion and stochastic asymptotic expansion.

Asymptotic expansions typically arise in the following way. We are interested in a sequence of functions $\{f_n(x)\}$, indexed by n, and write

$$f_n(x) = \gamma_0(x)b_{0,n} + \gamma_1(x)b_{1,n} + \gamma_2(x)b_{2,n} + \cdots + \gamma_k(x)b_{k,n} + o(b_{k,n}),$$

as $n \to \infty$, where $\{b_{r,n}\}_{r=0}^k$ is a sequence, such as $\{1, n^{-1/2}, n^{-1}, \ldots, n^{-k/2}\}$ or $\{1, n^{-1}, n^{-2}, \ldots, n^{-k}\}$. An essential condition is that $b_{r+1,n} = o(b_{r,n})$ as $n \to \infty$, for each $r = 0, 1, \ldots, k - 1$.

Often the function of interest $f_n(x)$ will be the exact density or distribution function of a statistic based on a sample of size n at a fixed x, and $\gamma_0(x)$ will be some simple first-order approximation, such as the normal density or distribution function. One important feature of asymptotic expansions is that they are not in general convergent series for $f_n(x)$ for any fixed x: taking successively more terms, letting $k \to \infty$ for fixed n, will not necessarily improve the approximation to $f_n(x)$.

We will concentrate here on asymptotic expansions for densities, but describe some of the key formulae in distribution function estimation.

For a sequence of random variables $\{Y_n\}$, a *stochastic asymptotic expansion* is expressed as

$$Y_n = X_0 b_{0,n} + X_1 b_{1,n} + \cdots + X_k b_{k,n} + o_p(b_{k,n}),$$

where $\{b_{k,n}\}$ is a given set of sequences and $\{X_0, X_1, \ldots\}$ are random variables having distributions not depending on n.

There are several examples of the use of stochastic asymptotic expansions in the literature, but they are not as well defined as asymptotic expansions, as there is usually considerable arbitrariness in the choice of the coefficient random variables $\{X_0, X_1, \ldots\}$, and it is often convenient to use instead of $X_0, X_1, \ldots$ random variables for which only the asymptotic distribution is free of n. A simple application of stochastic asymptotic expansion is the proof of asymptotic normality of the maximum likelihood estimator, as sketched in Chapter 8: we have

$$\sqrt{i(\theta)}(\widehat{\theta} - \theta) = \left\{ \frac{U(\theta)}{\sqrt{i(\theta)}} \right\} + O_p(n^{-1/2}),$$

in terms of the score $U(\theta)$ and Fisher information $i(\theta)$. The quantity $U(\theta)/\sqrt{i(\theta)}$ plays the role of X_0. By the CLT we can write

$$\frac{U(\theta)}{\sqrt{i(\theta)}} = X_0 + O_p(n^{-1/2}),$$

where X_0 is $N(0, 1)$.

9.2 Parameter orthogonality

We work now with a multidimensional parameter θ. As we saw in Chapter 8, typically θ may be partitioned as $\theta = (\psi, \chi)$, where ψ is the parameter of interest and χ is a nuisance parameter. When making inference about ψ, we may seek to minimise the effect of our lack of knowledge about the nuisance parameter using the notion of *parameter orthogonality*.

9.2.1 Definition

Suppose that θ is partitioned into components $\theta = (\theta_1, \ldots, \theta_{d_1}; \theta_{d_1+1}, \ldots, \theta_d) = (\theta_{(1)}, \theta_{(2)})$. Suppose that the Fisher information matrix $i(\theta) \equiv [i_{rs}(\theta)]$ has $i_{rs}(\theta) = 0$ for all $r = 1, \ldots, d_1$, $s = d_1 + 1, \ldots, d$, for all $\theta \in \Theta$, so that $i(\theta)$ is block diagonal. We then say that $\theta_{(1)}$ is *orthogonal* to $\theta_{(2)}$.

9.2.2 An immediate consequence

Orthogonality implies that the corresponding components of the score $U(\theta)$ are uncorrelated, since $U(\theta)$ has covariance matrix $i(\theta)$.

9.2.3 The case $d_1 = 1$

Suppose we have $\theta = (\psi, \chi_1, \ldots, \chi_q)$, with ψ the parameter of interest and $\chi_1, \ldots, \chi_q$, $q = d - 1$, nuisance parameters. Then it is always possible to find an *interest-respecting* reparametrisation under which the parameter of interest ψ and the nuisance parameters are orthogonal, that is we can find $\lambda_1, \ldots, \lambda_q$ as functions of $(\psi, \chi_1, \ldots, \chi_q)$ such that ψ is orthogonal to $(\lambda_1, \ldots, \lambda_q)$. We discuss obtaining such a reparametrisation for the case $q = 1$, though the same argument applies for $q > 1$.

Let $\tilde{l}$ and $\tilde{i}$ be the log-likelihood and information matrix in terms of (ψ, χ) and write $\chi = \chi(\psi, \lambda)$. Then

$$l(\psi, \lambda) = \tilde{l}\{\psi, \chi(\psi, \lambda)\}$$

and

$$\frac{\partial^2 l}{\partial \psi \partial \lambda} = \frac{\partial^2 \tilde{l}}{\partial \psi \partial \chi} \frac{\partial \chi}{\partial \lambda} + \frac{\partial^2 \tilde{l}}{\partial \chi^2} \frac{\partial \chi}{\partial \lambda} \frac{\partial \chi}{\partial \psi} + \frac{\partial \tilde{l}}{\partial \chi} \frac{\partial^2 \chi}{\partial \psi \partial \lambda},$$

on differentiating twice using the chain rule.

Now take expectations. The final term vanishes and orthogonality of ψ and λ then requires

$$\frac{\partial \chi}{\partial \lambda} \left(\tilde{i}_{\psi\chi} + \tilde{i}_{\chi\chi} \frac{\partial \chi}{\partial \psi} \right) = 0.$$

Assuming that $\frac{\partial \chi}{\partial \lambda}$ is non-zero, this is equivalent to

$$\tilde{i}_{\chi\chi} \frac{\partial \chi}{\partial \psi} + \tilde{i}_{\psi\chi} = 0. \tag{9.1}$$

This partial differential equation determines the dependence of λ on ψ and χ, and is solvable in general. However, the dependence is not determined uniquely and there remains considerable arbitrariness in the choice of λ.

9.2.4 An example

Let (Y_1, Y_2) be independent, exponentially distributed with means $(\chi, \psi\chi)$. Then $q = 1$ and equation (9.1) becomes

$$2\chi^{-2} \frac{\partial \chi}{\partial \psi} = -(\psi\chi)^{-1},$$

the solution of which is $\chi = g(\lambda)/\psi^{1/2}$, where $g(\lambda)$ is an arbitrary function of λ. A convenient choice is $g(\lambda) \equiv \lambda$, so that in the orthogonal parametrisation the means are $\lambda/\psi^{1/2}$ and $\lambda\psi^{1/2}$.

9.2.5 The case $d_1 > 1$

When $\dim(\psi) > 1$ there is no guarantee that a λ may be found so that ψ and λ are orthogonal.

If, for example, there were two components ψ_1 and ψ_2 for which it was required to satisfy (9.1), there would in general be no guarantee that the values of $\partial \chi_r / \partial \psi_1$ and $\partial \chi_r / \partial \psi_2$ so obtained would satisfy the compatibility condition

$$\frac{\partial^2 \chi_r}{\partial \psi_1 \partial \psi_2} = \frac{\partial^2 \chi_r}{\partial \psi_2 \partial \psi_1}.$$

9.2.6 Further remarks

For a fixed value ψ_0 of ψ it is possible to determine λ so that $i_{\psi\lambda}(\psi_0, \lambda) = 0$ identically in λ.

If λ is orthogonal to ψ, then any one-to-one smooth function of ψ is orthogonal to any one-to-one smooth function of λ.

9.2.7 Effects of parameter orthogonality

Assume that it is possible to make the parameter of interest ψ and the nuisance parameter, now denoted by λ, orthogonal. We have seen that this is always possible if ψ is one-dimensional. Any transformation from, say, (ψ, χ) to (ψ, λ) necessary to achieve this leaves the profile log-likelihood unchanged: see Problem 9.6.

Now the matrices $i(\psi, \lambda)$ and $i^{-1}(\psi, \lambda)$ are block diagonal. Therefore, $\widehat{\psi}$ and $\widehat{\lambda}$ are asymptotically independent and the asymptotic variance of $\widehat{\psi}$ where λ is unknown is the same as that where λ is known. A related property is that $\widehat{\lambda}_\psi$, the MLE of λ for specified ψ, varies only slowly in ψ in the neighbourhood of $\widehat{\psi}$, and that there is a corresponding slow variation of $\widehat{\psi}_\lambda$ with λ. More precisely, if $\psi - \widehat{\psi} = O_p(n^{-1/2})$, then $\widehat{\lambda}_\psi - \widehat{\lambda} = O_p(n^{-1})$. For a non-orthogonal nuisance parameter χ, we would have $\widehat{\chi}_\psi - \widehat{\chi} = O_p(n^{-1/2})$.

We sketch a proof of this result for the case where both the parameter of interest and the nuisance parameter are scalar. If $\psi - \widehat{\psi} = O_p(n^{-1/2})$, $\chi - \widehat{\chi} = O_p(n^{-1/2})$, we have

$$l(\psi, \chi)$$
$$= l(\widehat{\psi}, \widehat{\chi}) - \tfrac{1}{2}\{\widehat{j}_{\psi\psi}(\psi - \widehat{\psi})^2 + 2\widehat{j}_{\psi\chi}(\psi - \widehat{\psi})(\chi - \widehat{\chi}) + \widehat{j}_{\chi\chi}(\chi - \widehat{\chi})^2\} + O_p(n^{-1/2}),$$

where, for instance, $\widehat{j}_{\psi\psi}$ denotes the $\psi\psi$-block of the observed information evaluated at $(\widehat{\psi}, \widehat{\chi})$. Recalling that $\widehat{\chi}_\psi$ maximises $l(\psi, \chi)$ for fixed ψ, it then follows that

$$\widehat{\chi}_\psi - \widehat{\chi} = \frac{-\widehat{j}_{\psi\chi}}{\widehat{j}_{\chi\chi}}(\psi - \widehat{\psi}) + O_p(n^{-1})$$
$$= \frac{-i_{\psi\chi}}{i_{\chi\chi}}(\psi - \widehat{\psi}) + O_p(n^{-1}).$$

Then, because $\psi - \widehat{\psi} = O_p(n^{-1/2})$, $\widehat{\chi}_\psi - \widehat{\chi} = O_p(n^{-1/2})$ unless $i_{\psi\chi} = 0$, the orthogonal case, when the difference is $O_p(n^{-1})$.

Note also that, so far as asymptotic theory is concerned, we can have $\widehat{\chi}_\psi = \widehat{\chi}$ independently of ψ only if χ and ψ are orthogonal. In this special case we can write $l_p(\psi) = l(\psi, \widehat{\chi})$. In the general orthogonal case, $l_p(\psi) = l(\psi, \widehat{\chi}) + o_p(1)$, so that a first-order theory could use $l_p^*(\psi) = l(\psi, \widehat{\chi})$ instead of $l(\psi, \widehat{\chi}_\psi)$.

9.3 Pseudo-likelihoods

As we have discussed, typically we consider a model parametrised by a parameter θ which may be written as $\theta = (\psi, \lambda)$, where ψ is the parameter of interest and λ is a nuisance parameter, not necessarily orthogonal. In order to draw inferences about the parameter of interest, we must deal with the nuisance parameter.

Ideally, we would like to construct a likelihood function for ψ alone. The simplest method for doing so is to construct a likelihood function based on a statistic T such that the distribution of T depends only on ψ. In this case, we may form a genuine likelihood function for ψ based on the density function of T; this is called a *marginal likelihood*, since it is based on the marginal distribution of T.

Another approach is available whenever there exists a statistic S such that the conditional distribution of the data X given $S = s$ depends only on ψ. In this case, we may form a likelihood function for ψ based on the conditional density function of X given $S = s$; this is called a *conditional likelihood* function. The drawback of this approach is that we discard the part of the likelihood function based on the marginal distribution of S, which may contain information about ψ.

Conditional and marginal likelihoods are particular instances of *pseudo-likelihood functions*. The term pseudo-likelihood is used to indicate any function of the data which depends only on the parameter of interest and which behaves, in some respects, as if it were a genuine likelihood (so that the score has zero null expectation, the maximum likelihood estimator has an asymptotic normal distribution etc.).

Formally, suppose that there exists a statistic T such that the density of the data X may be written as

$$f_X(x; \psi, \lambda) = f_T(t; \psi) f_{X|T}(x|t; \psi, \lambda).$$

Inference can be based on the marginal distribution of T, which does not depend on λ. The marginal likelihood function based on t is given by

$$L(\psi; t) = f_T(t; \psi).$$

The drawback of this approach is that we lose the information about ψ contained in the conditional density of X given T. It may, of course, also be difficult to find such a statistic T.

To define formally a conditional log-likelihood, suppose that there exists a statistic S such that

$$f_X(x; \psi, \lambda) = f_{X|S}(x|s; \psi) f_S(s; \psi, \lambda).$$

The statistic S is sufficient in the model with ψ held fixed. A conditional likelihood function for ψ may be based on $f_{X|S}(x|s; \psi)$, which does not depend on λ. The conditional log-likelihood function may be calculated as

$$l(\psi; x \mid s) = l(\theta) - l(\theta; s),$$

where $l(\theta; s)$ denotes the log-likelihood function based on the marginal distribution of S and $l(\theta)$ is the log-likelihood based on the full data X. Note that we make two assumptions here about S. The first is that S is not sufficient in the general model with parameters (ψ, λ), for, if it was, the conditional likelihood would not depend on either ψ or λ. The other is that S, the sufficient statistic when ψ is fixed, is the same for all ψ; S does not depend on ψ.

Note that factorisations of the kind that we have assumed in the definitions of conditional and marginal likelihoods arise essentially only in exponential families and transformation families. Outside these cases more general notions of pseudo-likelihood must be found.

9.4 Parametrisation invariance

Note that the score $u(\theta; x)$ and the Fisher information $i(\theta)$ depend, not only on the value of the parameter θ, but also on the parametrisation. If we change from θ to $\psi = \psi(\theta)$ by a smooth one-to-one transformation, with inverse $\theta = \theta(\psi)$, and calculate the score and information in terms of ψ, then different values will be obtained.

Write $l^{(\theta)}$, $u^{(\theta)}$, $i^{(\theta)}$ and $l^{(\psi)}$, $u^{(\psi)}$, $i^{(\psi)}$ for the log-likelihood, score and Fisher information in the θ- and ψ-parametrisation, respectively. Then

$$l^{(\psi)}(\psi) = l^{(\theta)}\{\theta(\psi)\},$$

and, for $a = 1, \ldots, d$,

$$u_a^{(\psi)}(\psi; x) = \frac{\partial l^{(\theta)}\{\theta(\psi); x\}}{\partial \psi_a}$$

$$= \sum_{r=1}^{d} \frac{\partial l^{(\theta)}\{\theta(\psi); x\}}{\partial \theta_r} \frac{\partial \theta_r}{\partial \psi_a}$$

$$= \sum_{r=1}^{d} u_r^{(\theta)}\{\theta(\psi); x\} \frac{\partial \theta_r}{\partial \psi_a},$$

or

$$u^{(\psi)}(\psi; x) = \left[\frac{\partial \theta}{\partial \psi}\right]^T u^{(\theta)}\{\theta(\psi); x\},$$

where $\partial \theta / \partial \psi$ is the Jacobian of the transformation from θ to ψ, with (r, a) element $\partial \theta_r / \partial \psi_a$. Similarly,

$$i_{ab}^{(\psi)}(\psi) = \sum_{r=1}^{d} \sum_{s=1}^{d} \frac{\partial \theta_r}{\partial \psi_a} \frac{\partial \theta_s}{\partial \psi_b} i_{rs}^{(\theta)}\{\theta(\psi)\},$$

or

$$i^{(\psi)}(\psi) = \left[\frac{\partial \theta}{\partial \psi}\right]^T i^{(\theta)}\{\theta(\psi)\} \left[\frac{\partial \theta}{\partial \psi}\right].$$

The principle of *parametrisation invariance* is a valuable basis for choosing between different inferential procedures. Invariance requires that the conclusions of a statistical analysis be unchanged by reformulation in terms of ψ, for any reasonably smooth one-to-one function of θ.

Formally, if θ and ψ are two alternative parametrisations for the model function in question and $\pi(\cdot) : \mathcal{X} \to \mathcal{A}$, with $\mathcal{A}$ denoting as before some action space, is an inference procedure, and C_θ and C_ψ are the conclusions that $\pi(\cdot)$ leads to, expressed in the two parametrisations, then the same conclusion C_ψ should be reached by *both* application of $\pi(\cdot)$ in the ψ parametrisation *and* translation into the ψ parametrisation of the conclusion C_θ.

Consider, for example, the exponential distribution with density $\rho e^{-\rho x}$. It would for many purposes be reasonable to take the mean $1/\rho$ or, say, $\log \rho$ as the parameter of interest. Parametrisation invariance would require, for example, the same conclusions to be reached by: (i) formulation in terms of ρ, application of a method of analysis and drawing conclusions about ρ; (ii) formulation in terms of $1/\rho$, application of a method of analysis, drawing conclusions about $1/\rho$, then taking the reciprocal.

A particular use of the principle of parametrisation invariance is to decide between different test procedures. For example, of the three test procedures based on likelihood quantities that we have described, the likelihood ratio test and the score test are parametrisation invariant, while the Wald test is not.

9.5 Edgeworth expansion

In this section and in Section 9.6 we assume, for simplicity, the case of univariate, continuous random variables. Extensions to the multivariate and discrete cases are straightforward and are summarised, for example, by Severini (2000: Chapter 2).

Let $X_1, X_2, \ldots, X_n$ be independent, identically distributed random variables with cumulant generating function $K_X(t)$ and cumulants κ_r. Let $S_n = \sum_1^n X_i$, $S_n^* = (S_n - n\mu)/\sqrt{n}\sigma$, where $\mu \equiv \kappa_1 = \mathbb{E}X_1, \sigma^2 \equiv \kappa_2 = \mathrm{var}X_1$. Define the rth standardised cumulant by $\rho_r = \kappa_r/\kappa_2^{r/2}$.

The *Edgeworth expansion* for the density of the standardised sample mean S_n^* can be expressed as:

$$f_{S_n^*}(x) = \phi(x)\left\{1 + \frac{\rho_3}{6\sqrt{n}}\,H_3(x) + \frac{1}{n}\left[\frac{\rho_4 H_4(x)}{24} + \frac{\rho_3^2 H_6(x)}{72}\right]\right\} + O(n^{-3/2}). \quad (9.2)$$

Here $\phi(x)$ is the standard normal density and $H_r(x)$ is the rth-degree Hermite polynomial defined, say, by

$$H_r(x) = (-1)^r \frac{d^r \phi(x)}{dx^r}\bigg/\phi(x)$$
$$= (-1)^r \phi^{(r)}(x)/\phi(x).$$

We have $H_3(x) = x^3 - 3x$, $H_4(x) = x^4 - 6x^2 + 3$ and $H_6(x) = x^6 - 15x^4 + 45x^2 - 15$. The asymptotic expansion (9.2) holds uniformly for $x \in \mathbb{R}$.

The leading term in the Edgeworth expansion is the standard normal density, as is appropriate from CLT. The remaining terms may be considered as higher-order correction terms. The $n^{-1/2}$ term is an adjustment for the main effect of the skewness of the true density, via the standardised skewness ρ_3, and the n^{-1} term is a simultaneous adjustment for skewness and kurtosis. If the density of X_1 is symmetric, $\rho_3 = 0$ and a normal approximation to the density of S_n^* is accurate to order n^{-1}, rather than the usual $n^{-1/2}$ for $\rho_3 \neq 0$. The accuracy of the Edgeworth approximation, say

$$f_{S_n^*}(x) \doteq \phi(x)\left\{1 + \frac{\rho_3}{6\sqrt{n}}\,H_3(x) + \frac{1}{n}\left[\frac{\rho_4 H_4(x)}{24} + \frac{\rho_3^2 H_6(x)}{72}\right]\right\},$$

will depend on the value of x. In particular, Edgeworth approximations tend to be poor, and may even be negative, in the tails of the distribution, as $|x|$ increases.

Integrating the Edgeworth expansion (9.2) term by term (which requires a non-trivial justification), using the properties of the Hermite polynomials, we obtain an expansion for the distribution function of S_n^*:

$$F_{S_n^*}(x) = \Phi(x) - \phi(x)\left\{\frac{\rho_3}{6\sqrt{n}} H_2(x) + \frac{\rho_4}{24n} H_3(x) + \frac{\rho_3^2}{72n} H_5(x)\right\} + O(n^{-3/2}).$$

Also, if T_n is a sufficiently smooth function of S_n^*, then a formal Edgeworth expansion can be obtained for the density of T_n. Further details and references are given by Severini (2000: Chapter 2).

When studying the coverage probability of confidence intervals, for example, it is often convenient to be able to determine x as x_α say, so that $F_{S_n^*}(x_\alpha) = \alpha$, to the order considered in the Edgeworth approximation to the distribution function of S_n^*. The solution is known as the *Cornish–Fisher expansion* and the formula is

$$x_\alpha = z_\alpha + \frac{1}{6\sqrt{n}}(z_\alpha^2 - 1)\rho_3 + \frac{1}{24n}(z_\alpha^3 - 3z_\alpha)\rho_4 - \frac{1}{36n}(2z_\alpha^3 - 5z_\alpha)\rho_3^2 + O(n^{-3/2}),$$

where $\Phi(z_\alpha) = \alpha$.

The derivation of the Edgeworth expansion stems from the result that the density of a random variable can be obtained by inversion of its characteristic function. A form of this inversion result useful for our discussion here is that the density for $\bar{X}$, the mean of a set of independent, identically distributed random variables $X_1, \ldots, X_n$, can be obtained as

$$f_{\bar{X}}(\bar{x}) = \frac{n}{2\pi i} \int_{\tau-i\infty}^{\tau+i\infty} \exp[n\{K(\phi) - \phi\bar{x}\}]d\phi, \tag{9.3}$$

where K is the cumulant generating function of X, and τ is any point in the open interval around 0 in which the moment generating function M exists. For details, see Feller (1971: Chapter 16). In essence, the Edgeworth expansion (9.2) is obtained by expanding the cumulant generating function in a Taylor series around 0, exponentiating and inverting term by term. Details are given in Barndorff-Nielsen and Cox (1989: Chapter 4): see also Problem 9.7.

9.6 Saddlepoint expansion

The *saddlepoint expansion* for the density of S_n is

$$f_{S_n}(s) = \frac{1}{\sqrt{2\pi}} \frac{1}{\{nK_X''(\hat{\phi})\}^{1/2}} \times \exp\{nK_X(\hat{\phi}) - \hat{\phi}s\}\{1 + O(n^{-1})\}, \tag{9.4}$$

where $\hat{\phi} \equiv \hat{\phi}(s)$ satisfies $nK_X'(\hat{\phi}) = s$. Though we would be interested primarily in applying the expansion in the so-called *moderate deviation region* of the form $|s - n\mu| \le cn^{1/2}$, for fixed c, it is actually valid in a *large deviation region* of the form $|s - n\mu| \le bn$, for fixed b, and in some cases even for *all* s.

A detailed analysis shows that the $O(n^{-1})$ term is actually $(3\hat{\rho}_4 - 5\hat{\rho}_3^2)/(24n)$, where $\hat{\rho}_j \equiv \hat{\rho}_j(\hat{\phi}) = K_X^{(j)}(\hat{\phi})/\{K_X''(\hat{\phi})\}^{j/2}$ is the jth standardised derivative of the cumulant generating function for X_1 evaluated at $\hat{\phi}$.

A simple change of variable in (9.4) gives a saddlepoint expansion for the density of $\bar{X}_n = S_n/n$:

$$f_{\bar{X}_n}(x) = (2\pi)^{-1/2}\{n/K_X''(\hat{\phi})\}^{1/2} \times \exp\{n[K_X(\hat{\phi}) - \hat{\phi}x]\}(1 + O(n^{-1})), \qquad (9.5)$$

where $K_X'(\hat{\phi}) = x$. This expansion is valid for x in a large deviation region $|x - \mu| \le b$, for fixed b.

The saddlepoint expansion is quite different in form from the Edgeworth expansion. In order to use the former to approximate $f_{\bar{X}_n}(x)$ with either the leading term, or the leading term plus n^{-1} correction, it is necessary to know the whole cumulant generating function, not just the first four cumulants. It is also necessary to solve the equation $K_X'(\hat{\phi}) = x$ for each value of x. The leading term in (9.5) is not the normal (or any other) density; in fact it will not usually integrate to 1, although it can be renormalised to do so. The saddlepoint expansion is an asymptotic expansion in powers of n^{-1}, rather than $n^{-1/2}$, as in the Edgeworth expansion. This suggests that the main correction for skewness has been absorbed by the leading term, which is in fact the case.

Observe that, crucially, the saddlepoint expansion is stated with a *relative* error, while the Edgeworth expansion is stated with an *absolute* error.

The approximation obtained from the leading term of (9.5), ignoring the $O(n^{-1})$ correction term, is generally very accurate. In particular, the saddlepoint approximation tends to be much more accurate than an Edgeworth approximation in the tails of the distribution. In distributions that differ from the normal density in terms of asymmetry, such as the Gamma distribution, the saddlepoint approximation is extremely accurate throughout the range of x. It is customary to use as an approximation to $f_{\bar{X}_n}(x)$ a renormalised version of (9.5):

$$f_{\bar{X}_n}(x) \doteq c_n\{n/K_X''(\hat{\phi})\}^{1/2} \exp[n\{K_X(\hat{\phi}) - \hat{\phi}x\}], \qquad (9.6)$$

where c_n is determined, usually numerically, so that the right-hand side of (9.6) integrates to 1. If the $O(n^{-1})$ correction term is constant in x, (9.6) will be exact. For scalar random variables this happens only in the case of the normal, Gamma and inverse Gaussian distributions. The latter will be considered in Chapter 11. In general, the n^{-1} correction term $\{3\hat{\rho}_4(\hat{\phi}) - 5\hat{\rho}_3^2(\hat{\phi})\}/24$ varies only slowly with x and the relative error in the renormalised approximation (9.6) is $O(n^{-3/2})$, but only for x in a moderate deviation region $|x - \mu| \le cn^{-1/2}$, for fixed c.

The saddlepoint approximation is usually derived by one of two methods. The first (Daniels, 1954) uses the inversion formula (9.3) and contour integration, choosing the contour of integration to pass through the saddlepoint of the integrand on the line of steepest descent. We sketch instead a more statistical derivation, as described by Barndorff-Nielsen and Cox (1979).

We associate with the density $f(x)$ for X_1 an exponential family density $f(x; \phi)$ defined by

$$f(x; \phi) = \exp\{x\phi - K_X(\phi)\}f(x),$$

where K_X is the cumulant generating function of X_1, under $f(x)$. Then it is straightforward to check that the sum $S_n = X_1 + \cdots + X_n$ has associated density

$$f_{S_n}(s; \phi) = \exp\{s\phi - nK_X(\phi)\}f_{S_n}(s)$$

from which

$$f_{S_n}(s) = \exp\{nK_X(\phi) - s\phi\}f_{S_n}(s;\phi). \tag{9.7}$$

Now use the Edgeworth expansion to obtain an approximation to the density $f_{S_n}(s;\phi)$, remembering that cumulants all must refer to cumulants computed under the tilted density $f(x;\phi)$. Since ϕ is arbitrary, it is chosen so that the Edgeworth expansion for the tilted density is evaluated at its mean, where the $n^{-1/2}$ term in the expansion is zero. This value is defined by $nK'_X(\hat{\phi}) = s$ and (9.7) becomes

$$f_{S_n}(s) \doteq \exp\{nK_X(\hat{\phi}) - \hat{\phi}s\}\{2\pi nK''_X(\hat{\phi})\}^{-1/2}, \tag{9.8}$$

which is the approximation deriving from (9.4). The factor $\{2\pi nK''_X(\hat{\phi})\}^{-1/2}$ comes from the normal density evaluated at its mean.

A case of special interest is when $f(x)$ is itself in the exponential family, $f(x;\theta) = \exp\{x\theta - c(\theta) - h(x)\}$. Then, since $K_X(t) = c(\theta + t) - c(\theta)$, it follows that $\hat{\phi} = \hat{\theta} - \theta$, where $\hat{\theta}$ is the MLE based on $s = x_1 + \cdots + x_n$. Then (9.8) is

$$f_{S_n}(s;\theta) \doteq \exp\left[n\{c(\hat{\theta}) - c(\theta)\} - (\hat{\theta} - \theta)s\right]\{2\pi nc''(\hat{\theta})\}^{-1/2},$$

which can be expressed as

$$c \exp\{l(\theta) - l(\hat{\theta})\}|j(\hat{\theta})|^{-1/2}, \tag{9.9}$$

where $l(\theta)$ is the log-likelihood function based on $(x_1, \ldots, x_n)$, or s, and $j(\hat{\theta})$ is the observed information. Since $\hat{\theta} = \hat{\theta}(s)$ is a one-to-one function of s, with Jacobian $|j(\hat{\theta})|$, (9.9) can be used to obtain an approximation to the density of $\hat{\theta}$

$$f_{\hat{\theta}}(\hat{\theta};\theta) \doteq c \exp\{l(\theta) - l(\hat{\theta})\}|j(\hat{\theta})|^{1/2}. \tag{9.10}$$

This latter approximation is a particular example of the p^* formula, considered in Section 9.8.

It is not easy to integrate the right-hand side of the saddlepoint approximation (9.4) to obtain an approximation to the distribution function of S_n: see Lugannani and Rice (1980). The *Lugannani–Rice approximation* is

$$F_{S_n}(s) \doteq \Phi(r_s) + \phi(r_s)\left(\frac{1}{r_s} - \frac{1}{v_s}\right),$$

where

$$r_s = \text{sgn}(\hat{\phi})\sqrt{2n\{\hat{\phi}K'_X(\hat{\phi}) - K_X(\hat{\phi})\}},$$
$$v_s = \hat{\phi}\sqrt{nK''_X(\hat{\phi})},$$

and $\hat{\phi} \equiv \hat{\phi}(s)$ is the saddlepoint satisfying $nK'_X(\hat{\phi}) = s$. The error is $O(n^{-1})$, uniformly in s, and the expansion can be expressed in the asymptotically equivalent form

$$F_{S_n}(s) = \Phi(r_s^*)\{1 + O(n^{-1})\},$$

with

$$r_s^* = r_s - \frac{1}{r_s} \log \frac{r_s}{v_s}.$$

The $O(n^{-1})$ relative error holds in the large deviation region of s, and is actually of order $O(n^{-3/2})$ in the moderate deviation region.

9.7 Laplace approximation of integrals

Suppose $g : \mathbb{R} \to \mathbb{R}$ is a smooth function, and that we wish to evaluate the integral

$$g_n = \int_a^b e^{-ng(y)} dy.$$

The main contribution to the integral, for large n, will come from values of y near the minimum of $g(y)$, which may occur at a or b, or in the interior of the interval (a, b). Assume that $g(y)$ is minimised at $\tilde{y} \in (a, b)$ and that $g'(\tilde{y}) = 0$, $g''(\tilde{y}) > 0$. The other cases may be treated in a similar manner. For a useful summary of Laplace approximation see Barndorff-Nielsen and Cox (1989: Chapter 3).

Then, using Taylor expansion, we can write

$$g_n = \int_a^b e^{-n\{g(\tilde{y}) + \frac{1}{2}(\tilde{y}-y)^2 g''(\tilde{y}) + \cdots\}} dy$$

$$\doteq e^{-ng(\tilde{y})} \int_a^b e^{-\frac{n}{2}(\tilde{y}-y)^2 g''(\tilde{y})} dy$$

$$\doteq e^{-ng(\tilde{y})} \sqrt{\frac{2\pi}{ng''(\tilde{y})}} \int_{-\infty}^{\infty} \phi\left(y - \tilde{y}; \frac{1}{ng''(\tilde{y})}\right) dy,$$

where $\phi(y - \mu; \sigma^2)$ is the density of $N(\mu, \sigma^2)$. Since ϕ integrates to 1,

$$g_n \doteq e^{-ng(\tilde{y})} \sqrt{\frac{2\pi}{ng''(\tilde{y})}}. \tag{9.11}$$

A more detailed analysis gives

$$g_n = e^{-ng(\tilde{y})} \sqrt{\frac{2\pi}{ng''(\tilde{y})}} \left\{1 + \frac{5\tilde{\rho}_3^2 - 3\tilde{\rho}_4}{24n} + O(n^{-2})\right\},$$

where

$$\tilde{\rho}_3 = g^{(3)}(\tilde{y})/\{g''(\tilde{y})\}^{3/2},$$
$$\tilde{\rho}_4 = g^{(4)}(\tilde{y})/\{g''(\tilde{y})\}^2.$$

A similar analysis gives

$$\int_a^b h(y) e^{-ng(y)} dy = h(\tilde{y}) e^{-ng(\tilde{y})} \sqrt{\frac{2\pi}{ng''(\tilde{y})}} \{1 + O(n^{-1})\}. \tag{9.12}$$

A further refinement of the method, which allows $g(y)$ to depend weakly on n, gives

$$\int_a^b e^{-n\{g(y)-\frac{1}{n}\log h(y)\}}dy$$

$$= \int_a^b e^{-nq_n(y)}dy, \quad \text{say,}$$

$$= e^{-ng(y^*)}h(y^*)\sqrt{\frac{2\pi}{nq_n''(y^*)}}\{1 + (5\rho_3^{*2} - 3\rho_4^*)/(24n) + O(n^{-2})\}, \qquad (9.13)$$

where

$$q_n'(y^*) = 0, \ \rho_j^* = q_n^{(j)}(y^*)/\{q_n''(y^*)\}^{j/2}.$$

The Laplace approximations are particularly useful in Bayesian inference: see Section 9.12.

9.8 The p^* formula

9.8.1 Introduction

The log-likelihood is, except possibly for a term not depending on the parameter, a function of a sufficient statistic s and parameter θ. If the dimensions of s and θ are equal, the maximum likelihood estimator $\widehat{\theta}$ is usually a one-to-one function of s and then $\widehat{\theta}$ is minimal sufficient if and only if s is minimal sufficient. We can then take the log-likelihood as $l(\theta; \widehat{\theta})$, it being the same as if the data consisted solely of $\widehat{\theta}$ or s.

If $s = (t, a)$, where t has the dimension of θ and a is ancillary, then we can generally write the log-likelihood as $l(\theta; \widehat{\theta}, a)$. Recall the convention introduced in Chapter 7 that the minimal sufficient statistic based in data x can be re-expressed, by a one-to-one smooth transformation, as $(\widehat{\theta}, a)$, where a is ancillary, so that we can write the log-likelihood $l(\theta; x)$ as $l(\theta; \widehat{\theta}, a)$. Similarly, we can write the observed information $j(\theta) \equiv j(\theta; x) = j(\theta; \widehat{\theta}, a)$.

Under a transformation model, the maximal invariant statistic serves as the ancillary. In a full (m, m) exponential model the MLE is minimal sufficient and no ancillary is called for.

Example 9.1 We consider first the *location model*, which is the simplest example of a transformation model, the general theory of which was described in Chapter 5. We have $X_1, \ldots, X_n$ independent random variables with

$$X_j = \theta + \epsilon_j, \ j = 1, \ldots, n,$$

where $\epsilon_1, \ldots, \epsilon_n$ are independent random variables each having the known density function $\exp\{g(\cdot)\}$. The log-likelihood is given by

$$l(\theta) = \sum g(x_j - \theta).$$

Let $a = (a_1, \ldots, a_n)$, where $a_j = x_j - \widehat{\theta}$: it is readily shown that a is ancillary. We may write $x_j = a_j + \widehat{\theta}$, so that the log-likelihood may be written

$$l(\theta; \widehat{\theta}, a) = \sum g(a_j + \widehat{\theta} - \theta).$$

Example 9.2 As a further example, let $X_1, \ldots, X_n$ be an independent sample from a full (m, m) exponential density

$$\exp\{x^T\theta - k(\theta) + D(x)\}.$$

The log-likelihood is, ignoring an additive constant,

$$l(\theta) = \sum x_j^T\theta - nk(\theta).$$

Since $\widehat{\theta}$ satisfies the likelihood equation

$$\sum x_j - nk'(\theta) = 0,$$

the log-likelihood may be written

$$l(\theta; \widehat{\theta}) = nk'(\widehat{\theta})^T\theta - nk(\theta).$$

9.8.2 Approximate ancillaries

Outside full exponential family and transformation models it is often difficult to construct an appropriate ancillary a such that $(\widehat{\theta}, a)$ is minimal sufficient, and it is usually necessary to work with notions of *approximate ancillarity*. A statistic a is, broadly speaking, approximately ancillary if its asymptotic distribution does not depend on the parameter. Useful approximate ancillaries can often be constructed from signed log-likelihood ratios or from score statistics.

Severini (2000: Section 6.6) gives a summary of techniques for construction of approximate ancillaries. One particularly important approximate ancillary is the *Efron–Hinkley ancillary* (Efron and Hinkley, 1978). Consider the case of a scalar parameter θ and let, as before, i and j be the expected and observed information and let $l_\theta = \frac{\partial l}{\partial \theta}$, $l_{\theta\theta} = \frac{\partial^2 l}{\partial \theta^2}$ etc. Use the notation $v_{2,1} = \mathbb{E}(l_{\theta\theta}l_\theta; \theta)$, $v_{2,2} = \mathbb{E}(l_{\theta\theta}l_{\theta\theta}; \theta)$, $v_2 = \mathbb{E}(l_{\theta\theta}; \theta)$. Define

$$\gamma = i^{-1}(v_{2,2} - v_2^2 - i^{-1}v_{2,1}^2)^{1/2},$$

and use circumflex to denote evaluation at $\widehat{\theta}$. Then the Efron–Hinkley ancillary is defined by

$$a = (\widehat{i\gamma})^{-1}(\widehat{j} - \widehat{i}).$$

A particularly powerful result, which we will not prove but which amplifies comments made in Chapter 8 about the use of observed rather than Fisher information being preferable, is the following. For a location model with θ as the location parameter, if $\widehat{i}$ and $\widehat{j}$ denote respectively the Fisher and observed information evaluated at $\widehat{\theta}$,

$$\frac{\text{var}(\widehat{\theta} \mid a) - \widehat{j}^{-1}}{\text{var}(\widehat{\theta} \mid a) - \widehat{i}^{-1}} = O_p(n^{-1/2}),$$

where a denotes the Efron–Hinkley ancillary: $\widehat{j}^{-1}$ provides a more accurate estimate of the conditional variance of $\widehat{\theta}$ given a.

A simple example of construction of this ancillary is provided by the *exponential hyperbola*. Under this model, $(X_1, Y_1), \ldots, (X_n, Y_n)$ denote independent pairs of independent exponential random variables, such that each X_j has mean $1/\theta$ and each Y_j has mean θ.

The minimal sufficent statistic for the model may be written as $(\widehat{\theta}, a)$, where $\widehat{\theta} = (\bar{y}/\bar{x})^{1/2}$ is the MLE and $a = (\bar{x}\bar{y})^{1/2}$ is an (exact) ancillary. Simple calculations show that the Efron–Hinkley ancillary is

$$\sqrt{2n}(\bar{y}/\widehat{\theta} - 1) = \sqrt{2n}\{(\bar{x}\bar{y})^{1/2} - 1\},$$

which is in fact also exactly ancillary.

9.8.3 The key formula

A striking result due to Barndorff-Nielsen (1983) is that the conditional density function $f(\widehat{\theta}; \theta \mid a)$ for the MLE $\widehat{\theta}$ given an ancillary statistic a is, in wide generality, exactly or approximately equal to

$$p^*(\widehat{\theta}; \theta \mid a) = c(\theta, a)|j(\widehat{\theta})|^{1/2} \exp\{l(\theta) - l(\widehat{\theta})\}, \qquad (9.14)$$

that is

$$f(\widehat{\theta}; \theta \mid a) \doteq p^*(\widehat{\theta}; \theta \mid a).$$

In (9.14), $c(\theta, a)$ is a normalising constant, determined, usually numerically, so that the integral of p^* with respect to $\widehat{\theta}$, for fixed a, equals 1.

Equation (9.14) gives the exact conditional distribution of the MLE for a considerable range of models. In particular, this is the case for virtually all transformation models, for which $c(\theta, a)$ is independent of θ. The location-scale model provides a prototypical example, with the configuration statistic as the ancillary. Among models for which (9.14) is exact, but which is not a transformation model, is the inverse Gaussian distribution. Under many of these models the norming constant c equals $(2\pi)^{-d/2}$ exactly, $d = \dim(\theta)$. In general, $c = c(\theta, a) = (2\pi)^{-d/2}\bar{c}$, where $\bar{c} = 1 + O(n^{-1})$. Outside the realm of exactness cases, (9.14) is quite generally accurate to relative error of order $O(n^{-1})$:

$$f(\widehat{\theta}; \theta \mid a) = p^*(\widehat{\theta}; \theta \mid a)(1 + O(n^{-1})),$$

for *any fixed* $\widehat{\theta}$. For $\widehat{\theta}$ of the form $\widehat{\theta} = \theta + O_p(n^{-1/2})$, which, in practice, is the situation we are primarily interested in, the approximation achieves higher accuracy, the relative error in fact being of order $O(n^{-3/2})$. Severini (2000: Section 6.5) provides an account of definitions of approximate ancillarity, which are strong enough for the relative error to be of order $O(n^{-1})$ for values of the argument $\widehat{\theta}$ of this latter form without a being exactly ancillary.

Comparing (9.10) with (9.14), we see that the p^* formula is equivalent to the saddlepoint approximation in exponential families, with θ the natural parameter.

9.8.4 The adjusted signed root likelihood ratio, r^*

Integration of the p^* formula in the case of scalar θ to obtain an approximation to the distribution function of the MLE is intricate: a very clear description is given by Barndorff-Nielsen (1990). Write

$$r_t \equiv r_t(\theta) = \text{sgn}(t - \theta)\sqrt{2(l(t; t, a) - l(\theta; t, a))},$$

and let

$$v_t \equiv v_t(\theta) = j(t; t, a)^{-1/2}\{l_{;\widehat{\theta}}(t; t, a) - l_{;\widehat{\theta}}(\theta; t, a)\},$$

in terms of the so-called sample space derivative $l_{;\widehat{\theta}}$ defined by

$$l_{;\widehat{\theta}}(\theta; \widehat{\theta}, a) = \frac{\partial}{\partial\widehat{\theta}}l(\theta; \widehat{\theta}, a),$$

and with j the observed information. Then

$$\mathrm{Pr}_\theta(\widehat{\theta} \le t \mid a) = \Phi\{r_t^*(\theta)\}\{1 + O(n^{-3/2})\},$$

where $r_t^*(\theta) = r_t + r_t^{-1}\log\{v_t/r_t\}$, for $t = \theta + O(n^{-1/2})$.

The random quantity $r^*(\theta)$ corresponding to $r_t^*(\theta)$ is an approximate pivot, conditional on the ancillary, in the sense that its distribution is close to normal. We may view $r^*(\theta)$ as an adjusted form (the 'r^* statistic') of the signed root likelihood ratio statistic

$$r(\theta) = \mathrm{sgn}(\widehat{\theta} - \theta)\, [2\{l(\widehat{\theta}; \widehat{\theta}, a) - l(\theta; \widehat{\theta}, a)\}]^{1/2}$$

which improves the accuracy of the normal approximation.

To define $r^*(\theta)$ formally,

$$r^*(\theta) = r(\theta) + r(\theta)^{-1}\log\{v(\theta)/r(\theta)\},$$

where

$$v(\theta) = \widehat{j}^{-1/2}\{l_{;\widehat{\theta}}(\widehat{\theta}; \widehat{\theta}, a) - l_{;\widehat{\theta}}(\theta; \widehat{\theta}, a)\}, \tag{9.15}$$

with $\widehat{j}$ denoting evaluation of the observed information at $\widehat{\theta}$.

We have that $r^*(\theta)$ is distributed as $N(0, 1)$ to (relative) error of order $O(n^{-3/2})$:

$$\mathrm{Pr}_\theta\{r^*(\theta) \le t \mid a\} = \Phi(t)\{1 + O(n^{-3/2})\},$$

for $t = O(1)$.

The limits of an approximate $(1 - 2\alpha)$ confidence interval for θ may be found as those θ such that $\Phi\{r^*(\theta)\} = \alpha, 1 - \alpha$.

The above is expressed in terms of a one-parameter model. Versions of the adjusted signed root likelihood ratio statistic $r^*(\theta)$ relevant to inference about a scalar parameter of interest in the presence of a nuisance parameter are more complicated. To present just the key formula, suppose that the model depends on a multidimensional parameter $\theta = (\psi, \lambda)$, with ψ a scalar parameter of interest, with λ nuisance. Then the $N(0, 1)$ approximation to the distribution of the signed root likelihood ratio statistic $r_\mathrm{p} = \mathrm{sgn}(\widehat{\psi} - \psi)\, [2\{l_\mathrm{p}(\widehat{\psi}) - l_\mathrm{p}(\psi)\}]^{1/2}$ is improved by analytically adjusted versions of the form

$$r_\mathrm{a}(\psi) = r_\mathrm{p}(\psi) + r_\mathrm{p}(\psi)^{-1}\log(v_\mathrm{p}(\psi)/r_\mathrm{p}(\psi)),$$

that are distributed as $N(0, 1)$, conditionally on a (and hence unconditionally), to error of order $O(n^{-3/2})$. (We adopt the notation $r_\mathrm{a}(\psi)$ in preference to $r^*(\psi)$ to avoid notational conflict in Chapter 10).

Now the statistic v_p is defined (Barndorff-Nielsen, 1986) by

$$v_p(\psi) = \begin{vmatrix} l_{;\widehat{\theta}}(\widehat{\theta}) - l_{;\widehat{\theta}}(\psi, \widehat{\lambda}_\psi) \\ l_{\psi;\widehat{\theta}}(\psi, \widehat{\lambda}_\psi) \end{vmatrix} / \{|j_{\psi\psi}(\psi, \widehat{\lambda}_\psi)|^{1/2} |j(\widehat{\theta})|^{1/2}\}. \tag{9.16}$$

Here, as previously, the log-likelihood function has been written as $l(\theta; \widehat{\theta}, a)$, with $(\widehat{\theta}, a)$ minimal sufficient and a ancillary, $\widehat{\lambda}_\psi$ denotes the MLE of λ for given ψ, and

$$l_{;\widehat{\theta}}(\theta) \equiv l_{;\widehat{\theta}}(\theta; \widehat{\theta}, a) = \frac{\partial}{\partial \widehat{\theta}} l(\theta; \widehat{\theta}, a), \quad l_{\psi;\widehat{\theta}}(\theta) \equiv l_{\psi;\widehat{\theta}}(\theta; \widehat{\theta}, a) = \frac{\partial^2}{\partial \psi \partial \widehat{\theta}} l(\theta; \widehat{\theta}, a).$$

Again, j denotes the observed information matrix and $j_{\psi\psi}$ denotes its (ψ, ψ) component.

A key drawback to use of $r_a(\psi)$ (the same comment is true of $r^*(\theta)$) is the need to calculate sample space derivatives, which necessitates explicit specification of the ancillary a. We have commented that this is difficult in general, outside full exponential family and transformation models. Several methods of approximation to $r_a(\psi)$ which avoid this by approximating to the sample space derivatives have been developed. A computationally attractive approximation based on orthogonal parameters is described by DiCiccio and Martin (1993): recall that in the case we are assuming here of a scalar parameter of interest it is always possible to find a parametrisation in which the interest parameter ψ and the nuisance parameters λ are orthogonal. The DiCiccio and Martin (1993) approximation replaces $v_p(\psi)$ by

$$\tilde{v}_p(\psi) = l_\psi(\psi, \widehat{\lambda}_\psi) \frac{|j_{\lambda\lambda}(\psi, \widehat{\lambda}_\psi)|^{1/2} i_{\psi\psi}(\widehat{\theta})^{1/2}}{|j(\widehat{\theta})|^{1/2} i_{\psi\psi}(\psi, \widehat{\lambda}_\psi)^{1/2}}, \tag{9.17}$$

with the usual partitioning of the observed information j and the Fisher information i, and with l_ψ denoting, as before, the derivative of the log-likelihood l with respect to the parameter of interest. The corresponding adjusted version of the signed root likelihood ratio statistic,

$$\tilde{r}_a(\psi) = r_p(\psi) + r_p(\psi)^{-1} \log(\tilde{v}_p(\psi)/r_p(\psi)),$$

is distributed as $N(0, 1)$ to error of order $O(n^{-1})$, rather than order $O(n^{-3/2})$ for $r_a(\theta)$. A further point should be noted, that r_a is parametrisation invariant, with respect to interest-respecting reparametrisation, while $\tilde{r}_a$ depends on the orthogonal parametrisation adopted. Other approximations to r_a, due to various authors and with the same property of being distributed as $N(0, 1)$ to error of order $O(n^{-1})$, are detailed by Severini (2000: Chapter 7).

9.8.5 An example: Normal distribution with known coefficient of variation

Let $X_1, \ldots, X_n$ denote independent normally distributed random variables each with mean θ and standard deviation $r\theta$, where $\theta > 0$ and the coefficient of variation r is known; for simplicity take $r = 1$. This distribution is widely assumed in many biological and agricultural problems. The minimal sufficient statistic for the model may be written $(\widehat{\theta}, a)$, where

$$a = \sqrt{n} \frac{(\sum x_j^2)^{1/2}}{\sum x_j}$$

is easily seen to be an exactly ancillary statistic and

$$\widehat{\theta} = \frac{(\sum x_j^2)^{1/2}}{\sqrt{n}} \frac{2|a|}{(1+4a^2)^{1/2} + \text{sgn}(a)}$$

is the maximum likelihood estimator of θ. Assume that $a > 0$, which occurs with probability rapidly approaching 1 as $n \to \infty$.

The log-likelihood function may be written

$$l(\theta; \widehat{\theta}, a) = -\frac{n}{2\theta^2}\left[q^2\widehat{\theta}^2 - \frac{2q\theta\widehat{\theta}}{a}\right] - n \log \theta,$$

where

$$q = \frac{(1+4a^2)^{1/2} + 1}{2a}.$$

It follows that

$$p^*(\widehat{\theta}; \theta \mid a) = \frac{\sqrt{n\bar{c}}}{\sqrt{(2\pi)}} \left(\frac{\widehat{\theta}}{\theta}\right)^{n-1} \frac{1}{\theta}(1+q^2)^{1/2}$$

$$\times \exp\left\{-\frac{n}{2}\left[\frac{q^2}{\theta^2}(\widehat{\theta}^2 - \theta^2) - \frac{2q}{a\theta}(\widehat{\theta} - \theta)\right]\right\}.$$

This expression may be rewritten as

$$p^*(\widehat{\theta}; \theta \mid a) = \frac{\sqrt{n\bar{c}}}{\sqrt{(2\pi)}} \exp\left\{\frac{n}{2}(q - 1/a)^2\right\}(1+q^2)^{1/2}\left(\frac{\widehat{\theta}}{\theta}\right)^{n-1}\frac{1}{\theta}$$

$$\times \exp\left\{-\frac{n}{2}q^2(\widehat{\theta}/\theta - 1/(aq))^2\right\}.$$

It may be shown that the exact conditional density of $\widehat{\theta}$ given a is of the form

$$p(\widehat{\theta}; \theta \mid a) = b(a)\left(\frac{\widehat{\theta}}{\theta}\right)^{n-1}\frac{1}{\theta}\exp\left\{-\frac{n}{2}q^2(\widehat{\theta}/\theta - 1/(aq))^2\right\}, \qquad (9.18)$$

where $b(a)$ is a normalising constant depending on a. Hence, the conditional density approximation is exact for this model. An $N(0, 1)$ approximation to the conditional distribution of $r^*(\theta)$ is not exact, but highly accurate, as we shall see when we consider this model further in Chapter 11.

9.8.6 The score function

We now consider the application of the p^* formula to the score function $U(\theta)$. Given an ancillary a, the MLE $\widehat{\theta}$ and the score function will in general be in one-to-one correspondence for a region of values of $\widehat{\theta}$ around the true parameter value θ, and this region will carry all the probability mass, except for an asymptotically negligible amount. The Jacobian of the transformation from $\widehat{\theta}$ to the vector of derivatives l_r of the log-likelihood, $l_r \equiv \frac{\partial}{\partial\theta_r}l(\theta; \widehat{\theta}, a)$, the components of the score function, is the matrix $l_; = [l_{r;s}]$ of second-order log-likelihood derivatives

$$l_{r;s} = l_{r;s}(\theta; \widehat{\theta}, a) = \frac{\partial}{\partial\theta_r}\frac{\partial}{\partial\theta_s}l(\theta; \widehat{\theta}, a).$$

From (9.14) an approximation of high accuracy to the conditional distribution of the score vector given a is provided by

$$p(u; \theta \mid a) \doteq p^*(u; \theta \mid a),$$

where

$$p^*(u; \theta \mid a) = c(\theta, a) |\widehat{j}|^{1/2} |l_{;}|^{-1} e^{l - \widehat{l}}.$$

As an example of calculation of the derivatives required by this approximation, and as used also, for example, in (9.16), consider the *location model*. We saw previously that

$$l(\theta; \widehat{\theta}, a) = \sum g(a_j + \widehat{\theta} - \theta).$$

Then

$$l_{;} \equiv l_{\theta; \widehat{\theta}} = -\sum g''(a_j + \widehat{\theta} - \theta).$$

Note than an Edgeworth or saddlepoint approximation to the marginal distribution of U is easy to obtain in the case when U is a sum of independent, identically distributed variates.

9.9 Conditional inference in exponential families

A particularly important inference problem to which ideas of this chapter apply concerns inference about the natural parameter of an exponential family model.

Suppose that $X_1, \ldots, X_n$ are independent, identically distributed from the exponential family density

$$f(x; \psi, \lambda) = \exp\{\psi \tau_1(x) + \lambda \tau_2(x) - d(\psi, \lambda) - Q(x)\},$$

where we will suppose for simplicity that the parameter of interest ψ and the nuisance parameter λ are both scalar.

The natural statistics are $T = n^{-1} \sum \tau_1(x_i)$ and $S = n^{-1} \sum \tau_2(x_i)$. We know from the general properties of exponential families (Chapter 5) that the conditional distribution of $X = (X_1, \ldots, X_n)$ given $S = s$ depends only on ψ, so that inference about ψ may be derived from a conditional likelihood, given s.

The log-likelihood based on the full data $x_1, \ldots, x_n$ is

$$n\psi t + n\lambda s - nd(\psi, \lambda),$$

ignoring terms not involving ψ and λ, and the conditional log-likelihood function is the full log-likelihood minus the log-likelihood function based on the marginal distribution of S. We consider an approximation to the marginal distribution of S, based on a saddlepoint approximation to the density of S, evaluated at its observed value s.

The cumulant generating function of $\tau_2(X_i)$ is given by

$$K(z) = d(\psi, \lambda + z) - d(\psi, \lambda).$$

Write $d_\lambda(\psi, \lambda) = \frac{\partial}{\partial \lambda} d(\psi, \lambda)$ and $d_{\lambda\lambda}(\psi, \lambda) = \frac{\partial^2}{\partial \lambda^2} d(\psi, \lambda)$. The saddlepoint equation is then given by

$$d_\lambda(\psi, \lambda + \widehat{z}) = s.$$

With s the observed value of S, the likelihood equation for the model with ψ held fixed is

$$ns - nd_\lambda(\psi, \widehat{\lambda}_\psi) = 0,$$

so that $\lambda + \hat{z} = \widehat{\lambda}_\psi$, where $\widehat{\lambda}_\psi$ denotes the maximum likelihood estimator of λ for fixed ψ. Applying the saddlepoint approximation, ignoring constants, we therefore approximate the marginal likelihood function based on S as

$$|d_{\lambda\lambda}(\psi, \widehat{\lambda}_\psi)|^{-1/2} \exp\{n[d(\psi, \widehat{\lambda}_\psi) - d(\psi, \lambda) - (\widehat{\lambda}_\psi - \lambda)s]\};$$

the resulting approximation to the conditional log-likelihood function is given by

$$n\psi t + n\widehat{\lambda}_\psi^T s - nd(\psi, \widehat{\lambda}_\psi) + \frac{1}{2}\log|d_{\lambda\lambda}(\psi, \widehat{\lambda}_\psi)|$$

$$\equiv l(\psi, \widehat{\lambda}_\psi) + \frac{1}{2}\log|d_{\lambda\lambda}(\psi, \widehat{\lambda}_\psi)|.$$

The form of this conditional log-likelihood indicates that, instead of just using the profile log-likelihood of ψ, an adjustment term should be added. This notion is developed in detail in Section 9.11 below.

9.10 Bartlett correction

The first-order χ^2 approximation to the distribution of the likelihood ratio statistic $w(\psi)$ can be expressed as

$$\Pr_\theta\{w(\psi) \le \omega^\circ\} = \Pr\{\chi_q^2 \le \omega^\circ\}\{1 + O(n^{-1})\},$$

where q is the dimension of ψ and the full parameter vector is $\theta = (\psi, \lambda)$, with λ nuisance. The χ^2 approximation has relative error of order $O(n^{-1})$.

In the case of independent, identically distributed sampling, it can be shown that

$$\mathbb{E}_\theta w(\psi) = q\{1 + b(\theta)/n + O(n^{-2})\},$$

and so $\mathbb{E}_\theta w'(\psi) = q\{1 + O(n^{-2})\}$, where $w' = w/\{1 + b(\theta)/n\}$.

This adjustment procedure, of replacing w by w', is known as *Bartlett correction*. In spite of its simplicity, this device yields remarkably good results under continuous models, the reason being that division by $\{1 + b(\theta)/n\}$ adjusts, in fact, not only the mean but simultaneously all the cumulants – and hence the whole distribution – of w towards those of χ_q^2. It can be shown that

$$\Pr_\theta\{w'(\psi) \le \omega^\circ\} = \Pr\{\chi_q^2 \le \omega^\circ\}\{1 + O(n^{-2})\}.$$

In practice, because of the (possible) presence of an unknown nuisance parameter λ, $b(\theta)$ may be unknown. If $b(\theta)$ is replaced by $b(\psi, \widehat{\lambda}_\psi)$, the above result still holds, even to $O(n^{-2})$. An explicit expression for $b(\theta)$ is given by Barndorff-Nielsen and Cox (1994: Chapter 6).

Note that the effect of the Bartlett correction is due to the special character of the likelihood ratio statistic, and the same device applied to, for instance, the score test does not have a similar effect. Also, under discrete models this type of adjustment does not generally lead to an improved χ^2 approximation.

9.11 Modified profile likelihood

We noted in Section 8.6 that the profile likelihood $L_p(\psi)$ for a parameter of interest ψ can largely be thought of as if it were a genuine likelihood. However, this amounts to behaving as if the nuisance parameter χ over which the maximisation has been carried out were known. Inference on ψ based on treating $L_p(\psi)$ as a proper likelihood may therefore be grossly misleading if the data contain insufficient information about χ, as is likely to happen, for instance, if the dimension of χ is large. Modified profile likelihood is intended as a remedy for this type of problem.

The modified profile likelihood $\tilde{L}_p(\psi)$ for a parameter of interest ψ, with nuisance parameter χ, due to Barndorff-Nielsen (1983), is defined by

$$\tilde{L}_p(\psi) = M(\psi)L_p(\psi), \tag{9.19}$$

where M is a modifying factor

$$M(\psi) = \left| \frac{\partial \widehat{\chi}}{\partial \widehat{\chi}_\psi} \right| |\widehat{j}_\psi|^{-1/2}.$$

Here $|\cdot|$ denotes the absolute value of a matrix determinant, and $\partial \widehat{\chi}/\partial \widehat{\chi}_\psi$ is the matrix of partial derivatives of $\widehat{\chi}$ with respect to $\widehat{\chi}_\psi$, where $\widehat{\chi}$ is considered as a function of $(\widehat{\psi}, \widehat{\chi}_\psi, a)$. Also, $\widehat{j}_\psi = j_{\chi\chi}(\psi, \widehat{\chi}_\psi)$, the observed information on χ assuming ψ is known. An instructive example to look at to grasp the notation is the case of $X_1, \ldots, X_n$ independent, identically distributed $N(\mu, \sigma^2)$. Here we see that $\widehat{\sigma}_\mu^2 = \frac{1}{n}\sum(X_j - \mu)^2 = \widehat{\sigma}^2 + (\widehat{\mu} - \mu)^2$.

The modified profile likelihood $\tilde{L}_p$ is, like L_p, parametrisation invariant. An alternative expression for the modifying factor M is

$$M(\psi) = |l_{\chi;\widehat{\chi}}(\psi, \widehat{\chi}_\psi; \widehat{\psi}, \widehat{\chi}, a)|^{-1} \times |j_{\chi\chi}(\psi, \widehat{\chi}_\psi; \widehat{\psi}, \widehat{\chi}, a)|^{1/2}. \tag{9.20}$$

Identity (9.20) follows from the likelihood equation for $\widehat{\chi}_\psi$:

$$l_\chi(\psi, \widehat{\chi}_\psi(\widehat{\psi}, \widehat{\chi}, a); \widehat{\psi}, \widehat{\chi}, a) = 0.$$

Differentiation with respect to $\widehat{\chi}$ yields

$$l_{\chi\chi}(\psi, \widehat{\chi}_\psi; \widehat{\psi}, \widehat{\chi}, a)\frac{\partial \widehat{\chi}_\psi}{\partial \widehat{\chi}} + l_{\chi;\widehat{\chi}}(\psi, \widehat{\chi}_\psi; \widehat{\psi}, \widehat{\chi}, a) = 0,$$

from which (9.20) follows.

Asymptotically, $\tilde{L}_p$ and L_p are equivalent to first order. A justification for using $\tilde{L}_p$ rather than L_p is that (9.19) arises as a higher-order approximation to a marginal likelihood for ψ when such a marginal likelihood function is available, and to a conditional likelihood for ψ when this is available.

Specifically, suppose that the density $f(\widehat{\psi}, \widehat{\chi}; \psi, \chi \mid a)$ factorises, either as

$$f(\widehat{\psi}, \widehat{\chi}; \psi, \chi \mid a) = f(\widehat{\psi}; \psi \mid a)f(\widehat{\chi}; \psi, \chi \mid \widehat{\psi}, a) \tag{9.21}$$

or as

$$f(\widehat{\psi}, \widehat{\chi}; \psi, \chi \mid a) = f(\widehat{\chi}; \psi, \chi \mid a)f(\widehat{\psi}; \psi \mid \widehat{\chi}, a). \tag{9.22}$$

In the case (9.21), (9.19) can be obtained as an approximation (using the p^* formula) to the marginal likelihood for ψ based on $\widehat{\psi}$ and conditional on a, that is to the likelihood for ψ determined by $f(\widehat{\psi}; \psi \mid a)$. Similarly, under (9.22) the same expression (9.19) is obtained as an approximation to the conditional likelihood for ψ given $\widehat{\chi}$ and a, that is to the likelihood for ψ obtained from $f(\widehat{\psi}; \psi \mid \widehat{\chi}, a)$. Proofs of both results are given by Barndorff-Nielsen and Cox (1994: Chapter 8).

Sometimes the joint conditional distribution of $\widehat{\psi}$ and $\widehat{\chi}_\psi$ may be factorised as

$$f(\widehat{\psi}, \widehat{\chi}_\psi; \psi, \chi \mid a) = f(\widehat{\chi}_\psi; \psi, \chi \mid a) f(\widehat{\psi}; \psi \mid \widehat{\chi}_\psi, a),$$

while (9.22) does not hold. In this case, (9.19) may be obtained as an approximation to $f(\widehat{\psi}; \psi \mid \widehat{\chi}_\psi, a)$, considered as a pseudo-likelihood for ψ.

Note that, if $\widehat{\chi}_\psi$ does not depend on ψ,

$$\widehat{\chi}_\psi \equiv \widehat{\chi}, \tag{9.23}$$

then

$$\tilde{L}_p(\psi) = |\widehat{j}_\psi|^{-1/2} L_p(\psi). \tag{9.24}$$

In the case that ψ and χ are orthogonal, which is a weaker assumption than (9.23), we have that (9.23) holds to order $O(n^{-1})$, as does (9.24).

The version of modified profile likelihood defined by (9.24) was first presented by Cox and Reid (1987). It is easy to construct and seems to give reasonable results in applications. It is easier to compute than (9.19), but is not invariant with respect to one-to-one transformations of χ, which leave the parameter of interest fixed. A simple Bayesian motivation for (9.24) may be given. Let ψ and the nuisance parameter χ be orthogonal, and let the prior density of ψ and χ be $\pi(\psi, \chi)$. Then the posterior density of ψ is proportional to

$$\int \exp\{l(\psi, \chi)\} \pi(\psi, \chi) d\chi. \tag{9.25}$$

We consider this at a fixed value of ψ. As a function of χ, $l(\psi, \chi)$ achieves its maximum at $\chi = \widehat{\chi}_\psi$. Expanding about this point using Laplace's method, as given by (9.12), shows that (9.25) is approximately

$$(2\pi)^{d_\chi/2} \pi(\psi, \widehat{\chi}_\psi) \exp\{l(\psi, \widehat{\chi}_\psi)\} / |\widehat{j}_\psi|^{1/2},$$

with d_χ denoting the dimension of χ. Now argue as follows. As ψ varies in the range of interest, within $O(n^{-1/2})$ of $\widehat{\psi}$, $\widehat{\chi}_\psi$ varies by $O_p(n^{-1})$, by orthogonality, and therefore so too does the term involving the prior density. Because of its dependence on ψ, the factor involving the determinant varies by $O(n^{-1/2})$, while the part depending on the likelihood is $O(1)$. Therefore, ignoring an error of order $O(n^{-1})$, inference about ψ can be based on an effective log-likelihood of

$$l(\psi, \widehat{\chi}_\psi) - \frac{1}{2} \log |\widehat{j}_\psi|,$$

as given by (9.24).

9.12 Bayesian asymptotics

In Chapter 3, we discussed Monte Carlo techniques for simulating from a posterior distribution, or for evaluation of a posterior expectation of interest. In many circumstances, it will be adequate to have easily computed analytic approximations to these.

In this section we review briefly the asymptotic theory of Bayesian inference. The results provide demonstration of the application of asymptotic approximations discussed earlier, in particular Laplace approximations. Key references in such use of Laplace approximation in Bayesian asymptotics include Tierney and Kadane (1986) and Tierney, Kass and Kadane (1989).

The key result is that the posterior distribution given data x is asymptotically normal. Write

$$\pi_n(\theta \mid x) = f(x; \theta)\pi(\theta) / \int f(x; \theta)\pi(\theta)d\theta$$

for the posterior density. Denote by $\widehat{\theta}$ the MLE.

For θ in a neighbourhood of $\widehat{\theta}$ we have, by Taylor expansion,

$$\log\left\{\frac{f(x; \theta)}{f(x; \widehat{\theta})}\right\} \doteq -\tfrac{1}{2}(\theta - \widehat{\theta})^T j(\widehat{\theta})(\theta - \widehat{\theta}).$$

Provided the likelihood dominates the prior, we can approximate $\pi(\theta)$ in a neighbourhood of $\widehat{\theta}$ by $\pi(\widehat{\theta})$. Then we have

$$f(x; \theta)\pi(\theta) \doteq f(x; \widehat{\theta})\pi(\widehat{\theta})\exp\{-\tfrac{1}{2}(\theta - \widehat{\theta})^T j(\widehat{\theta})(\theta - \widehat{\theta})\},$$

so that, to first order,

$$\pi_n(\theta \mid x) \sim N\left(\widehat{\theta}, j^{-1}(\widehat{\theta})\right).$$

A more natural approximation to the posterior distribution when the likelihood does not dominate the prior is obtained if we expand about the posterior mode $\widehat{\theta}_\pi$, which maximises $f(x; \theta)\pi(\theta)$. An analysis similar to the above then gives

$$\pi_n(\theta \mid x) \sim N\left(\widehat{\theta}_\pi, j_\pi^{-1}(\widehat{\theta}_\pi)\right),$$

where j_π is minus the matrix of second derivatives of $\log f(x; \theta)\pi(\theta)$.

A more accurate approximation to the posterior is provided by the following. We have

$$\begin{aligned}
\pi_n(\theta \mid x) &= f(x; \theta)\pi(\theta) / \int f(x; \theta)\pi(\theta)d\theta \\
&\doteq \frac{c\exp\{l(\theta; x)\}\pi(\theta)}{\exp\{l(\widehat{\theta}; x)\}|j(\widehat{\theta})|^{-1/2}\pi(\widehat{\theta})},
\end{aligned}$$

by Laplace approximation of the denominator. We can rewrite as

$$\pi_n(\theta \mid x) \doteq c|j(\widehat{\theta})|^{1/2}\exp\{l(\theta) - l(\widehat{\theta})\} \times \{\pi(\theta)/\pi(\widehat{\theta})\};$$

note the similarity to the density approximation (9.14) for $\widehat{\theta}$.

Finally, we consider use of the Laplace approximation to approximate to the posterior expectation of a function $g(\theta)$ of interest,

$$\mathbb{E}\{g(\theta) \mid x\} = \frac{\int g(\theta)e^{n\bar{l}_n(\theta)}\pi(\theta)d\theta}{\int e^{n\bar{l}_n(\theta)}\pi(\theta)d\theta} ,$$

where $\bar{l}_n = n^{-1}\sum_{i=1}^{n}\log f(x_i;\theta)$ is the average log-likelihood function. Recall that such expectations arise as the solutions to Bayes decision problems. It turns out to be more effective to rewrite the integrals as

$$\mathbb{E}\{g(\theta) \mid x\} = \frac{\int e^{n\{\bar{l}_n(\theta)+q(\theta)/n\}}d\theta}{\int e^{n\{\bar{l}_n(\theta)+p(\theta)/n\}}d\theta}$$

and to use the version (9.13) of the Laplace approximation. Applying this to the numerator and denominator gives

$$E\{g(\theta) \mid x\} \doteq \frac{e^{n\bar{l}_n(\theta^*)+q(\theta^*)}}{e^{n\bar{l}_n(\tilde{\theta})+p(\tilde{\theta})}} \times \frac{\{-n\bar{l}_n''(\tilde{\theta}) - p''(\tilde{\theta})\}^{1/2}}{\{-n\bar{l}_n''(\theta^*) - q''(\theta^*)\}^{1/2}} \frac{\{1+O(n^{-1})\}}{\{1+O(n^{-1})\}},$$

where θ^* maximises $n\bar{l}_n(\theta) + \log g(\theta) + \log\pi(\theta)$ and $\tilde{\theta}$ maximises $n\bar{l}_n(\theta) + \log\pi(\theta)$. Further detailed analysis shows that the relative error is, in fact, $O(n^{-2})$. If the integrals are approximated in their unmodified form the result is not as accurate.

9.13 Problems

9.1 Suppose that $(y_1, \ldots, y_n)$ are generated by a stationary first-order Gaussian autoregression with correlation parameter ρ, mean μ and innovation variance τ. That is, $Y_1 \sim N(\mu, \tau/(1-\rho^2))$ and, for $j = 2, \ldots, n$,

$$Y_j = \mu + \rho(Y_{j-1} - \mu) + \epsilon_j,$$

where $(\epsilon_1, \ldots, \epsilon_n)$ are independent, identically distributed $N(0, \tau)$.

Find the log-likelihood function. Show that, if μ is known to be zero, the log-likelihood has $(3, 2)$ exponential family form, and find the natural statistics.

9.2 Let $Y_1, \ldots, Y_n$ be independent Poisson (θ). Find the score function and the expected and observed information.

Consider the new parametrisation $\psi = \psi(\theta) = e^{-\theta}$. Compute the score function and the expected and observed information in the ψ-parametrisation.

9.3 Show that, if the parameters ψ and χ are orthogonal, any one-to-one smooth function of ψ is orthogonal to any one-to-one smooth function of χ.

9.4 Suppose that Y is distributed according to a density of the form

$$p(y;\theta) = \exp\{s(y)^T c(\theta) - k(\theta) + D(y)\}.$$

Suppose that θ may be written $\theta = (\psi, \lambda)$, where ψ denotes the parameter of interest, possibly vector valued, and that $c(\theta) = (c_1(\psi), c_2(\theta))^T$, for functions c_1, c_2, where $c_1(\cdot)$ is a one-to-one function of ψ. Then, writing $s(y) = (s_1(y), s_2(y))^T$, the log-likelihood function is of the form

$$l(\psi, \lambda) = s_1(y)^T c_1(\psi) + s_2(y)^T c_2(\theta) - k(\theta).$$

Let ϕ be the *complementary mean parameter* given by

$$\phi \equiv \phi(\theta) = \mathbb{E}\{s_2(Y); \theta\}.$$

Show that ψ and ϕ are orthogonal parameters.

Let Y have a Gamma distribution with shape parameter ψ and scale parameter ϕ, and density

$$f(y; \psi, \phi) = \phi^{-\psi} y^{\psi-1} \exp(-y/\phi) / \Gamma(\psi).$$

Show that $\psi\phi$ is orthogonal to ψ.

9.5 Let $Y_1, \dots, Y_n$ be independent random variables such that Y_j has a Poisson distribution with mean $\exp\{\lambda + \psi x_j\}$, where $x_1, \dots, x_n$ are known constants.

Show that the conditional distribution of $Y_1, \dots, Y_n$ given $S = \sum Y_j$ does not depend on λ. Find the conditional log-likelihood function for ψ, and verify that it is equivalent to the profile log-likelihood.

9.6 Verify that in general the likelihood ratio and score tests are invariant under a reparametrisation $\psi = \psi(\theta)$, but that the Wald test is not.

Write $\theta = (\theta_1, \theta_2)$, where θ_1 is the parameter of interest. Suppose $\psi = \psi(\theta) = (\psi_1, \psi_2)$ is an interest respecting transformation, with $\psi_1 \equiv \psi_1(\theta) = \theta_1$. Show that the profile log-likelihood is invariant under this reparametrisation.

9.7 Verify that the rth degree Hermite polynomial H_r satisfies the identity

$$\int_{-\infty}^{\infty} e^{ty} H_r(y) \phi(y) dy = t^r e^{\frac{1}{2}t^2}.$$

Verify that the moment generating function of S_n^* has the expansion

$$M_{S_n^*}(t) = \exp\{K_{S_n^*}(t)\}$$

$$= e^{\frac{1}{2}t^2} \exp\left\{\frac{1}{6\sqrt{n}} \rho_3 t^3 + \frac{1}{24n} \rho_4 t^4 + O(n^{-3/2})\right\}$$

$$= e^{\frac{1}{2}t^2} \left\{1 + \frac{\rho_3}{6\sqrt{n}} t^3 + \frac{\rho_4}{24n} t^4 + \frac{\rho_3^2}{72n} t^6 + O(n^{-3/2})\right\}.$$

On using the above identity, this latter expansion may be written

$$M_{S_n^*}(t) = \int_{-\infty}^{\infty} e^{ty} \left\{1 + \frac{1}{6\sqrt{n}} \rho_3 H_3(y)\right.$$

$$\left. + \frac{1}{24n} \rho_4 H_4(y) + \frac{1}{72n} \rho_3^2 H_6(y) + O(n^{-3/2})\right\} \phi(y) dy.$$

Compare with the definition

$$M_{S_n^*}(t) = \int_{-\infty}^{\infty} e^{ty} f_{S_n^*}(y) dy,$$

to provide a heuristic justification for the Edgeworth expansion.

9.8 Let $Y_1, \ldots, Y_n$ be independent, identically distributed $N(\mu, \sigma^2)$. Obtain the saddle-point approximation to the density of $S_n = \sum_{i=1}^n Y_i$, and comment on its exactness.

9.9 Let $Y_1, \ldots, Y_n$ be independent, identically distributed exponential random variables with density function $f(y) = e^{-y}$. Obtain the saddlepoint approximation to the density of $S_n = \sum_{i=1}^n Y_i$, and show that it matches the exact density except for the normalising constant.

9.10 Let $Y_1, \ldots, Y_n$ be independent, identically distributed exponential random variables of mean μ. Verify that the p^* formula for the density of $\widehat{\mu}$ is exact.

9.11 Let $y_1, \ldots, y_n$ be independent realisations of a continuous random variable Y with density belonging to a location-scale family,

$$p(y; \mu, \sigma) = \frac{1}{\sigma} p_0 \left(\frac{y - \mu}{\sigma} \right),$$

$(y - \mu)/\sigma \in \mathcal{X}, \mu \in \mathbb{R}, \sigma > 0$. Assume that the maximum likelihood estimate $(\widehat{\mu}, \widehat{\sigma})$ of (μ, σ) based on $y = (y_1, \ldots, y_n)$ exists and is finite and that p_0 is suitably differentiable. Define the sample configuration a by

$$a = \left(\frac{y_1 - \widehat{\mu}}{\widehat{\sigma}}, \ldots, \frac{y_n - \widehat{\mu}}{\widehat{\sigma}} \right).$$

Show that the p^* formula for the conditional density of $(\widehat{\mu}, \widehat{\sigma})$ given a is

$$p^*(\widehat{\mu}, \widehat{\sigma}; \mu, \sigma \mid a) = c(\mu, \sigma, a) \frac{\widehat{\sigma}^{n-2}}{\sigma^n} \prod_{i=1}^n p_0 \left(\frac{\widehat{\sigma}}{\sigma} a_i + \frac{\widehat{\mu} - \mu}{\sigma} \right),$$

and is exact.

9.12 Let $Y_1, \ldots, Y_n$ be independent, identically distributed $N(\mu, \sigma^2)$, and suppose the parameter of interest is the variance σ^2.

Obtain the form of the profile log-likelihood. Show that the profile score has an expectation which is non-zero.

Find the modified profile log-likelihood for σ^2 and examine the expectation of the modified profile score.

9.13 Let $Y_1, \ldots, Y_n$ be independent exponential random variables, such that Y_j has mean $\lambda \exp(\psi x_j)$, where $x_1, \ldots, x_n$ are known scalar constants and ψ and λ are unknown parameters.

In this model the maximum likelihood estimators are not sufficient and an ancillary statistic is needed. Let

$$a_j = \log Y_j - \log \widehat{\lambda} - \widehat{\psi} x_j,$$

$j = 1, \ldots, n$. Show that $a = (a_1, \ldots, a_n)$ is ancillary.

Find the form of the profile log-likelihood function and that of the modified profile log-likelihood function for ψ.

9.14 Let $Y_1, \ldots, Y_n$ be independent, identically distributed $N(\mu, \sigma^2)$ and consider testing $H_0 : \mu = \mu_0$. Show that the likelihood ratio statistic for testing H_0 may be expressed as

$$w = n \log\{1 + t^2/(n - 1)\},$$

where t is the usual Student's t statistic.

Show directly that

$$\mathbb{E}w = 1 + \frac{3}{2n} + O(n^{-2})$$

in this case, so that the Bartlett correction factor $b \equiv 3/2$.

Examine numerically the adequacy of the χ_1^2 approximation to w and to $w' = w/(1 + 3/2n)$.

9.15 Let $(X_1, Y_1), \ldots, (X_n, Y_n)$ be independent pairs of independently normally distributed random variables such that, for each j, X_j and Y_j each have mean μ_j and variance σ^2.

Find the maximum likelihood estimator of σ^2 and show that it is not consistent.

Find the form of the modified profile log-likelihood function for σ^2 and examine the estimator of σ^2 obtained by its maximisation.

Let $S = \sum_{i=1}^{n}(X_i - Y_i)^2$. What is the distribution of S? Find the form of the marginal log-likelihood for σ^2 obtained from S and compare it with the modified profile likelihood.

(This is the 'Neyman–Scott problem', which typifies situations with large numbers of nuisance parameters. Note, however, that the model falls outside the general framework that we have been considering, in that the dimension of the parameter $(\mu_1, \ldots, \mu_n, \sigma^2)$ depends on the sample size, and tends to ∞ as $n \to \infty$.)

9.16 Consider a multinomial distribution with four cells, the probabilities for which are

$$\pi_1(\theta) = \frac{1}{6}(1 - \theta), \pi_2(\theta) = \frac{1}{6}(1 + \theta),$$

$$\pi_3(\theta) = \frac{1}{6}(2 - \theta), \pi_4(\theta) = \frac{1}{6}(2 + \theta),$$

where θ is unknown, $|\theta| < 1$.

What is the minimal sufficient statistic? Show that $A' = (N_1 + N_2, N_3 + N_4)$ and $A'' = (N_1 + N_4, N_2 + N_3)$ are both ancillary.

If A is ancillary, we may write

$$f_X(x; \theta) = f_{X|A}(x \mid a; \theta) f_A(a).$$

The conditional expected information for θ given $A = a$ is

$$i_A(\theta \mid a) = \mathbb{E}_\theta \left\{ \frac{-\partial^2 \log f_{X|A}(X \mid a, \theta)}{\partial \theta^2} \bigg| A = a \right\}$$

$$= \mathbb{E}_\theta \left\{ \frac{-\partial^2 \log f_X(X; \theta)}{\partial \theta^2} \bigg| A = a \right\}.$$

If we take expectations over the distribution of A:

$$\mathbb{E}\{i_A(\theta \mid A)\} = i(\theta).$$

With two ancillaries competing,

$$\mathbb{E}\{i_{A'}(\theta \mid A')\} = \mathbb{E}\{i_{A''}(\theta \mid A'')\},$$

so that expected conditional information is no basis for choice between them. Suggest why, to discriminate between them, A' might be considered preferable to A'' if

$$\text{var}\{i_{A'}(\theta \mid A')\} > \text{var}\{i_{A''}(\theta \mid A'')\}.$$

Show that in the current multinomial example A' is preferable to A'' in these terms.

Predictive inference

The focus of our discussion so far has been inference for the unknown *parameter* of the probability distribution assumed to have generated the sample data. Sometimes, interest lies instead in assessing the values of future, unobserved values from the same probability distribution, typically the next observation. We saw in Section 3.9 that in a Bayesian approach such prediction is easily accommodated, since there *all* unknowns are regarded as random variables, so that the distinction between an unknown constant (parameter) and a future observation (random variable) disappears. However, a variety of other approaches to prediction have been proposed.

The prediction problem is as follows. The data x are the observed value of a random variable X with density $f(x; \theta)$, and we wish to predict the value of a random variable Z, which, conditionally on $X = x$, has distribution function $G(z \mid x; \theta)$, depending on θ.

As a simple case, we might have X formed from independent and identically distributed random variables $X_1, \ldots, X_n$, and Z is a further, independent, observation from the same distribution. A more complicated example is that of *time series prediction*, where the observations are correlated and prediction of a future value depends directly on the observed value as well as on any unknown parameters that have to be estimated. Example 10.2 is a simple case of time series prediction.

Apart from the fully Bayesian approach of Section 3.9, it is possible to identify at least five approaches to predictive inference, which we classify as (a) exact methods, (b) decision theory approaches, (c) methods based on predictive likelihood, (d) asymptotic approaches, (e) bootstrap methods. In this chapter we provide brief outlines of each of these methods.

Book-length treatments of predictive inference are due to Aitchison and Dunsmore (1975) and Geisser (1993). Both focus primarily on the Bayesian approach.

10.1 Exact methods

By an exact method, we mean that, for any $\alpha \in (0, 1)$, we can construct a 'prediction set' $S_\alpha(X_1, \ldots, X_n)$ such that $\Pr\{Z \in S_\alpha(X_1, \ldots, X_n); \theta\} = 1 - \alpha$ for any value of the unknown parameter θ. Typically, if Z is a scalar, then S_α will be an interval, but in principle we can allow more general sets.

Cox (1975) and Barndorff-Nielsen and Cox (1996) described two general approaches for constructing an exact prediction set. The first idea, due to Guttman (1970), uses similar tests. Suppose $X \sim f(x; \theta)$, $Z \sim g(z; \theta^*)$, where we do not initially assume $\theta^* = \theta$. Suppose we can find a similar test of size α for the hypothesis $H_0 : \theta^* = \theta$. Such a test would lead to an acceptance region A_α such that

$$\Pr\{(X_1, \ldots, X_n, Z) \in A_\alpha; \theta^* = \theta\} = 1 - \alpha \text{ for all } \theta. \tag{10.1}$$

For given $X_1, \ldots, X_n$, define $S_\alpha(X_1, \ldots, X_n)$ to be the set of all values of Z for which $(X_1, \ldots, X_n, Z) \in A_\alpha$. Because of (10.1), this has exact coverage probability $1 - \alpha$, whatever the true value of θ.

A second method uses pivotal statistics. Suppose $T = T(X_1, \ldots, X_n, Z)$ is pivotal, that is the distribution of T does not depend on θ. Suppose R_α is a set such that $\Pr\{T \in R_\alpha\} = 1 - \alpha$. Then, for given $X_1, \ldots, X_n$, the set $S = \{Z : T(X_1, \ldots, X_n, Z) \in R_\alpha\}$ defines a $100(1 - \alpha)\%$ prediction set for Z.

Example 10.1 Suppose for instance that we wish to predict a new observation Z from $N(\mu, \sigma^2)$, with known variance σ^2 but unknown mean μ, on the basis of a set $X_1, \ldots, X_n$ of independent identically distributed observations, with mean $\bar{X}$, from the same distribution. Then

$$(Z - \bar{X})/\{\sigma\sqrt{1 + 1/n}\}$$

is pivotal, being distributed as $N(0, 1)$, and can be used to construct a prediction interval for Z: details are worked out in Problem 10.1.

Example 10.2 (Barndorff-Nielsen and Cox, 1996) Suppose $X_1, \ldots, X_{n+1}$ are from a first-order autoregressive process

$$X_1 \sim N\left(\mu, \frac{\sigma^2}{1 - \rho^2}\right), \tag{10.2}$$

$$X_{i+1} = \mu + \rho(X_i - \mu) + \epsilon_{i+1},$$

where $|\rho| < 1$ and $\epsilon_2, \epsilon_3, \ldots$ are independent (of each other and of X_1) with distribution $N(0, \sigma^2)$. It is readily verified that, provided (10.2) is satisfied, then $X_i \sim N\left(\mu, \frac{\sigma^2}{1-\rho^2}\right)$ for all $i \geq 1$; for this reason (10.2) is called the stationary distribution of the process. The problem considered here is to predict the value of X_{n+1} given observations $X_1, \ldots, X_n$. Specifically, we consider the case where μ is unknown and ρ and σ^2 are known.

We first calculate the maximum likelihood estimator (MLE) of μ. The likelihood function for μ is derived by writing the joint density of $X_1, \ldots, X_n$ in the form

$$f(X_1) \prod_{i=1}^{n-1} f(X_{i+1} \mid X_i) = (2\pi\sigma^2)^{-n/2}(1 - \rho^2)^{1/2} \times$$

$$\exp\left[-\frac{1}{2}\left\{(1 - \rho^2)(X_1 - \mu)^2 + \sum_{i=1}^{n-1}(X_{i+1} - \rho X_i - (1 - \rho)\mu)^2\right\}\right]. \tag{10.3}$$

The MLE $\widehat{\mu}$ minimises

$$(1 - \rho^2)(X_1 - \mu)^2 + \sum_{i=1}^{n-1}\{X_{i+1} - \rho X_i - (1 - \rho)\mu\}^2$$

and is therefore given by

$$\widehat{\mu} = \frac{X_1 + (1 - \rho)(X_2 + \ldots + X_{n-1}) + X_n}{n - n\rho + 2\rho}.$$

It follows that $\mathbb{E}(\widehat{\mu}) = \mu$ and hence that

$$T = X_{n+1} - \rho X_n - (1 - \rho)\widehat{\mu}$$

has a normal distribution with mean 0 and a variance that also does not depend on μ; in other words, T is a pivotal quantity.

To calculate the variance of T, write

$$\widehat{\mu} - \mu = \frac{(1 + \rho)(X_1 - \mu) + \sum_{i=1}^{n-1} \epsilon_{i+1}}{n - n\rho + 2\rho},$$

$$T = \epsilon_{n+1} - \frac{1 - \rho}{n - n\rho + 2\rho}\{(1 + \rho)(X_1 - \mu) + \sum_{i=1}^{n-1} \epsilon_{i+1}\},$$

which is a sum of independent random variables, so

$$\mathrm{var}(T) = \sigma^2 \left\{ 1 + \left(\frac{1 - \rho}{n - n\rho + 2\rho} \right)^2 \left(\frac{1 + \rho}{1 - \rho} + n - 1 \right) \right\}$$

$$= \sigma^2 r_n^2(\rho) \quad \text{say}.$$

With probability $1 - \alpha$, we have $|T| < z_{\alpha/2}\sigma r_n(\rho)$ (recall that, for arbitrary β, z_β is the solution of $\Phi(z_\beta) = 1 - \beta$, Φ being the standard normal distribution function) and, therefore, an exact $100(1 - \alpha)\%$ prediction interval for X_{n+1} is given by

$$\left(\rho X_n + (1 - \rho)\widehat{\mu} - z_{\alpha/2}\sigma r_n(\rho), \ \rho X_n + (1 - \rho)\widehat{\mu} + z_{\alpha/2}\sigma r_n(\rho) \right). \tag{10.4}$$

As with the case of a simple normal problem (Example 10.1), the more practical problem is one in which ρ and σ^2 are also unknown. They may be estimated (along with μ) by jointly maximising the likelihood (10.3) with respect to all three unknown parameters; the solution of this is left as an exercise. In that case, a reasonable approximation to a $100(1 - \alpha)\%$ prediction interval would be to substitute the maximum likelihood estimators $\widehat{\rho}$ and $\widehat{\sigma}$ into (10.4). However, for similar reasons to those for Example 10.1 (see also Problem 10.1), such an approximation will not give an exact interval. In this case, we are not aware of any procedure that gives exact coverage probability $1 - \alpha$, but asymptotic methods (Section 10.4) are a potential alternative approach.

Example 10.3 (Barndorff-Nielsen and Cox, 1996) This concerns a different viewpoint of empirical Bayes analysis (Chapter 3) formulated as a prediction problem.

Suppose $Z_1, \ldots, Z_n$ are independent $N(\mu, \sigma^2)$ and, conditionally on $(Z_1, \ldots, Z_n)$, $X_1, \ldots, X_n$ are independent with $X_i \sim N(Z_i, \tau^2)$. We observe $X_1, \ldots, X_n$ and would like to predict Z_1. This differs from the formulation of Chapter 3 in that the means of $X_1, \ldots, X_n$ are random variables with a known distribution, as opposed to simply being unknown parameters $\mu_1, \ldots, \mu_n$. In this case, admissibility results, such as the James–Stein paradox, no longer apply, but there is an analogous result, that the estimation of a particular mean

Z_1 should depend not only on X_1 but on the entire sample $X_1, \ldots, X_n$ through some process of 'shrinking towards the mean' (see equation (10.5) below). The specific formulation considered here assumes σ^2 and τ^2 are known but the overall mean μ is unknown.

In this case the obvious estimator of μ (which is also the MLE) is $\widehat{\mu} = \bar{X} = n^{-1} \sum_{i=1}^{n} X_i$, so it is natural to consider pivotal statistics of the form

$$T_\lambda = Z_1 - \lambda X_1 - (1 - \lambda)\bar{X}, \qquad (10.5)$$

where $0 \leq \lambda \leq 1$, $\lambda = 1$ being the limiting case of no shrinkage.

We can directly calculate the value of λ, say λ^*, that minimises the variance of T_λ; we then use T_{λ^*} directly to compute an exact prediction interval for Z_1. Details are in Problem 10.3.

10.2 Decision theory approaches

The discussion so far has focussed on procedures for constructing a prediction interval or a prediction set that guarantees a specified coverage probability. However, it stands to reason that this cannot be the only criterion for comparing two predictive procedures. For example, there may be two different ways of constructing a prediction interval, both having exactly the desired coverage probability, but the first of which always results in a shorter interval than the second. In this case it seems obvious that we would prefer the first method, but our discussion so far does not incorporate this as a criterion.

An alternative viewpoint is to define a loss function between the true and predicted probability densities, and apply the standard concepts of decision theory that we have discussed in Chapter 3. Suppose we are trying to estimate $g(z; \theta)$, the density of a random variable we are trying to predict, based on an observation X from a density $f(x; \theta)$ that depends on the same parameter θ. Typically X will be a random sample $(X_1, \ldots, X_n)$ from the same density g, so $f(x; \theta) = \prod_{i=1}^{n} g(x_i; \theta)$, but it is not necessary that $X_1, \ldots, X_n, Z$ all have the same density, so long as they all depend on the same θ.

Suppose, then, we are considering a density $\tilde{g}(z|x)$ as an estimate of the true density $g(z; \theta)$. A common measure of discrepancy between g and $\tilde{g}$ is the Kullback–Leibler divergence:

$$L(g, \tilde{g}) = \int \log \left\{ \frac{g(z; \theta)}{\tilde{g}(z|x)} \right\} g(z; \theta) dz. \qquad (10.6)$$

Elementary application of Jensen's inequality shows

$$L(g, \tilde{g}) \geq 0 \qquad (10.7)$$

with equality only if $\tilde{g} = g$. Thus if we know θ we should always set $\tilde{g} = g$, which is a natural property for a loss function to have, though there are many other reasons why (10.6) is considered a good measure of discrepancy.

If we are comparing two predictors $\tilde{g}_1$ and $\tilde{g}_2$, according to (10.6) we would prefer $\tilde{g}_1$ if $L(g, \tilde{g}_1) < L(g, \tilde{g}_2)$, or in other words if

$$\int \log \left\{ \frac{\tilde{g}_1(z|x)}{\tilde{g}_2(z|x)} \right\} g(z; \theta) dz > 0. \qquad (10.8)$$

However, (10.8) depends on both x and θ. To convert (10.8) from a comparison of loss functions to a comparison of risk functions, we integrate with respect to x:

$$\int \int \log \left\{ \frac{\tilde{g}_1(z|x)}{\tilde{g}_2(z|x)} \right\} g(z; \theta) dz f(x; \theta) dx > 0. \tag{10.9}$$

If, further, θ has a prior density $\pi(\theta)$, we may integrate (10.9) with respect to $\pi(\theta) d\theta$: $\tilde{g}_1$ has smaller Bayes risk than $\tilde{g}_2$ if

$$\int \int \int \log \left\{ \frac{\tilde{g}_1(z|x)}{\tilde{g}_2(z|x)} \right\} g(z; \theta) dz f(x; \theta) dx \pi(\theta) d\theta > 0. \tag{10.10}$$

However, if we rewrite $f(x; \theta) \pi(\theta) = f(x) \pi(x|\theta)$, where $f(x) = \int f(x; \theta) \pi(\theta) d\theta$ is the marginal density and $\pi(\theta|x)$ is the posterior density of θ given x, then an interchange of the order of integration in (10.10) leads to

$$\int \int \log \left\{ \frac{\tilde{g}_1(z|x)}{\tilde{g}_2(z|x)} \right\} \hat{g}(z|x) dz f(x) dx > 0, \tag{10.11}$$

where $\hat{g}(z|x) = \int g(z; \theta) \pi(\theta|x) d\theta$ is the Bayesian predictive density of Z as defined in Section 3.9. Note that we are using $\hat{g}$ specifically to denote a Bayesian predictive density (depending on some prior π which is not explicitly represented in the notation), whereas $\tilde{g}$ refers to a general method of constructing an estimate of the density g. However, a comparison of (10.11) with (10.7) shows that (10.11) is always satisfied if $\tilde{g}_1(z|x) = \hat{g}(z|x)$. In other words, if our decision criterion is Bayes risk with respect to prior $\pi(\theta)$, then the optimal predictive density is always the Bayesian predictive density with respect to the same $\pi(\theta)$. Of course this could have been anticipated from the very general properties of Bayesian decision procedures given in Chapter 3, but much of the interest in this topic is that Bayesian predictive densities often have desirable properties even when evaluated from a non-Bayesian perspective. In particular, the point of Aitchison's (1975) paper was to point out that with $\tilde{g}_1 = \hat{g}$ and $\tilde{g}_2$ lying in a wide class of natural alternatives, (10.9) is often satisfied for all θ, a much stronger result than the integrated form (10.10).

Aitchison compared two approaches: if $\tilde{g}_2(z|x) = g(z; \tilde{\theta})$, where $\tilde{\theta} = \tilde{\theta}(x)$ is some point estimate of θ given x, then $\tilde{g}_2$ is called an 'estimative density'. This is to be compared with the 'predictive density' $\tilde{g}_1 = \hat{g}$. An example where (10.9) holds is given next.

Example 10.4 Suppose g is the Gamma(α, θ) density with known shape parameter α:

$$g(z; \theta) = \frac{\theta^\alpha z^{\alpha-1} e^{-\theta z}}{\Gamma(\alpha)}. \tag{10.12}$$

We may assume $X_1, \ldots, X_n$ are independent from the same density, but, since $\sum X_i$ is sufficient for θ, it is equivalent to define $X = \sum X_i$, in which case the density of X is

$$f(x; \theta) = \frac{\theta^k x^{k-1} e^{-\theta x}}{\Gamma(k)}, \tag{10.13}$$

where $k = \alpha n$.

The obvious 'estimative' approach is to set $\tilde{\theta} = \frac{k}{x}$, the maximum likelihood estimator, which leads to

$$\tilde{g}_2(z|x) = \frac{k^\alpha z^{\alpha-1}}{x^\alpha \Gamma(\alpha)} e^{-kz/x}. \tag{10.14}$$

For the 'predictive' approach, we assume θ has a Gamma prior with parameters (a, b). The posterior is then of the same form with parameters (A, B), where $A = a + k$, $B = b + x$. Subsequently we assume $b = 0$, which leads to the predictive density

$$\tilde{g}_1(z|x) = \int \frac{\theta^\alpha z^{\alpha-1} e^{-\theta z}}{\Gamma(\alpha)} \cdot \frac{x^A \theta^{A-1} e^{-x\theta}}{\Gamma(A)} d\theta$$

$$= \frac{\Gamma(A+\alpha)}{\Gamma(A)\Gamma(\alpha)} \frac{z^{\alpha-1} x^A}{(z+x)^{\alpha+A}}. \tag{10.15}$$

From (10.14) and (10.15) we have

$$\log\left\{\frac{\tilde{g}_1(z|x)}{\tilde{g}_2(z|x)}\right\} = \log\left\{\frac{\Gamma(A+\alpha)}{\Gamma(A)}\right\} - \alpha \log k + k\frac{z}{x}$$
$$- (\alpha + A) \log\left(1 + \frac{z}{x}\right). \tag{10.16}$$

To evaluate (10.9) based on (10.16), we must integrate with respect to both z and x, using (10.12) and (10.13). As a side calculation, if $Z \sim \text{Gamma}(\alpha, \theta)$, we have

$$\mathbb{E}(Z^r) = \frac{\theta^{-r} \Gamma(\alpha + r)}{\Gamma(\alpha)}$$

and on differentiating both sides with respect to r and setting $r = 0$,

$$\mathbb{E}(\log Z) = -\log \theta + \psi(\alpha),$$

where $\psi(t) = \frac{d}{dt}\{\log \Gamma(t)\}$ is known as the digamma function. Since we also have $X + Z \sim \text{Gamma}(k + \alpha, \theta)$ we calculate

$$\mathbb{E}\left(\frac{Z}{X}\right) = \frac{\alpha}{k-1},$$

$$\mathbb{E}\left\{\log\left(1 + \frac{Z}{X}\right)\right\} = \mathbb{E}\{\log(X + Z)\} - \mathbb{E}\{\log(X)\}$$
$$= \psi(\alpha + k) - \psi(k).$$

Hence the left-hand side of (10.9) becomes

$$\log\left\{\frac{\Gamma(A+\alpha)}{\Gamma(A)}\right\} - \alpha \log k + \frac{k\alpha}{k-1} - (\alpha + A)\{\psi(\alpha + k) - \psi(k)\}. \tag{10.17}$$

Aitchison gave a simple but non-rigorous argument why (10.17) is positive for all A, α, k but this is in any case confirmed by numerical experience. He also developed an example based on the multivariate normal distribution for which a similar result holds, that is that a suitable class of Bayesian predictive densities outperforms the estimative density when assessed by Kullback–Leibler distance to the true $g(z; \theta)$. Those two cases are apparently the only known examples when a strict ordering holds for any sample size n; however, it is widely suspected that the same result holds in practice for a much wider class of distributions. Harris (1989) developed an alternative approach related to the bootstrap (Chapter 11) and showed that this also improves asymptotically on the estimative approach in the case of exponential families; this approach is described further in Section 10.5. In a very broadly based asymptotic approach, Komaki (1996) derived a general construction for improving estimative approaches by means of a shift in a direction orthogonal to the model. He also defined general conditions under which this shift is achieved by a Bayesian predictive density. Unfortunately a full presentation of this approach lies well beyond the scope of the present discussion.

10.3 Methods based on predictive likelihood

The name 'predictive likelihood' was apparently first introduced by Hinkley (1979), though parallel ideas had been presented in an earlier paper by Lauritzen (1974), and there are precedents in much earlier work of Karl Pearson and R.A. Fisher. Other contributors include Mathiasen (1979), Butler (1986, 1989) and Davison (1986), and the whole topic was reviewed by Bjørnstad (1990), who identified 14 different versions of predictive likelihood, though adding that many of them are very similar. On the other hand, the concept has been criticised on the grounds that frequentist coverage probabilities of predictive likelihood-based prediction intervals or prediction sets do not necessarily improve on those of more naive procedures; we briefly discuss that aspect at the end of this section.

We do not attempt to review all the different methods, but concentrate on some of the leading developments in roughly chronological order.

10.3.1 Basic definitions

Suppose we have past data X from a density $f(x; \theta)$ defined by a finite-dimensional parameter θ, and we are interested in predicting some as yet unobserved random variable Z, whose distribution depends on θ and possibly also on X. A common situation, but by no means the only one to which these concepts have been applied, is that $X = (X_1, \ldots, X_n)$, $Z = (X_{n+1}, \ldots, X_{n+m})$, where $X_1, \ldots, X_{n+m}$ are independent and identically distributed from some parametric family of density functions with parameter θ. In general we assume there is a joint density $f_{X,Z}(x, z; \theta)$.

If θ has a prior density $\pi(\theta)$, the Bayesian predictive density (Section 3.9) for a value $Z = z$ given $X = x$ is defined by

$$\hat{g}(z|x) = \frac{\int f_{X,Z}(x, z; \theta)\pi(\theta)d\theta}{\int f_X(x; \theta)\pi(\theta)d\theta} = \int f_{Z|X}(z|x; \theta)\pi(\theta|x)d\theta. \tag{10.18}$$

Just as the ordinary likelihood may be viewed as an attempt to define the information in θ without requiring a prior density, so the predictive likelihood for Z may be viewed as an attempt to provide alternatives to (10.18) without specifying a prior $\pi(\theta)$. To distinguish predictive likelihood from its Bayesian relatives, we shall use the symbol $L(z|x)$. If z is regarded as the quantity of interest then θ is a nuisance parameter, and all predictive likelihood approaches may be viewed as attempts to remove θ from the conditional density of Z given X.

Bjørnstad (1990) remarked that, although there are many definitions of predictive likelihood, they all reduce to one of three main operations to eliminate θ: integration, maximisation or conditioning. The simplest is the 'profile predictive likelihood'

$$L(z|x) = \sup_{\theta} f_{X,Z}(x, z; \theta)$$

introduced by Mathiasen (1979). This differs from the 'estimative' approach of Section 10.2 in that θ is chosen to maximise the *joint* density of (X, Z) instead of X alone; nevertheless it suffers from a similar objection, that it does not adequately allow for the uncertainty in θ when defining predictive inference for Z.

Lauritzen (1974) and Hinkley (1979) defined an alternative approach by exploiting the properties of sufficient statistics. A simple version (Hinkley's Definition 1) applies when X and Z are independent. Suppose R, S, T are minimal sufficient statistics for θ based on (Y, Z), Y, Z respectively. An immediate consequence is that the conditional distribution of Z given T does not depend on θ; therefore, to define a predictive likelihood for Z, it suffices to do so for T. Because X and Z are independent, it follows that $R = r(S, T)$ is a function of S and T; Hinkley assumed this is invertible, that is that T is determined by knowledge of R and S. Then Hinkley's definition of the predictive likelihood for $T = t$ given $S = s$ is

$$L_T(t|s) = f_{S|R}(s|r(s, t)), \tag{10.19}$$

that is the conditional density of S given R. It follows that the predictive likelihood for $Z = z$ is

$$L_Z(z|s) = L_T(t(z)|s) \cdot f_{Z|T}(z|t(z)).$$

Example 10.5 Suppose $Y = (X_1, \ldots, X_n)$, $Z = (X_{n+1}, \ldots, X_{n+m})$, where $X_1, \ldots, X_{n+m}$ are independent Bernoulli random variables with $\Pr\{X_i = 1\} = 1 - \Pr\{X_i = 0\} = \theta$ for all i. Then $S = \sum_{i=1}^{n} X_i$, $T = \sum_{i=n+1}^{n+m} X_i$, $R = S + T$ and the conditional distribution of S given R is hypergeometric:

$$\Pr\{S = s|R = s + t\} = \frac{\binom{n}{s}\binom{m}{t}}{\binom{n+m}{s+t}} \tag{10.20}$$

wherever $0 \le s \le n$, $0 \le t \le m$. Thus (10.20), interpreted as a function of t for given s, is the predictive likelihood for T given S.

In more complicated situations, including ones where X and Z are dependent, Hinkley proposed the following (Definition 2):

Suppose R is minimal sufficient for (X, Z), S is minimal sufficient for X and let T be a function of (Z, S) such that: (i) R is determined by S and T, (ii) a minimal sufficient reduction of Z is determined by (S, T). Assume T is determined uniquely by (R, S). Then the predictive likelihood of T is again defined by (10.19) and the predictive likelihood of Z is given by

$$L_Z(z|s) = L_T(t(z, s)|s) \cdot f_{Z|(S,T)}(z|s, t(z, s)).$$

Example 10.6 Suppose $Y = (X_1, \ldots, X_n)$, $Z = X_{n+1}$, where $X_1, \ldots, X_{n+1}$ are independent from the uniform distribution on $(0, \theta)$ for some unknown $\theta > 0$.

Let $M_j = \max\{X_1, \ldots, X_j\}$. Then $R = M_{n+1}$, $S = M_n$, but Z is not necessarily determined by knowledge of R and S; therefore, Definition 1 does not apply. Hinkley defined

$$T = \begin{cases} 0 & \text{if } X_{n+1} \leq S, \\ X_{n+1} & \text{if } X_{n+1} > S. \end{cases}$$

The conditions of Definition 2 are now satisfied. When $t = 0$, the joint density of (S, T), evaluated at (s, t), is $ns^{n-1}\theta^{-n} \cdot s\theta^{-1}$ (the first factor is the marginal density of S and the second is the conditional probability that $X_{n+1} < S$ given $S = s$). When $t > s$, the joint density is $ns^{n-1}\theta^{-n} \cdot \theta^{-1}$ on $s < t < \theta$. The marginal density of R is $(n + 1)r^n\theta^{-n-1}$. Therefore, the predictive likelihood for T is

$$L_T(t|s) = \begin{cases} \frac{n}{n+1} & \text{if } t = 0, \ r = s, \\ \frac{ns^{n-1}}{(n+1)t^n} & \text{if } t = r > s, \end{cases}$$

and the predictive likelihood for Z is

$$L_Z(z|s) = \begin{cases} \frac{n}{(n+1)s} & \text{if } 0 \leq z \leq s, \\ \frac{ns^{n-1}}{(n+1)z^n} & \text{if } z > s. \end{cases}$$

10.3.2 Butler's predictive likelihood

Butler (1986) argued that Hinkley's predictive likelihood is too restrictive to be applied to a wide range of problems and proposed the following 'conditional' approach.

Suppose (X, Z) is transformed to (R, U), where R is minimal sufficient and the components of U are locally orthogonal (to each other and to R). This is equivalent to assuming $K^T K = I$ and $K^T J = 0$, where $K = \frac{\partial u}{\partial(x,z)}$ and $J = \frac{\partial r}{\partial(x,z)}$ are matrices of first-order partial derivatives. Then the Jacobian of the transformation from (x, z) to (r, u) has the form

$$\left| \frac{\partial(r, u)}{\partial(x, z)} \right| = \left| \begin{matrix} J \\ K \end{matrix} \right| = \left| \begin{pmatrix} J \\ K \end{pmatrix} \begin{pmatrix} J^T & K^T \end{pmatrix} \right|^{1/2} = \left| \begin{matrix} JJ^T & JK^T \\ KJ^T & KK^T \end{matrix} \right|^{1/2} = |JJ^T|^{1/2}$$

and hence the likelihood is rewritten

$$f_{X,Z}(x, z; \theta)dxdz = \frac{f_{X,Z}(x, z; \theta)}{f_R(r; \theta)} |JJ^T|^{-1/2}du \cdot f_R(r; \theta)dr. \qquad (10.21)$$

Butler defined the predictive likelihood as

$$L(z|x) = \frac{f_{X,Z}(x, z; \theta)}{f_R(r; \theta)|JJ^T|^{1/2}} \qquad (10.22)$$

noting that this is 'the largest portion of (10.21) which may be used to infer z without a prior distribution for θ'.

Under the conditions where Hinkley's Definition 1 holds, Butler remarked that (10.22) is the same as Hinkley's predictive likelihood except for the $|JJ^T|^{1/2}$ factor. For other cases, including those where X and Z are dependent, Butler suggested that (10.22) is a simpler and more widely applicable approach than Hinkley's Definition 2.

10.3.3 Approximate predictive likelihood

The predictive likelihoods defined so far all suffer from one objection: they require the existence of a minimal sufficient statistic for (X, Z), and are only effective when this represents a genuine reduction of the data; for instance, if R was simply the vector of ordered values of X and Z, (10.22) would be meaningless. For situations in which no non-trivial minimal sufficient statistic exists, various approximations to predictive likelihood were suggested by Leonard (1982), Davison (1986) and Butler (1989). We follow Davison's development here, which has the virtue of being defined for a general parametric family without any specification of sufficient or ancillary statistics.

The basis of Davison's formula is an approximation to the Bayesian predictive density (10.18). Suppose θ is p-dimensional and $f_X(x; \theta)\pi(\theta)$ has a unique maximum with respect to θ at $\theta = \theta^*$. We assume that $-\log\{f_X(x; \theta)\pi(\theta)\}$ is at least twice continuously differentiable in a neighbourhood of $\theta = \theta^*$ and let $I(\theta^*)$ denote the matrix of second-order partial derivatives, evaluated at $\theta = \theta^*$. Then a multidimensional version of Laplace's integral formula (Section 9.7) shows that

$$\int f_X(x; \theta)\pi(\theta)d\theta = (2\pi)^{p/2}f_X(x; \theta^*)\pi(\theta^*)|I(\theta^*)|^{-1/2}\left\{1 + O_p(n^{-1})\right\}. \qquad (10.23)$$

Note that, if the contribution of $\pi(\theta)$ is ignored, θ^* is the maximum likelihood estimator of θ, and $I(\theta^*)$ is the observed information matrix.

Davison applied (10.23) to both the numerator and the denominator of (10.18) to deduce the approximation

$$\hat{g}(z|x) = \frac{f_{X,Z}(x, z; \theta^*(z))\pi(\theta^*(z))|I(\theta^*)|^{1/2}}{f_X(x; \theta^*)\pi(\theta^*)|J(\theta^*(z))|^{1/2}}\left\{1 + O_p(n^{-1})\right\}, \qquad (10.24)$$

where $\theta^*(z)$ is the value of θ that maximises $f_{X,Z}(x, z; \theta)\pi(\theta)$, and $J(\theta^*(z))$ is the corresponding matrix of second-order derivatives of $-\log\{f_{X,Z}(x, z; \theta)\pi(\theta)\}$ at $\theta = \theta^*(z)$. In many cases, Davison pointed out, the approximation error is actually $O_p(n^{-2})$ rather than $O_p(n^{-1})$ – this happens, essentially, because the leading error terms to $O_p(n^{-1})$ cancel in the numerator and denominator. The approximation (10.24) is similar to the approximation to posterior densities defined by Tierney and Kadane (1986), though Tierney and Kadane did not apply their method to the calculation of predictive densities.

In cases where π is not specified, Davison proposed (10.24) (omitting the $\pi(\cdot)$ terms and the $O_p(n^{-1})$ errors) as an 'approximate predictive likelihood'. This is equivalent to approximating a Bayesian predictive density under a flat prior.

The appearance of Jacobian terms in the numerator and denominator of (10.24) is reminiscent of the p^* formula for the distribution of the MLE (Section 9.8), so there are clear connections with conditional inference. On the other hand, there is no direct justification for a uniform prior; for example, the formula (10.24) (with π omitted) is not invariant with respect to transformations of the parameter. This is because, if ψ is a non-linear 1–1 transformation of θ, a uniform prior for θ does not transform into a uniform prior for ψ. In this respect, the approximate predictive likelihood suffers from the same objection as has often been voiced against the Bayesian predictive likelihood, namely, that there is no clear-cut basis for assigning a prior density and assuming a uniform prior does not resolve this difficulty.

10.3.4 Objections to predictive likelihood

Cox (1986) argued that prediction intervals constructed from predictive likelihood would need to be 'calibrated' to ensure accurate frequentist coverage probabilities. Hall, Peng and Tajvidi (1999) took this objection further, commenting that predictive likelihood methods 'do not adequately allow for the effects of curvature with respect to the parameter when used to construct prediction intervals and prediction limits. Therefore, they cannot be expected to correct for the dominant term in an expansion of coverage error of a prediction interval.' To counter these objections they proposed a bootstrap approach, which we describe in Section 10.5. On the other hand, Butler (1989, 1990) argued that conditional forms of calibration (conditioning on an exact ancillary statistic if one is available, or otherwise on an approximate ancillary) are more appropriate than unconditional calibration.

10.4 Asymptotic methods

Asymptotic methods for prediction have been developed by a number of authors but in particular Cox (1975) and Barndorff-Nielsen and Cox (1994, 1996). The central idea is as follows. Suppose we have some statistic T (a function of $X_1, \ldots, X_n$ and Z) for which $\Pr\{T \leq t\} = G(t; \theta)$. If G is independent of θ then T is pivotal and construction of a prediction interval with exact coverage probability follows as in Section 10.1. In most problems, however, no exact pivotal exists. In that case a natural approach is to estimate $G(t; \theta)$ in some way, which we write as $\tilde{G}(t)$. We proceed as if $\tilde{G}$ was the pivotal distribution of T. For given α, define $\tilde{t}_\alpha$ to satisfy $\tilde{G}(\tilde{t}_\alpha) = \alpha$. Then the relationship $\{T \leq \tilde{t}_\alpha\}$ (regarded as a statement about Z, for given $X_1, \ldots, X_n$) defines an approximate $100\alpha\%$ prediction interval for Z. In regular cases, the true coverage probability is of the form $\alpha + \frac{c}{n} + o\left(\frac{1}{n}\right)$, where c can in principle be explicitly calculated. Defining $\alpha_1 = \alpha - \frac{c}{n}$ and assuming some reasonable amount of continuity, the set $\{T \leq t_{\alpha_1}\}$ has coverage probability $\alpha + o\left(\frac{1}{n}\right)$. Thus, by adjusting the nominal coverage probability from α to α_1, we may correct for the coverage probability bias.

In this section we focus primarily on the α-quantile of the predictive distribution, and resulting one-sided prediction sets with coverage probability α. To translate the results into the more familiar setting of two-sided prediction intervals with coverage probability $1 - \alpha$ (Section 10.1), the usual procedure is to estimate two quantiles of the predictive distribution, corresponding to probabilities $\frac{\alpha}{2}$ and $1 - \frac{\alpha}{2}$, and to define the difference between the two quantiles as a prediction interval.

If the predictive distribution is normal, the correction from α to α_1 may be given an alternative interpretation, as follows. In this case t_α is typically of the form $A + Bz_\alpha$, where A is the predictive mean of Z, B is the predictive standard deviation and z_α is the α-quantile of the standard normal distribution. Replacing α by $\alpha_1 = \alpha - \frac{c}{n}$ is equivalent to replacing z_α by $z_\alpha + \epsilon$ where, to first order,

$$\alpha - \frac{c}{n} = \Phi(z_\alpha + \epsilon) = \alpha + \epsilon\phi(z_\alpha).$$

Thus $\epsilon = -\frac{c}{n\phi(z_\alpha)}$ and hence $z_{\alpha_1} = z_\alpha \left(1 - \frac{c}{nz_\alpha\phi(z_\alpha)}\right)$. The correction thus amounts to replacing the predictive standard deviation B by $B\left(1 - \frac{c}{nz_\alpha\phi(z_\alpha)}\right)$ or, what is equivalent to the same order of approximation, replacing the predictive variance B^2 by $B^2 \left(1 - \frac{2c}{n z_\alpha \phi(z_\alpha)}\right)$. In general there is no guarantee that $\frac{c}{z_\alpha \phi(z_\alpha)}$ is independent of α, but in some cases it is, and, then, this interpretation is particularly appealing. Problem 10.5 provides an example.

In some cases it may be possible to partition the parameter vector, say $\theta = (\mu, \psi)$, so that, when ψ is known, T is exactly pivotal (in other words, the distribution of T depends on ψ but not on μ). Each of the Examples 10.1–10.3 is of this form. In that case, the argument may be simplified by performing the asymptotic calculation with respect to ψ alone, instead of the whole of θ. See Example 10.7 below.

Another issue is that in cases of genuine stochastic dependence (for example, prediction in a time series), the conditional distribution of T given $X = (X_1, \ldots, X_n)$ may itself depend on X; thus, we should replace $G(t; \theta)$ by $G(t; x, \theta)$, where x is the numerical value of the conditioning variable X. To simplify the notation, we do not indicate the possible dependence on x, but the following arguments (as far as (10.28)) essentially continue to hold in that case.

There are various approaches to constructing an estimator $\tilde{G}(t)$. An obvious approach is to substitute an estimator $\tilde{\theta}$ for θ; possibly, but not necessarily, the maximum likelihood estimator. Thus $\tilde{G}(t) = G(t; \tilde{\theta})$. This is the estimative approach defined in Section 10.2. An alternative method is Bayesian: $\tilde{G}(t) = \int G(t; \theta) \pi(\theta|X) d\theta$, where $\pi(\theta|X)$ is the posterior distribution of θ given $X_1, \ldots, X_n$. In this case the constant c depends on the prior density $\pi(\theta)$. Sometimes it is possible to choose π so that $c = 0$. In this case π is called a matching prior (Datta *et al.*, 2000). The implication is that, if we perform Bayesian prediction with a matching prior, the resulting prediction interval has coverage probability very close to the nominal value. However, even in the absence of a matching prior, asymptotic calculations and/or simulations often show that the Bayesian prediction interval is superior to the estimative approach.

Full details of the asymptotic calculations use second-order asymptotics (Chapter 9), which lie beyond the scope of the present discussion. In cases where there is a closed-form formula for $\tilde{G}(t)$ however, it is often possible to calculate the needed asymptotic terms directly, and we concentrate on that case here. The argument follows Smith (1997, 1999) and extends an earlier argument of Cox (1975), Barndorff-Nielsen and Cox (1996).

Our approach to asymptotics is 'formal': we make extensive use of the Central Limit Theorem and Taylor expansions, keeping track of the orders of magnitudes of the various terms involved, but not attempting to prove rigorously that the remainder terms are of the orders stated. We also assume that asymptotic expressions may be differentiated term by term without rigorously justifying the interchange of limit and derivative.

Recall that $\tilde{G}(\tilde{t}_\alpha) = G(t_\alpha; \theta) = \alpha$. For brevity we write t_α in place of $t_\alpha(\theta)$ and let primes denote derivatives with respect to t, for example $G'(t; \theta) = \frac{\partial}{\partial t}\{G(t; \theta)\}$.

Suppose $\tilde{G}(t)$ allows a stochastic asymptotic expansion

$$\tilde{G}(t) = G(t; \theta) + \frac{R}{\sqrt{n}} + \frac{S}{n} + \ldots,$$

where $R = R(t, \theta)$ and $S = S(t, \theta)$ are random functions of both t and θ. This kind of expansion typically holds with maximum likelihood estimators, Bayesian estimators, etc.

Also, since the principal error term is of $O_p\left(\frac{1}{\sqrt{n}}\right)$, it follows at once that $\tilde{t}_\alpha - t_\alpha = O_p\left(\frac{1}{\sqrt{n}}\right)$. Then

$$
\begin{aligned}
0 &= \tilde{G}(\tilde{t}_\alpha) - G(t_\alpha; \theta) \\
&= \tilde{G}(\tilde{t}_\alpha) - \tilde{G}(t_\alpha) + \tilde{G}(t_\alpha) - G(t_\alpha; \theta) \\
&= (\tilde{t}_\alpha - t_\alpha)\tilde{G}'(t_\alpha) + \frac{1}{2}(\tilde{t}_\alpha - t_\alpha)^2\tilde{G}''(t_\alpha) + \frac{R(t_\alpha, \theta)}{\sqrt{n}} + \frac{S(t_\alpha, \theta)}{n} + o_p\left(\frac{1}{n}\right) \\
&= (\tilde{t}_\alpha - t_\alpha)\left\{G'(t_\alpha; \theta) + \frac{R'(t_\alpha, \theta)}{\sqrt{n}}\right\} + \frac{1}{2}(\tilde{t}_\alpha - t_\alpha)^2 G''(t_\alpha; \theta) \\
&\quad + \frac{R(t_\alpha, \theta)}{\sqrt{n}} + \frac{S(t_\alpha, \theta)}{n} + o_p\left(\frac{1}{n}\right).
\end{aligned}
\tag{10.25}
$$

A first-order approximation (retaining only terms of $O_p\left(\frac{1}{\sqrt{n}}\right)$) shows that

$$
\tilde{t}_\alpha - t_\alpha = -\frac{R(t_\alpha, \theta)}{\sqrt{n}G'(t_\alpha, \theta)} + O_p\left(\frac{1}{n}\right).
\tag{10.26}
$$

Substituting from (10.26) back into (10.25) and collecting up the $O_p\left(\frac{1}{n}\right)$ terms, we get

$$
\tilde{t}_\alpha - t_\alpha = -\frac{R}{\sqrt{n}G'} + \frac{1}{n}\left(\frac{RR'}{G'^2} - \frac{R^2 G''}{2G'^3} - \frac{S}{G'}\right) + o_p\left(\frac{1}{n}\right).
\tag{10.27}
$$

Now write $G(\tilde{t}_\alpha; \theta) - G(t_\alpha; \theta) = (\tilde{t}_\alpha - t_\alpha)G' + \frac{1}{2}(\tilde{t}_\alpha - t_\alpha)^2 G'' + \ldots$ taking further Taylor expansion based on (10.27), to deduce

$$
G(\tilde{t}_\alpha; \theta) - G(t_\alpha; \theta) = -\frac{R}{\sqrt{n}} + \frac{1}{n}\left(\frac{RR'}{G'} - S\right) + \ldots.
$$

Finally, on taking expectations in this last expression,

$$
\Pr\{T \le \tilde{t}_\alpha\} - \alpha = -\frac{1}{\sqrt{n}}\mathbb{E}(R) + \frac{1}{n}\mathbb{E}\left(\frac{RR'}{G'} - S\right) + \ldots.
\tag{10.28}
$$

Typically $\mathbb{E}(R)$ is of $O\left(\frac{1}{\sqrt{n}}\right)$ or smaller, so that the leading term in (10.28) is $\frac{c}{n}$ for some constant c, as previously indicated.

In the case that $\Pr\{T \le t | X = x\} = G(t; x, \theta)$ depending on x, each of the expressions R, R', S and G' will also depend on x, but this does not change the essential form of the result; in all cases, the expectation in (10.28) is taken jointly with respect to X and Z.

Example 10.7 Suppose $X_1, \ldots, X_n, Z$ are independent $N(\mu, \sigma^2)$, where μ and σ^2 are both unknown. Recall from our earlier example 10.1 that, if σ^2 is known, then $T = \sqrt{\frac{n}{n+1}}(Z - \bar{X})$ is a pivotal statistic with distribution $N(0, \sigma^2)$, and write $\theta = \sigma^{-1}$. Then $G(t; \theta) = \Phi(t\theta)$, where Φ is the standard normal distribution function. If we define t_α so that $G(t_\alpha; \theta) = \alpha$, then we will have $t_\alpha = z_\alpha/\theta$, where z_α is the α-quantile of the standard normal distribution.

The usual estimator of σ^2 is $s^2 = \frac{1}{n-1}\sum_{i=1}^n (X_i - \bar{X})^2$, for which $(n-1)\frac{s^2}{\sigma^2} \sim \chi_{n-1}^2$. Thus we suggest the estimator $\tilde{\theta} = s^{-1}$ and consequently $\tilde{t}_\alpha = z_\alpha/\tilde{\theta}$. The next step is to calculate the approximate mean and variance of $\tilde{\theta}$.

If $W \sim \chi_\nu^2$, then W has mean ν and variance 2ν, so we may write $W = \nu\left(1 + \sqrt{\frac{2}{\nu}}\xi\right)$, where ξ has mean 0 and variance 1. By the Taylor expansion $(1 + x)^{-1/2} = 1 - \frac{1}{2}x + \frac{3}{8}x^2 + \dots$ as $x \to 0$, we have $W^{-1/2} = \nu^{-1/2}\left(1 - \frac{1}{\sqrt{2\nu}}\xi + \frac{3}{4\nu}\xi^2 + \dots\right)$ and hence

$$\mathbb{E}(W^{-1/2}) = \nu^{-1/2}\left(1 + \frac{3}{4\nu} + \dots\right).$$

By a similar Taylor expansion or else the exact result $\mathbb{E}(W^{-1}) = \frac{1}{\nu-2}$ for $\nu > 2$, we also have

$$\mathbb{E}(W^{-1}) = \nu^{-1}\left(1 + \frac{2}{\nu} + \dots\right).$$

Applying these results to the case $W = (n-1)\frac{s^2}{\sigma^2}$, where $\nu = n - 1$, and replacing ν by n, where it does not change the asymptotic expressions, we have $\tilde{\theta} = \theta \nu^{1/2} W^{-1/2}$, and hence

$$\mathbb{E}(\tilde{\theta} - \theta) \approx \frac{3\theta}{4n}, \tag{10.29}$$

$$\mathbb{E}\{(\tilde{\theta} - \theta)^2\} \approx \frac{\theta^2}{2n}. \tag{10.30}$$

Recalling $G(t;\theta) = \Phi(t\theta)$, let $\tilde{G}(t) = \Phi(t\tilde{\theta})$, so

$$\tilde{G}(t) - G(t;\theta) = \Phi(t\tilde{\theta}) - \Phi(t\theta)$$
$$\approx \left\{t(\tilde{\theta} - \theta) - \frac{1}{2}t^3\theta(\tilde{\theta} - \theta)^2\right\}\phi(t\theta),$$

where $\phi(x) = (2\pi)^{-1/2}e^{-x^2/2}$ is the standard normal density and we use the relation $\phi'(x) = -x\phi(x)$.

We apply the expansion (10.28), where we write $n^{-1/2}R = t(\tilde{\theta} - \theta)\phi(t\theta)$, $n^{-1}S = -\frac{1}{2}t^3\theta(\tilde{\theta} - \theta)^2\phi(t\theta)$, and hence also $n^{-1/2}R' = (1 - t^2\theta^2)(\tilde{\theta} - \theta)\phi(t\theta)$. Combining these with $G' = \theta\phi(t\theta)$ and applying (10.29), (10.30), we have

$$-n^{-1/2}\mathbb{E}(R) \approx -\frac{3t\theta\phi(t\theta)}{4n},$$

$$n^{-1}\mathbb{E}\left(\frac{RR'}{G'}\right) \approx \frac{t\theta(1 - t^2\theta^2)\phi(t\theta)}{2n},$$

$$-n^{-1}\mathbb{E}(S) \approx \frac{t^3\theta^3\phi(t\theta)}{4n}.$$

All of these expressions are evaluated at $t = t_\alpha = \frac{z_\alpha}{\theta}$, so, by (10.28), the final result is

$$\Pr\left\{T \le \frac{z_\alpha}{\tilde{\theta}}\right\} = \alpha - \frac{z_\alpha(1 + z_\alpha^2)\phi(z_\alpha)}{4n} + o\left(\frac{1}{n}\right). \tag{10.31}$$

Equation (10.31) shows that the prediction interval defined by $\{T \le \tilde{t}\theta\}$ has true coverage probability approximately $\alpha + \frac{c}{n}$, where $c = -\frac{1}{4}t\theta(1 + t^2\theta^2)\phi(t\theta)$. In practice we would replace c by an estimate $\tilde{c}$, calculated by substituting $\tilde{\theta}$ for θ.

Problem 10.4 provides an alternative interpretation of this result, showing that the result is (to the stated order of approximation) equivalent to the exact result obtained in Problem 10.1. Although in this case the asymptotic calculation is not needed to provide an accurate prediction interval, the example has been given here to provide an illustration of a very general approach.

10.5 Bootstrap methods

Bootstrap methods in statistics are described in detail in Chapter 11. The present section, however, describes some simple forms of (parametric) bootstrap applied to prediction, and can be read independently of Chapter 11.

Harris (1989) considered the following method to improve on a simple 'estimative' approach to prediction. Suppose $g(z; \theta)$ is the probability density function (or, in the case of discrete random variables, the probability mass function) of Z given parameter θ. We assume Z is independent of observed data X. Suppose also we have an estimate of θ, denoted $\widehat{\theta}$. In most cases this will be the maximum likelihood estimator, though the theory is not dependent on that particular choice of $\widehat{\theta}$. The estimative approach uses $g(z; \widehat{\theta})$ as a predictive density for Z. If the distribution function of $\widehat{\theta}$ were known as a function of the true θ, say $H(t; \theta) = \Pr\{\widehat{\theta} \leq t; \theta\}$, then we could improve on this by

$$g^*(z; \theta) = \int g(z; t) dH(t; \theta). \tag{10.32}$$

The notation in (10.32) is intended to allow for both discrete and continuous cases. If $\widehat{\theta}$ is continuous with density $h(\cdot; \theta)$, then $dH(t; \theta)$ may be replaced by $h(t; \theta)dt$. Alternatively, if $\widehat{\theta}$ is discrete and $\Pr\{\widehat{\theta} = t\} = h(t; \theta)$, then the right side of (10.32) may be replaced by $\sum_t g(z; t)h(t; \theta)$.

There are two potential difficulties with (10.32). The first is that $H(t; \theta)$ may not be readily available. Harris' paper was confined to the case of exponential families, for which, in many cases, $H(t; \theta)$ is computable analytically. However, even if this were not the case, it would be possible to approximate (10.32) by simulation. The more fundamental difficulty is that θ is unknown. Harris suggested resolving this problem by replacing θ by $\widehat{\theta}$ in (10.32), leading to the predictive density

$$\widehat{g}(z|x) = \int g(z; t) dH(t; \widehat{\theta}(x)). \tag{10.33}$$

This is different from either the estimative or the (Bayesian) predictive approach; since it effectively replaces $H(t; \theta)$ in (10.32) by an estimate constructed from the sample, Harris called it a parametric bootstrap approach.

Example 10.8 Suppose $X_1, \ldots, X_n, Z$ are independent from a Poisson distribution with common mean θ. The MLE is $\widehat{\theta} = \bar{X}$; since $n\bar{X}$ has a Poisson distribution with mean $n\theta$, the distribution of $\widehat{\theta}$ is

$$\Pr\{\widehat{\theta} = t\} = \frac{e^{-n\theta}(n\theta)^{nt}}{(nt)!}, \quad t = 0, \frac{1}{n}, \frac{2}{n}, \ldots$$

Therefore, (10.33) in this case becomes

$$\widehat{g}(z|x) = \sum_{nt=0}^{\infty} \frac{e^{-t}t^z}{z!} \cdot \frac{e^{-n\bar{x}}(n\bar{x})^{nt}}{(nt)!}. \tag{10.34}$$

In discussion of this approach, Harris compared (10.33) with both estimative and Bayesian approaches. Arguing from Aitchison's (1975) point of view of using Kullback–Leibler distance as a measure of fit, he showed for the Poisson case that (10.34) is superior to the estimative approach for all large n, and he gave a general argument to show that the same asymptotic result essentially holds for all exponential families.

An alternative approach given by Hall, Peng and Tajvidi (1999) aims explicitly to reduce the coverage probability bias in prediction intervals or more general prediction sets. This is related to the approach of Section 10.4, where we showed that the true coverage probability of a prediction set of nominal coverage probability α is typically of the form $\alpha + \frac{c}{n} + o\left(\frac{1}{n}\right)$, where n is the sample size (of X) and c is some non-zero constant. The emphasis in Section 10.4 was on removing the $\frac{c}{n}$ term by means of an analytic correction; in contrast, the idea behind bootstrap approaches is to achieve the same thing through simulation.

Suppose, once again, we are interested in predicting a random variable Z with density $g(z; \theta)$ based on some independent sample $X = (X_1, \ldots, X_n)$ from a density $f(x; \theta)$. Hall *et al.* defined ∇f to be the vector of first-order derivatives of f with respect to θ. They also defined $\widehat{\theta}$ and $\widehat{\theta}_z$ to be the MLEs of θ based on X and $(X, Z = z)$ respectively, and defined matrices $\widehat{J}(\theta), \widehat{J}_z(\theta)$ to be

$$\widehat{J}(\theta) = \sum_{i=1}^{n} \frac{\nabla f(X_i; \theta)\nabla f(X_i; \theta)^T}{f(X_i; \theta)^2}, \quad \widehat{J}_z(\theta) = J(\theta) + \frac{\nabla g(z; \theta)\nabla g(z; \theta)^T}{g(z; \theta)^2}. \tag{10.35}$$

They then defined

$$\widehat{g}(z|x) = \frac{g(z; \widehat{\theta}_z)|\widehat{J}(\widehat{\theta})|^{1/2}|\widehat{J}_z(\widehat{\theta}_z)|^{-1/2}}{\int g(z; \widehat{\theta}_z)|\widehat{J}(\widehat{\theta})|^{1/2}|\widehat{J}_z(\widehat{\theta}_z)|^{-1/2}dz}. \tag{10.36}$$

Equation (10.36) is similar to Davison's approximate predictive likelihood but with two differences, (i) the matrices $\widehat{J}(\theta)$ and $\widehat{J}_z(\theta)$ have been used in place of the observed information matrices in Davison's definition, (ii) the predictive likelihood has been normalised so that it integrates to 1. Hall, Peng and Tajvidi implied that (10.36) would improve on Davison's predictive likelihood, but not enough to provide adequate correction for the coverage probability bias.

To this end, they defined prediction sets for Z by the constructions

$$\widehat{S}_{\text{pr}\,\alpha} = \{z; \widehat{g}(z|x) > \widehat{c}_{\text{pr}\,\alpha}\},$$
$$\widehat{S}_\alpha = \{z; g(z; \widehat{\theta}) > \widehat{c}_\alpha\},$$
$$S_\alpha = \{z; g(z; \theta) > c_\alpha\},$$

where $\widehat{c}_{\text{pr }\alpha}$, $\widehat{c}_{\alpha}$ and c_{α} are defined so that

$$\int_{\widehat{S}_{\text{pr }\alpha}} \widehat{g}(z|x)dz = \int_{\widehat{S}_{\alpha}} g(z;\widehat{\theta})dz = \int_{S_{\alpha}} g(z;\theta)dz = \alpha.$$

Thus S_{α} could be regarded as an ideal prediction set, constructed under the assumption that θ is known and having exact coverage probability α, while $\widehat{S}_{\alpha}$ and $\widehat{S}_{\text{pr }\alpha}$ are approximations based on the estimative density and (10.36) respectively. Hall, Peng and Tajvidi argued that

$$\text{Pr}\{Z \in \widehat{S}_{\alpha}\} = \alpha + \frac{c_1}{n} + o\left(\frac{1}{n}\right), \quad \text{Pr}\{Z \in \widehat{S}_{\text{pr }\alpha}\} = \alpha + \frac{c_2}{n} + o\left(\frac{1}{n}\right),$$

with rather complicated expressions for c_1 and c_2; however their main point was that both c_1 and c_2 are functions of the second-order derivatives of f with respect to θ and therefore neither construction compensates for the 'curvature' in f with respect to θ; moreover, the alternative forms of (10.36) (such as leaving out the normalisation, or using Davison's form of $\widehat{J}$ and $\widehat{J}_z$ in place of (10.35)) may change the value of c_2 but will not eliminate this term. Thus, in this sense, the approximate predictive likelihood and its relatives do not adequately correct the coverage probability bias of the naïve prediction interval.

As an alternative approach, they suggested the following. After computing $\widehat{\theta}$, draw a synthetic sample $X^* = (X_1^*, \dots, X_n^*)$ and Z^* from the densities $f(\cdot; \widehat{\theta})$ and $g(\cdot; \widehat{\theta})$. Using these data, compute the corresponding estimator $\widehat{\theta}^*$ and hence the prediction sets $\widehat{S}_{\beta}^*$ and $\widehat{S}_{\text{pr }\beta}^*$ for several values of β near α. Repeating this experiment many times, estimate the empirical coverage probabilities

$$\widehat{p}(\beta) = \text{Pr}\{Z^* \in \widehat{S}_{\beta}^*|X\}, \quad \widehat{p}_{\text{pr}}(\beta) = \text{Pr}\{Z^* \in \widehat{S}_{\text{pr }\beta}^*|X\}. \tag{10.37}$$

These values are conditional on X in the sense that they depend on $\widehat{\theta}$, which is a function of X. By interpolating among the βs, the idea is to find estimates $\widehat{\beta}_{\alpha}$, $\widehat{\beta}_{\text{pr }\alpha}$ such that the conditional probabilities in (10.37) are as close as possible to α. Then the regions $\widehat{S}_{\widehat{\beta}_{\alpha}}$ and $\widehat{S}_{\text{pr }\widehat{\beta}_{\text{pr }\alpha}}$ are bootstrap-calibrated prediction sets for which they argued that the coverage probabilities are of the form $\alpha + O(n^{-2})$.

Thus, either version of the bootstrap method may be considered a computationally intensive, but conceptually straightforward, means of obtaining prediction sets whose coverage probability is correct with an error of $O(n^{-2})$. In practice, Hall *et al.* suggested, the additional optimisations required for the approximate predictive likelihood method make the calculations too computationally intensive in this case, but, since the bootstrap recalibration appears to apply equally well to the naïve estimative method, prediction sets of the form of $\widehat{S}_{\widehat{\beta}_{\alpha}}$ may be the most practical approach.

10.6 Conclusions and recommendations

The focus of this chapter has really been on alternatives to the direct Bayesian approach of Section 3.9. The Bayesian approach is very popular among practitioners because of its conceptual simplicity combined with ease of computation by Markov chain Monte Carlo methods. However its properties, when assessed from other points of view than that of Bayesian decision theory, remain uncertain. We have focussed here on two alternative criteria, one based on decision theory and the other on the exact or asymptotic coverage probabilities of prediction sets. Apart from Bayesian methods and the naive or estimative approach, alternative ways of constructing prediction sets include exact methods based on pivotals (Section 10.1), predictive likelihood (Section 10.3), analytic methods for correcting the coverage probability bias (Section 10.4) and bootstrap methods (Section 10.5). At the present time, bootstrap methods appear to be the most direct approach to obtaining prediction sets of asymptotically correct coverage probability, but there remains ample scope for further exploration of all of these methods.

10.7 Problems

10.1 Let $Z_1, \ldots, Z_n$ be independent, identically distributed $N(\mu, \sigma^2)$ random variables, with σ^2 known. Suppose that it is required to construct a prediction interval $I_{1-\alpha} \equiv I_{1-\alpha}(Z_1, \ldots, Z_n)$ for a future, independent random variable Z_0 with the same $N(\mu, \sigma^2)$ distribution, such that

$$\Pr(Z_0 \in I_{1-\alpha}) = 1 - \alpha,$$

with the probability here being calculated from the *joint* distribution of $Z_0, Z_1, \ldots, Z_n$. Let

$$I_{1-\alpha}(Z_1, \ldots, Z_n; \sigma^2) = \left[\bar{Z}_n + z_{\alpha/2}\sigma\sqrt{1 + 1/n}, \bar{Z}_n - z_{\alpha/2}\sigma\sqrt{1 + 1/n} \right],$$

where $\bar{Z}_n = n^{-1}\sum_{i=1}^n Z_i$, and $\Phi(z_\beta) = \beta$, with Φ the distribution function of $N(0, 1)$.
Show that $\Pr\{Z_0 \in I_{1-\alpha}(Z_1, \ldots, Z_n; \sigma^2)\} = 1 - \alpha$.

Now suppose that σ^2 is unknown. Let $\widehat{\sigma}^2 = (n-1)^{-1}\sum_{i=1}^n (Z_i - \bar{Z}_n)^2$. By considering the distribution of $(Z_0 - \bar{Z}_n)/(\widehat{\sigma}\sqrt{\frac{n+1}{n-1}})$, show that

$$\Pr\{Z_0 \in I_{1-\alpha}(Z_1, \ldots, Z_n; \widehat{\sigma}^2)\} < 1 - \alpha.$$

(So, if σ^2 is known, a prediction interval of exactly the desired property is easily constructed. However, in the case where σ^2 is unknown, substitution of the maximum likelihood estimator reduces coverage.)
Show how to construct an interval $I_{1-\gamma}(Z_1, \ldots, Z_n; \widehat{\sigma}^2)$ with

$$\Pr\{Z_0 \in I_{1-\gamma}(Z_1, \ldots, Z_n; \widehat{\sigma}^2)\} = 1 - \alpha.$$

(So, we can *recalibrate* the nominal coverage, from $1 - \alpha$ to $1 - \gamma$, to allow for estimation of the variance σ^2, leaving a prediction interval of exactly the desired property.)

10.2 Suppose $X_1, \ldots, X_n, Z$ are independent exponentially distributed with mean μ. We assume $X_1, \ldots, X_n$ are observed and we would like to calculate a prediction interval for Z. Show that $\frac{Z}{\bar{X}}$ is a pivotal quantity and find its distribution. Hence calculate an exact $100(1 - \alpha)\%$ prediction interval for Z, as a function of $\bar{X}$.

10.3 Following the notation of Example 10.3, consider $T_\lambda = Z_1 - \lambda X_1 - (1 - \lambda)\bar{X}$, where $0 \leq \lambda \leq 1$.
Show that

$$\text{var}(T_\lambda) = \frac{n-1}{n}(1 - \lambda)^2\sigma^2 + \left(\lambda^2 + \frac{1 - \lambda^2}{n}\right)\tau^2.$$

Hence find λ^*, the value of λ that minimises the variance of T_λ, and for any $\alpha \in (0, 1)$ construct an exact $100(1 - \alpha)\%$ prediction interval for Z_1 based on $X_1, \ldots, X_n$.

10.4 Let us return to Example 3.3 from Chapter 3, where we set prior parameters $\alpha = \beta = k = \nu = 0$. Example 3.3 showed how to derive the posterior distribution of (τ, μ), given $X = (X_1, \ldots, X_n)$, and Problem 3.15 showed how to extend this to the predictive distribution of a new $Z \sim N(\mu, \tau^{-1})$, given X. Let us call the resulting predictive density $\tilde{g}_1(z|x)$, where x is the numerical value of X.

A competing candidate for $\tilde{g}$ is the estimative density $\tilde{g}_2(z|x)$, obtained by substituting sample estimates $\hat{\mu} = \bar{X} = \frac{1}{n}\sum_1^n X_i$ and $\hat{\tau}^{-1} = s^2 = \frac{1}{n-1}\sum_1^n (X_i - \bar{X})^2$ in the true normal density for Z, which we denote $g(z; \mu, \tau)$.

Show that $\tilde{g}_1$ is closer to g than is $\tilde{g}_2$, where the measure of closeness is the Kullback–Leibler distance.

10.5* Now consider Example 3.4 from Chapter 3, where (V, μ) have a prior distribution defined by parameters $m = k = \Psi = \nu = 0$. Suppose we have $X_1, \ldots, X_n$, Z independent $N_p(\mu, V^{-1})$ and wish to predict Z given $X_1, \ldots, X_n$. Two choices here are: (i) the predictive Bayesian density $\tilde{g}_1$ that you calculated in Problem 3.16, (ii) the estimative density $\tilde{g}_2$ obtained by substituting $\bar{X}$ for μ and $S = \frac{1}{n-1}\sum(X_i - \bar{X})(X_i - \bar{X})^T$ for V^{-1}. Show that $\tilde{g}_1$ is superior to $\tilde{g}_2$ when evaluated by Kullback–Leibler distance.

Remark: This result is due to Aitchison. Problem 10.4 represents the easier case when $p = 1$. The result depends on certain properties of the Wishart distribution that depend on the definitions in Example 3.4, but also use the following that you may quote without proof: (i) if $D = \sum_{j=1}^m (Z_j - \bar{Z})(Z_j - \bar{Z})^T$, where $Z_1, \ldots, Z_m$ are independent $N_p(0, A)$ and $\bar{Z} = \frac{1}{m}\sum_{j=1}^m Z_j$, then D and $\bar{Z}$ are independent and $D \sim W_p(A, m - 1)$; (ii) if $D \sim W_p(A, m)$, then $\mathbb{E}\{D^{-1}\} = A^{-1}/(m - p - 1)$.

10.6 Suppose T' has a student's t distribution with ν degrees of freedom. Show that for large ν,

$$\Pr\{T' \le z\} = \Phi(z) - \frac{1}{4\nu}(z + z^3)\phi(z) + o\left(\frac{1}{\nu}\right).$$

Hence show that the result (10.31) is equivalent to the statement that, within an error of $o_p\left(\frac{1}{n}\right)$, the distribution of $T\tilde{\theta}$ is t_n (or t_{n-1}).

Use this result to show that the asymptotic solution derived from (10.31) is equivalent to the exact result obtained in the last part of Problem 10.1.

10.7 Suppose $Z = (Z_1, \ldots, Z_n)$ are independent $N(\mu, \sigma^2)$ and, conditionally on Z, $\{X_{ij} : 1 \le i \le n, 1 \le j \le m\}$ are independent, $X_{ij}|Z_i \sim N(Z_i, \tau^2)$. This extends Problem 10.3 in the sense that, when $m > 1$, there are additional degrees of freedom within each subgroup that make it possible to estimate τ^2 and σ^2, and hence to solve the prediction problem when all three of μ, σ^2 and τ^2 are unknown. This model is also known as the (one-way) random effects analysis of variance model. As in Example 10.3 and Problem 10.3, we assume the objective is to predict a specified component of Z, which without loss of generality we take to be Z_1. Note that, if τ^2 is known, by sufficiency all predictions will be based on $\bar{X}_i = \frac{1}{m}\sum_j X_{ij}$ for $i = 1, \ldots, n$. This is the same as Example 10.3 except that $\text{var}(\bar{X}_i) = \frac{\tau^2}{m} = \tau_m^2$, say, instead of τ^2. All asymptotic calculations will be as $n \to \infty$ for fixed m.

(i) Consider first the case where μ, σ^2 and τ^2 are all known. In this case, no X_{ij} for $i \geq 2$ contributes any information about Z_1, so the prediction is based on the conditional distribution of Z_1 given $\bar{X}_1$.

The following result is standard from the theory of the bivariate normal distribution. Suppose (X, Z) is bivariate normal with means (μ_X, μ_Z), variances (σ_X^2, σ_Z^2) and correlation ρ. Then the conditional distribution of Z given $X = x$ is normal with mean $\mu_Z + \rho \frac{\sigma_Z}{\sigma_X}(x - \mu_X)$ and variance $\sigma_Z^2(1 - \rho^2)$.

Assuming this, show that an exact one-sided prediction interval with coverage probability α may be defined as

$$\left(-\infty, \; \frac{\sigma^2}{\sigma^2 + \tau_m^2} \bar{X}_1 + \frac{\tau_m^2}{\sigma^2 + \tau_m^2} \mu + \frac{\sigma \tau_m}{\sqrt{\sigma^2 + \tau_m^2}} z_\alpha \right).$$

(ii) Now suppose μ is unknown but σ^2, τ^2 are known. Define

$$G(t; x, \mu) = \Pr\{Z_1 \leq t | \bar{X}_1 = x; \mu\}$$

$$= \Phi\left\{ \frac{\sqrt{\sigma^2 + \tau_m^2}}{\sigma \tau_m} \left(t - \frac{\sigma^2}{\sigma^2 + \tau_m^2} x - \frac{\tau_m^2}{\sigma^2 + \tau_m^2} \mu \right) \right\}$$

$$= \Phi(At + Bx + C\mu) \quad \text{say,}$$

where $A = \frac{\sqrt{\sigma^2 + \tau_m^2}}{\sigma \tau_m}$, $B = -\frac{\sigma}{\tau_m \sqrt{\sigma^2 + \tau_m^2}}$, $C = -\frac{\tau_m}{\sigma \sqrt{\sigma^2 + \tau_m^2}}$ are all known constants.

Let $\tilde{G}(t; x) = \Phi(At + Bx + C\bar{X})$, where $\bar{X} = \frac{1}{mn} \sum_i \sum_j X_{ij}$. If we ignored the distinction between $\bar{X}$ and μ, the one-sided interval

$$Z \leq \frac{z_\alpha - Bx - C\bar{X}}{A} \tag{10.38}$$

would have coverage probability α.

Use equation (10.31) to show that the true coverage probability of (10.38) is of the form

$$\alpha + \frac{c}{n} + o\left(\frac{1}{n}\right),$$

where $c = -\frac{1}{2} \frac{\tau_m^2}{\sigma^2} z_\alpha \phi(z_\alpha)$.

(iii) According to the theory of Section 10.4, if we define $\alpha_1 = \alpha - \frac{c}{n}$, the interval

$$Z \leq \frac{z_{\alpha_1} - Bx - C\bar{X}}{A}$$

has coverage probability $\alpha + o\left(\frac{1}{n}\right)$.

Show that this is asymptotically equivalent to the method of Problem 10.3, with error $o\left(\frac{1}{n}\right)$.

(iv) Finally, we consider the case where all three of μ, σ^2, τ^2 are unknown. We again estimate μ by $\bar{X}$, τ_m^2 by $\tilde{\tau}_m^2 = \frac{1}{nm(m-1)} \sum_i (\sum_j X_{ij} - \bar{X}_i)^2$, σ^2 by $\tilde{\sigma}^2 = \frac{1}{n-1} \sum_i (\bar{X}_i - \bar{X})^2 - \tilde{\tau}_m^2$. Under this procedure there is a slight possibility that $\tilde{\sigma}^2 < 0$, but the probability of that is negligible when n is large, and we ignore it. Note that $\bar{X}$, $n(m-1)\frac{\tilde{\tau}_m^2}{\tau_m^2}$ and $(n-1)\frac{\tilde{\sigma}^2 + \tilde{\tau}_m^2}{\sigma^2 + \tau_m^2}$ are independent random variables with respective distributions $N\left(\mu, \frac{\sigma^2+\tau_m^2}{n}\right)$, $\chi^2_{n(m-1)}$ and χ^2_{n-1}. The constants A, B, C of part (ii) are now unknown but may be estimated by replacing σ^2, τ_m^2 by $\tilde{\sigma}^2$, $\tilde{\tau}_m^2$; the resulting estimates will be denoted $\tilde{A}$, $\tilde{B}$, $\tilde{C}$. The approximate one-sided prediction interval given by (10.38) may thus be amended to

$$Z \leq \frac{z_\alpha - \tilde{B}x - \tilde{C}\bar{X}}{\tilde{A}}. \tag{10.39}$$

Show that the interval (10.39) has approximate coverage probability $\alpha + \frac{c}{n}$ as $n \to \infty$ for fixed m, and show how to calculate c.

11

Bootstrap methods

Since its introduction by Efron (1979), the bootstrap has become a method of choice for empirical assessment of errors and related quantities in a vast range of problems of statistical estimation. It offers highly accurate inference in many settings. Together with Markov chain Monte Carlo methods, which are now routinely applied in Bayesian analysis of complex statistical models, bootstrap methods stand among the most significant methodological developments in statistics of the late twentieth century.

Bootstrap methodology encompasses a whole body of ideas, but principal among them are: (1) the *substitution principle*, of replacement in frequentist inference of an unknown probability distribution F by an estimate $\tilde{F}$ constructed from the sample data; and (2) replacement of analytic calculation, which may be intractable or just awkward, by *Monte Carlo simulation* from $\tilde{F}$.

In the most straightforward formulations of bootstrapping, $\tilde{F}$ has a simple form. In non-parametric inference, $\tilde{F}$ is the empirical distribution function $\widehat{F}$ of an observed random sample $Y = \{Y_1, \ldots, Y_n\}$, the distribution which places an equal probability mass, $1/n$, on each observed data point Y_i. Monte Carlo simulation from $\widehat{F}$ then amounts to independent sampling, with replacement, from $\{Y_1, \ldots, Y_n\}$. In a parametric context, a parametric model $F(y; \eta)$ with a parameter η of fixed dimension is replaced by its maximum likelihood estimate $F(y; \widehat{\eta})$. In both cases, the frequentist inference is performed treating $\tilde{F}$ as if it were the true distribution underlying the data sample Y. So, for example, the variance of a statistic $T(Y)$ under sampling from F is estimated by its variance under sampling from $\tilde{F}$, or the bias of $T(Y)$ as an estimator of a population quantity $\theta(F)$ is estimated by the bias under sampling from $\tilde{F}$ of $T(Y)$ as an estimator of $\theta(\tilde{F})$.

The range of applications of bootstrap methods is enormous, and so extensive is the literature on the topic that even book-length treatments such as Davison and Hinkley (1997), Shao and Tu (1995), Efron and Tibshirani (1993) or Hall (1992) treat only certain aspects. A recent review is given by Davison *et al.* (2003). In this chapter we provide a brief, general overview of the bootstrap, focussing on the key conceptual ideas. Since we have been concerned predominantly in the rest of the book with procedures of statistical inference operating under the assumption of a specified parametric model, we will concentrate mainly on parametric forms of bootstrap. Of key interest is analysis of how the levels of accuracy obtained by bootstrapping compare with those provided by more sophisticated analytic constructions, such as those considered in Chapter 9. As we shall see, in the development of bootstrap methodology primary emphasis has been placed on the frequentist desire for accurate repeated sampling properties, but there are important connections to the Fisherian ideas that we have also considered in detail.

Section 11.1 describes a general framework for an inference problem concerning an unknown scalar parameter of interest, and presents an illustration of how common parametric and non-parametric inference procedures may be placed within that framework. In Section 11.2 we elucidate the conceptual basis of the bootstrap, through the notion of 'prepivoting', due to Beran (1987, 1988) We detail the operation of simple forms of bootstrapping from that perspective, and describe, from this same perspective, how better bootstrap procedures can be obtained, for both parametric and non-parametric problems, by simple modifications. A series of numerical illustrations are presented in Section 11.3 and Section 11.4. Finally, in Section 11.5, we consider bootstrapping from a Fisherian perspective.

11.1 An inference problem

Suppose that $Y = \{Y_1, \ldots, Y_n\}$ is a random sample from an unknown underlying distribution F, and let $\eta(F)$ be an $\mathbb{R}^d$-valued functional of F. We suppose that inference is required for a scalar quantity $\theta = g\{\eta(F)\} \equiv \theta(F)$, for a suitably smooth injective function $g : \mathbb{R}^d \to \mathbb{R}$. In a parametric setting, we consider a parametric family of distributions indexed by the d-dimensional parameter η, and typically we will have $\eta = (\theta, \xi)$, so that inference is required for the scalar interest parameter θ, in the presence of the nuisance parameter ξ. In a non-parametric framework, inference is required for the quantity $\theta(F)$, which might be, for example, the variance of F.

Let $u(Y, \theta)$ be a pivot, a function of the data sample Y and the unknown parameter θ, such that a confidence set of nominal coverage $1 - \alpha$ for θ is

$$\mathcal{I}(Y) \equiv \mathcal{I}_{1-\alpha}(Y) = \{\psi : u(Y, \psi) \leq 1 - \alpha\}. \tag{11.1}$$

We have that, under repeated sampling of Y from the underlying distribution F, the random set $\mathcal{I}(Y)$ contains the true θ a proportion approximately equal to $1 - \alpha$ of the time:

$$\Pr_\theta\{\theta \in \mathcal{I}(Y)\} \approx 1 - \alpha.$$

We define the *coverage error* of the confidence set $\mathcal{I}(Y)$ to be

$$\Pr_\theta\{\theta \in \mathcal{I}(Y)\} - (1 - \alpha).$$

As a very simple example, suppose that $Y_1, \ldots, Y_n$ are independent, identically distributed $N(\theta, 1)$. Then $\bar{Y} = n^{-1} \sum_{i=1}^n Y_i$ is distributed as $N(\theta, 1/n)$, so a confidence set of *exact* coverage $1 - \alpha$ is obtained from (11.1) using the pivot

$$u(Y, \psi) = \Phi\{\sqrt{n}(\bar{Y} - \psi)\},$$

in terms of the distribution function $\Phi(\cdot)$ of $N(0, 1)$.

Typically, $u(Y, \psi)$ will be monotonic in ψ, so that the confidence set is a semi-infinite *interval* of the form $(\widehat{\theta}_l(Y), \infty)$ or $(-\infty, \widehat{\theta}_u(Y))$, say. We will speak of $u(Y, \theta)$ as a 'confidence set root'. A notational point is of importance here. In our development, we will denote by θ the *true* parameter value, with ψ denoting a generic point in the parameter space, a 'candidate value' for inclusion in the confidence set.

If two-sided inference is required, an 'equi-tailed' two-sided confidence set $\mathcal{J}_{1-\alpha}(Y)$ of nominal coverage $1 - \alpha$ may be obtained by taking the set difference of two one-sided sets of the form (11.1) as

$$\mathcal{J}_{1-\alpha}(Y) = \mathcal{I}_{1-\alpha/2}(Y) \backslash \mathcal{I}_{\alpha/2}(Y). \tag{11.2}$$

As noted by Hall (1992: Chapter 3), coverage properties under repeated sampling of two-sided confidence sets of the form (11.2) are rather different from those of one-sided sets of the form (11.1). Typically, even if the coverage error of the interval (11.1) is of order $O(n^{-1/2})$ in the sample size, the two-sided set (11.2) has coverage error of order $O(n^{-1})$. We note that results we describe later relating to reduction of the order of magnitude of the coverage error by different bootstrap schemes refer specifically to confidence sets of the form (11.1): the coverage properties of two-sided sets are rather more subtle. In any case, as noted by Efron (2003b), a two-sided confidence set might come very close to the desired coverage $1 - \alpha$ in a very lopsided fashion, failing, say, to cover the true parameter value much more than $\alpha/2$ on the left and much less than $\alpha/2$ on the right. The purpose of a two-sided confidence interval should be accurate inference in both directions, and a detailed discussion of the properties of one-sided confidence sets is more relevant.

We now provide two examples, the first parametric and the second relating to non-parametric inference about θ.

Example 11.1 Signed root likelihood ratio statistic

Suppose, as above, that it may be assumed that Y has probability density $f_Y(y; \eta)$ belonging to a specified parametric family, depending on an unknown parameter η.

Assume first of all that η is in fact scalar, so that there is no nuisance parameter and $\theta \equiv \eta$. We saw in Chapter 8 that inference about θ may be based on the likelihood ratio statistic $w(\theta) = 2\{l(\widehat{\theta}) - l(\theta)\}$, with $l(\theta) = \log f_Y(y; \theta)$ the log-likelihood and $\widehat{\theta}$ the maximum likelihood estimator of θ. As θ is scalar, inference is conveniently based on the signed root likelihood ratio statistic, $r(\theta) = \mathrm{sgn}(\widehat{\theta} - \theta)w(\theta)^{1/2}$, which is distributed as $N(0, 1)$ to error of order $O(n^{-1/2})$. Therefore a confidence set of nominal coverage $1 - \alpha$ for θ is $\{\psi : u(Y, \psi) \leq 1 - \alpha\}$, with

$$u(Y, \psi) = \Phi\{r(\psi)\}.$$

It is easily seen that $u(Y, \psi)$ is monotonic decreasing in ψ, so that the confidence set is of the form $(\widehat{\theta}_l, \infty)$, where the lower confidence limit $\widehat{\theta}_l$ is obtained by solving $\Phi\{r(\psi)\} = 1 - \alpha$. The coverage error of the confidence set is of order $O(n^{-1/2})$, but can perhaps be reduced by bootstrapping.

More typically, we will have $\eta = (\theta, \xi)$, with nuisance parameter ξ. Now, as described in Chapter 8, inference about θ may be based on the profile log-likelihood $l_p(\theta) = l(\theta, \widehat{\xi}_\theta)$, and the associated likelihood ratio statistic $w_p(\theta) = 2\{l_p(\widehat{\theta}) - l_p(\theta)\}$, with $l(\theta, \xi) = \log f_Y(y; \theta, \xi)$ the log-likelihood, $\widehat{\eta} = (\widehat{\theta}, \widehat{\xi})$ the overall maximum likelihood estimator of η and $\widehat{\xi}_\theta$ the constrained maximum likelihood estimator of ξ, for fixed θ. The corresponding signed root likelihood ratio statistic is $r_p(\theta) = \mathrm{sgn}(\widehat{\theta} - \theta)w_p(\theta)^{1/2}$. Again, we have that r_p is distributed as $N(0, 1)$ to error of order $O(n^{-1/2})$, and therefore, following the same approach as in the no-nuisance-parameter case, a confidence set of nominal coverage $1 - \alpha$ for θ is $\{\psi : u(Y, \psi) \leq 1 - \alpha\}$, now with

$$u(Y, \psi) = \Phi\{r_p(\psi)\}.$$

The coverage error of the confidence set is again of order $O(n^{-1/2})$.

We saw in Chapter 9 how the error associated with the $N(0, 1)$ approximation may be reduced to order $O(n^{-3/2})$ by analytically adjusted versions of r_p of the form

$$r_a = r_p + r_p^{-1} \log(v_p/r_p),$$

that are distributed as $N(0, 1)$ to error of order $O(n^{-3/2})$. Here the statistic v_p depends on specification of an ancillary statistic, and is defined by (9.16).

The DiCiccio and Martin (1993) approximation to $v_p(\theta)$ based on orthogonal parameters is defined by (9.17). This is used in our numerical illustrations of Section 11.4.

Alternative forms of confidence set root $u(Y, \theta)$ may be based on other forms of asymptotically $N(0, 1)$ pivot, such as Wald and score statistics.

Example 11.2 Studentised parameter estimate
Suppose now that we are unable, or unwilling, to specify any parametric form for the underlying distribution $F(\eta)$, but that we have available a non-parametric estimator $\widehat{\theta}$ of θ, of finite variance, asymptotically normally distributed and with estimated variance $\widehat{\sigma}^2$. A simple special case relates to the mean θ of F, which may be estimated by the sample mean $\widehat{\theta} = \bar{Y} = n^{-1} \sum_{i=1}^{n} Y_i$, which has variance which may be estimated by $\widehat{\sigma}^2 = n^{-1}s^2$, where $s^2 = n^{-1} \sum_{i=1}^{n}(Y_i - \bar{Y})^2$ is the (biased) sample variance. A non-parametric confidence set of nominal coverage $1 - \alpha$ may be defined as above by

$$u(Y, \psi) = \Phi\{(\widehat{\theta} - \psi)/\widehat{\sigma}\}.$$

Again, the confidence set is of one-sided form $(\widehat{\theta}_l, \infty)$, with $u(Y, \widehat{\theta}_l) = 1 - \alpha$. Typically, coverage error is again $O(n^{-1/2})$.

A general strategy for construction of an appropriate initial confidence set root $u(Y, \theta)$ is as follows: see Beran (1987). Let $s_n(\theta) \equiv s_n(Y, \theta)$ be a pivot, such as $\sqrt{n}(\widehat{\theta} - \theta)$. Let $H_n(\cdot; F)$ denote the distribution function of $s_n(\theta)$: $H_n(x; F) = \Pr\{s_n(Y, \theta) \leq x \mid F\}$. Suppose further that asymptotically $s_n(Y, \theta)$ has distribution function $H(\cdot; F)$. We know from Chapter 8, for example, that $\sqrt{n}(\widehat{\theta} - \theta)$ is quite generally asymptotically distributed as $N(0, 1/i_1(\theta))$, in terms of the Fisher information $i_1(\theta)$, and in many circumstances the finite sample distribution $H_n(\cdot; F)$ will also be known. Let $\widehat{F}_n$ be a consistent estimate of F, in the sense that $d(\widehat{F}_n, F)$ converges to zero in probability, where d is a metric on an appropriate family of distribution functions. In a parametric situation, for instance, $\widehat{F}_n$ would be the fitted distribution, obtained by maximum likelihood. Then, let $\widehat{H} = H(\cdot; \widehat{F}_n)$ and $\widehat{H}_n = H_n(\cdot; \widehat{F}_n)$. Possible confidence set roots are given by

$$u(Y, \psi) = \widehat{H}\{s_n(\psi)\}, \tag{11.3}$$

and

$$u(Y, \psi) = \widehat{H}_n\{s_n(\psi)\}. \tag{11.4}$$

Example 11.3 Mean of normal distribution

Let $Y_1, \ldots, Y_n$ be a random sample of size n from the normal distribution $N(\mu, \sigma^2)$, with both μ and σ^2 unknown, and suppose inference is required for μ. We have seen that the maximum likelihood estimators are $\widehat{\mu} = \bar{Y} = n^{-1} \sum_{i=1}^n Y_i$ and $\widehat{\sigma}^2 = n^{-1} \sum_{i=1}^n (Y_i - \bar{Y})^2$ respectively.

The exact and asymptotic distributions of $s_n(\mu) = \sqrt{n}(\widehat{\mu} - \mu)$ are identical, $N(0, \sigma^2)$. Therefore, $H_n(x) \equiv H(x) = \Phi(x/\sigma)$, so that $\widehat{H_n}(x) \equiv \widehat{H}(x) = \Phi(x/\widehat{\sigma})$ and (11.3) and (11.4) both yield the confidence set root

$$u(Y, \psi) = \Phi \left\{ \frac{\sqrt{n}(\widehat{\mu} - \psi)}{\widehat{\sigma}} \right\}. \tag{11.5}$$

Of course, in this example inference would more naturally be made using the distributional result that

$$\frac{\sqrt{n-1}(\widehat{\mu} - \mu)}{\widehat{\sigma}} \sim t_{n-1},$$

the t-distribution on $n-1$ degrees of freedom, so that a more appropriate confidence set root is

$$u(Y, \psi) = T \left\{ \frac{\sqrt{n-1}(\widehat{\mu} - \psi)}{\widehat{\sigma}} \right\}, \tag{11.6}$$

where T denotes the distribution function of t_{n-1}. Confidence sets constructed from the confidence set root (11.6) have exactly the desired coverage level, while use of (11.5) results in coverage error of order $O(n^{-1/2})$.

11.2 The prepivoting perspective

From the prepivoting perspective (Beran, 1987, 1988) the bootstrap may be viewed simply as a device by which we attempt to transform the confidence set root $U = u(Y, \theta)$ into a Uniform(0, 1) random variable.

The underlying notion is that, if U were exactly distributed as Uniform(0, 1), the confidence set would have coverage exactly equal to $1 - \alpha$: $\Pr_\eta(\theta \in \mathcal{I}) = \Pr_\eta\{u(Y, \theta) \leq 1 - \alpha\} = \Pr\{\text{Un}(0, 1) \leq 1 - \alpha\} = 1 - \alpha$. But U is typically *not* uniformly distributed, so the coverage error of $\mathcal{I}$ is non-zero: $\Pr_\eta(\theta \in \mathcal{I}) - (1 - \alpha) \neq 0$.

By bootstrapping, we hope to produce a new confidence set root u_1 so that the associated confidence set $\{\psi : u_1(Y, \psi) \leq 1 - \alpha\}$ has lower coverage error for θ. The error properties of different bootstrap schemes can be assessed by measuring how close to uniformity is the distribution of $U_1 = u_1(Y, \theta)$.

In the conventional bootstrap approach, the distribution function $G(x; \psi)$ of $u(Y, \psi)$ is estimated by

$$\widehat{G}(x) = \Pr^*\{u(Y^*, \widehat{\theta}) \leq x\}, \tag{11.7}$$

where $\widehat{\theta}$ denotes the estimator of θ (non-parametric or maximum likelihood, depending on the framework) constructed from the data sample Y, and we define the conventional prepivoted root by

$$\widehat{u}_1(Y, \psi) = \widehat{G}\{u(Y, \psi)\},$$

for *each* candidate parameter value ψ.

In a parametric problem, Pr^* denotes the probability under the drawing of bootstrap samples Y^* from the fitted maximum likelihood model $f_Y(y; \widehat{\eta})$.

In the non-parametric setting, Pr^* denotes the probability under the drawing of boot-strap samples Y^* from the empirical distribution function $\widehat{F}$: recall that such a sam-ple is obtained by independently sampling, with replacement, from $\{Y_1, \ldots, Y_n\}$. Note that in this case we can write down the bootstrap distribution explicitly. Let $u_{r_1 \ldots r_n} = u(Y_1, \ldots, Y_1, \ldots, Y_n, \ldots Y_n, \psi)$ denote the value of the pivot obtained from a dataset consisting of r_1 copies of Y_1, r_2 copies of $Y_2, \ldots, r_n$ copies of Y_n, $r_1 + \ldots + r_n = n$. Then

$$\text{Pr}^*\{u(Y^*, \psi) = u_{r_1 \ldots r_n}\} = \frac{n!}{r_1! \ldots r_n!} (\frac{1}{n})^n.$$

In practice, in both contexts, the prepivoting will in general be carried out by performing a Monte Carlo simulation, involving the drawing of an actual series of, say, R bootstrap samples, rather than analytically: the ease of doing so, instead of carrying out a mathematical calculation, represents a considerable part of the appeal of the bootstrap! A rule of thumb would suggest taking R to be of the order of a few thousands.

The basic idea here is that if the bootstrap estimated the sampling distribution exactly, so that $\widehat{G}$ was the true (continuous) distribution function G of $u(Y, \theta)$, then $\widehat{u}_1(Y, \theta)$ would be *exactly* Uniform$(0, 1)$ in distribution, as a consequence of the so-called 'probability integral transform': if Z is a random variable with continuous distribution function $H(\cdot)$, then $H(Z)$ is distributed as Uniform$(0, 1)$. Therefore the confidence set $\{\psi : \widehat{u}_1(Y, \psi) \leq 1 - \alpha\}$ would have exactly the desired coverage. Use of $\widehat{G}$ in place of G incurs an error, though in general the error associated with $\widehat{u}_1(Y, \psi)$ is smaller in magnitude than that obtained from $u(Y, \psi)$: see, for example, DiCiccio *et al.* (2001), Lee and Young (2004).

Consider the nuisance parameter case of Example 11.1 above. Conventional bootstrap-ping amounts to replacing the asymptotic $N(0, 1)$ distribution of r_p by its distribution when the true parameter value is $\widehat{\eta} = (\theta, \widehat{\xi})$. The bootstrap confidence set is of the form (θ^*, ∞), where $r_p(\theta^*) = \widehat{c}_{1-\alpha}$, with $\widehat{c}_{1-\alpha}$ denoting the $1 - \alpha$ quantile of $r_p(\theta)$ under sam-pling from the specified model with parameter value $(\theta, \widehat{\xi})$. In general, this reduces the order of the coverage error of the confidence set to $O(n^{-1})$. That the conventional boot-strap approximates the true distribution of r_p to error of order $O(n^{-1})$ was established by DiCiccio and Romano (1995).

In Example 11.2, prepivoting by the same technique amounts to replacing the asymptotic $N(0, 1)$ distribution of $(\widehat{\theta} - \theta)/\widehat{\sigma}$ by the distribution of $(\widehat{\theta}^* - \widehat{\theta})/\widehat{\sigma}^*$, with $\widehat{\theta}^*$ and $\widehat{\sigma}^*$ denoting the estimator and its standard error estimator respectively for a bootstrap sample obtained by uniform resampling from $\{Y_1, \ldots, Y_n\}$, and $\widehat{\theta}$, fixed under the bootstrap sampling, denoting the value of the estimator for the actual data sample Y. The confidence set is again of the form $(\widehat{\theta}_l^*, \infty)$, where $\widehat{u}_1(Y, \widehat{\theta}_l^*) = 1 - \alpha$. In general, the bootstrapping again reduces the order of the coverage error to $O(n^{-1})$.

In Example 11.3, it is a simple exercise to verify that prepivoting the confidence set root (11.5) produces the root (11.6), so that the bootstrapping produces confidence sets of exactly the desired coverage.

Conventional bootstrapping of the kind outlined above enjoys an appealing simplicity. A well-specified empirical sampling model is used in place of the unknown underlying distribution $F(\eta)$, either $\widehat{F}$ in the non-parametric context, or $F(\widehat{\eta})$ in the parametric framework. Yet, the results can be spectacular, as the following illustration demonstrates.

Illustration 11.1 We consider bootstrap inference for a normal distribution with known coefficient of variation, as considered in Section 9.5.2. This distribution is widely used in many agricultural and biological applications. We have $Y_1, \ldots, Y_n$ independent, identically distributed $N(\theta, \theta^2)$. For this model, the signed root likelihood ratio statistic is, in terms of the maximum likelihood estimator $\widehat{\theta}$, given by

$$r(\theta) = \text{sgn}(\widehat{\theta} - \theta)n^{1/2}[q^2\{(\widehat{\theta}/\theta)^2 - 1\} - 2(q/a)(\widehat{\theta}/\theta - 1) - 2\log(\widehat{\theta}/\theta)]^{1/2}, \quad (11.8)$$

with $q = \{(1 + 4a^2)^{1/2} + 1\}/(2a)$, in terms of the ancillary statistic $a = n^{1/2}(\sum Y_i^2)^{1/2}/\sum Y_i$. It is easily seen that $r(\theta)$ is *exactly* pivotal: the distribution does not depend on the parameter value θ. Thus, the parametric bootstrap procedure based on the fitted parametric $N(\widehat{\theta}, \widehat{\theta}^2)$ model will estimate the sampling distribution of $r(\theta)$ exactly, and therefore produce confidence sets of zero coverage error, ignoring any error that arises from use of a finite Monte Carlo simulation.

In general, the bootstrap approach will not completely eliminate error. This is likely to be especially true in non-parametric applications and in parametric situations involving nuisance parameters.

Recently, considerable attention has focussed on modifications of the basic bootstrap approach outlined above, which are aimed at reducing the levels of error remaining after prepivoting.

A basic device by which we may, in principle, obtain better bootstrap inference is to change the distribution from which bootstrap samples are drawn. Specifically, instead of using a single distribution $\tilde{F}$ as the basis for prepivoting, we utilise a family of models, explicitly constrained to depend on the candidate value ψ of the parameter of interest.

Constrained bootstrap procedures of this kind encompass a variety of statistical methods and are closely related in the non-parametric setting to empirical and other forms of non-parametric likelihood (Owen, 1988; DiCiccio and Romano, 1990) Besides the inference problem of confidence set construction (and the associated hypothesis testing problem) considered here, applications of weighted bootstrap ideas are numerous, and include variance stabilisation, non-parametric curve estimation, non-parametric sensitivity analysis etc.: see Hall and Presnell (1999a,b,c).

In detail, in our prepivoting formulation of bootstrapping, we replace $\widehat{u}_1(Y, \psi)$ by the constrained or weighted prepivoted root

$$\tilde{u}_1(Y, \psi) = \tilde{G}\{u(Y, \psi); \psi\},$$

with

$$\tilde{G}(x; \psi) = \text{Pr}^{\dagger}\{u(Y^{\dagger}, \psi) \leq x\}. \quad (11.9)$$

Now, in the parametric setting, $\mathrm{Pr}^\dagger$ denotes the probability under the drawing of bootstrap samples $Y^\dagger$ from a constrained fitted model $F(\widehat{\eta}_\psi)$, which depends on the candidate value ψ. Focussing on a formulation of the inference problem in which $\eta = (\theta, \xi)$, the particular proposal of DiCiccio *et al.* (2001) is to take $\widehat{\eta}_\psi = (\psi, \widehat{\xi}_\psi)$, in which the nuisance parameter is replaced by its constrained maximum likelihood estimator for any specified value of the interest parameter. It turns out that this proposal succeeds quite generally (Lee and Young, 2004) in reducing the error of the confidence set: compared with use of the original confidence set root u, the conventionally prepivoted root $\widehat{u}_1$ reduces error by a factor of order $O(n^{-1/2})$, while constrained prepivoting to obtain $\tilde{u}_1$ reduces error by a factor of order $O(n^{-1})$. A natural alternative in which $\widehat{\eta}_\psi = (\psi, \widehat{\xi})$, so that the nuisance parameter is fixed at its overall maximum likelihood value, typically only reduces error by order $O(n^{-1/2})$. In the no-nuisance-parameter case of Illustration 11.1, we have $\widehat{\eta}_\psi = \psi$.

In a non-parametric problem, $\mathrm{Pr}^\dagger$ denotes the probability under the drawing of bootstrap samples $Y^\dagger$ from the distribution $\widehat{F}_p$, which places probability mass p_i on Y_i, where $p \equiv p(\psi) = (p_1, \ldots, p_n)$ is chosen to minimise (say) the Kullback–Leibler distance

$$-n^{-1} \sum_{i=1}^{n} \log(np_i)$$

between $\widehat{F}_p$ and $\widehat{F}$, subject to $\theta(\widehat{F}_p) = \psi$. Thus, in this context, constrained prepivoting replaces sampling from $\{Y_1, \ldots, Y_n\}$ with *equal* probabilities $1/n$ on each Y_i, by sampling with *unequal* probabilities p_i.

Illustration 11.2 A second illustration concerns inference for the inverse Gaussian distribution, $IG(\theta, \lambda)$. This distribution, first obtained by the physicist Schrödinger, is related to the distribution of the random time at which the trajectory of a Brownian motion reaches a preassigned positive level for the first time, and has important applications in mathematical finance. Specifically, the $IG(\nu^2/\sigma^2, a^2/\sigma^2)$ density describes the density of the first hitting time of level a of a Brownian motion (see Feller, 1971: Chapter 10) with drift ν and diffusion constant σ^2, starting at 0.

Let $Y = \{Y_1, \ldots, Y_n\}$ be a random sample from the $IG(\theta, \lambda)$ distribution with density

$$f_Y(y; \theta, \lambda) = \frac{\sqrt{\lambda} e^{\sqrt{\theta\lambda}}}{\sqrt{2\pi y^3}} e^{-(\theta y + \lambda/y)/2},$$

and suppose that λ is known, so that again there is no nuisance parameter. We have $\sqrt{\lambda/\theta} = \bar{Y}$, and

$$r(\theta) = n^{1/2} \lambda^{1/4} \widehat{\theta}^{-1/4} (\widehat{\theta}^{1/2} - \theta^{1/2}).$$

Recall that in this situation an adjusted version $r_a(\theta)$ of the signed root statistic $r(\theta)$ is given by

$$r_a(\theta) = r(\theta) + \log\{v(\theta)/r(\theta)\}/r(\theta),$$

where $v(\theta)$ is defined by (9.15).

Table 11.1 *Coverages (%) of confidence sets for parameter θ of inverse Gaussian distribution $IG(\theta, \lambda)$, λ known, sample size $n = 10$*

Nominal	1.0	2.5	5.0	10.0	90.0	95.0	97.5	99.0
$\Phi(r)$	0.5	1.4	3.0	6.6	89.0	94.8	97.6	99.1
$\Phi(r_a)$	0.8	2.2	4.4	9.2	91.6	96.2	98.3	99.4
MLE bootstrap	1.1	2.7	5.4	10.4	89.7	94.8	97.4	99.0
Constrained bootstrap	1.0	2.4	4.9	9.9	90.0	95.0	97.5	99.0

It is easily verified that in the current inverse Gaussian case we have

$$v(\theta) = \frac{1}{2} n^{1/2} \lambda^{1/4} \widehat{\theta}^{-3/4} (\widehat{\theta} - \theta).$$

Notice that in a situation such as this, with no nuisance parameter, confidence sets based on the constrained prepivoted root $\tilde{u}_1(Y, \psi)$ will, again modulo simulation error, have exactly the nominal desired coverage.

We conducted a small simulation to compare the coverage properties of confidence sets derived from the confidence set roots $u(Y, \psi) = \Phi\{r(\psi)\}$, $u(Y, \psi) = \Phi\{r_a(\psi)\}$, and the two prepivoted confidence set roots $\widehat{u}_1(Y, \psi) = \widehat{G}[\Phi\{r(\psi)\}]$ (conventional MLE bootstrapping), and $\tilde{u}_1(Y, \psi) = \tilde{G}[\Phi\{r(\psi)\}; \psi]$ (constrained bootstrapping). Table 11.1 compares actual and nominal coverages provided by the four constructions, based on 100 000 simulated datasets Y. All bootstrap confidence sets were based on $R = 4999$ bootstrap samples, λ is assumed known equal to 1.0, the true $\theta = 1.0$, and the sample size is $n = 10$. The results demonstrate very clearly that poor levels of accuracy obtained by normal approximation to the distribution of r for small coverage levels can be improved by the analytic adjustment r_a. The coverage accuracy of confidence sets derived by conventional bootstrapping competes with that obtained from r_a, but, as expected, coverage figures for the constrained bootstrap approach are very close to the desired values.

A number of remarks are in order.

Remark 1 We note that, by contrast with the conventional bootstrap approach, in principle at least, a different fitted distribution is required by constrained bootstrapping for each candidate parameter value ψ. In the context of Example 11.1, for instance, the confidence set is $\{\psi : r_p(\psi) \leq c_{1-\alpha}(\psi, \widehat{\xi}_\psi)\}$, where now $c_{1-\alpha}(\psi, \widehat{\xi}_\psi)$ denotes the $1 - \alpha$ quantile of the sampling distribution of $r_p(\psi)$ when the true parameter value is $(\psi, \widehat{\xi}_\psi)$, so that a different bootstrap quantile is applied for each candidate ψ, rather than the single quantile $c_{1-\alpha}(\widehat{\theta}, \widehat{\xi})$.

Remark 2 However, computational shortcuts, which reduce the demands of weighted bootstrapping, are possible. These include the use of stochastic search procedures, which allow construction of the confidence set without a costly simulation at each candidate parameter value, such as the Robbins–Monro procedure (Garthwaite and Buckland, 1992; Carpenter, 1999) and, in the non-parametric case, approximation to the probability weights $p(\psi)$ (Davison and Hinkley, 1997), rather than explicit evaluation for each ψ.

The Robbins–Monro procedure operates as follows. Suppose we seek the value, $\bar{\psi}$ say, of a quantity ψ such that $p(\bar{\psi}) = 1 - \alpha$, where $p(\psi)$ denotes the probability of some event. We suppose that $p(\psi)$ decreases as ψ increases, as will be the case for our applications of interest. Here, $p(\psi)$ is either the conventional bootstrap confidence set root $\hat{u}_1(Y, \psi)$, or the constrained bootstrap confidence set root $\tilde{u}_1(Y, \psi)$. In the former case, the event in question is the event that the (conventional) bootstrap sample Y^* has $u(Y^*, \hat{\theta}) \leq u(Y, \psi)$, while, in the latter case, the event is that the (constrained) bootstrap sample $Y^\dagger$ has $u(Y^\dagger, \psi) \leq u(Y, \psi)$. The idea is to perform a sequence of trials, at each of a sequence of values of ψ, in our application a trial corresponding to generation of a single bootstrap sample. If the jth ψ value is ψ_j then,

$$\psi_{j+1} = \begin{cases} \psi_j + c\alpha/j & \text{if the event occurs,} \\ \psi_j - c(1 - \alpha)/j & \text{otherwise.} \end{cases}$$

Here $c > 0$ is some specified constant, and we note that the modification to the current ψ value gets smaller as the sequence of trials progresses. Also,

$$\mathbb{E}(\psi_{j+1}) = p(\psi_j)(\psi_j + c\alpha/j) + \{1 - p(\psi_j)\}(\psi_j - c(1 - \alpha)/j)$$
$$= \psi_j + c\{p(\psi_j) - (1 - \alpha)\}.$$

If $\psi_j > \bar{\psi}$, we have $p(\psi_j) < 1 - \alpha$, so that the new ψ value is expected to be smaller than ψ_j (a move in the right direction), while, if $\psi_j < \bar{\psi}$, then ψ_{j+1} is expected to be larger (also a move in the right direction). Strategies for implementation, in particular the choice of the constant c, are suggested by Garthwaite and Buckland (1992). Initialisation can utilise, in our context, the confidence limit provided by the initial (non-bootstrapped) root $u(Y, \psi)$. It is quite practical to determine a confidence limit by the Robbins–Monro procedure using a series of a few thousand trials, that is a few thousand bootstrap samples, as we have suggested is appropriate.

Remark 3 The theoretical effects of weighted bootstrapping in the nonparametric context are analysed for various classes of problem, including those involving robust estimators and regression estimation, as well as the smooth function model of Hall (1992), by Lee and Young (2003). The basic striking conclusion is as indicated above: if $u(Y, \theta)$ is uniform to order $O(n^{-j/2})$,

$$\Pr\{u(Y, \theta) \leq u\} = u + O(n^{-j/2}),$$

then, quite generally, $\hat{u}_1(Y, \theta)$ is uniform to order $O(n^{-(j+1)/2})$, while $\tilde{u}_1(Y, \theta)$ is uniform to the higher order $O(n^{-(j+2)/2})$. The result holds for confidence set roots of the kind described in Example 11.2, as well as more complicated roots. An example is the widely used bootstrap *percentile method* confidence set root

$$u(Y, \psi) = \Pr^*\{\hat{\theta}(Y^*) > \psi\}, \tag{11.10}$$

where Y^* denotes a bootstrap sample drawn from the empirical distribution function $\hat{F}$ of the data sample Y and $\Pr^*$ denotes probability under the drawing of such samples. We observe that $u(Y, \psi)$ is monotonically decreasing in ψ, so that the confidence set $\mathcal{I}$ is of the form $(\hat{\theta}_l, \infty)$, where it is easily seen that $\hat{\theta}_l \equiv \hat{\theta}_l(Y)$ is the α quantile of $\hat{\theta}(Y^*)$ under the drawing of bootstrap samples Y^*. As we have noted, in practice the confidence limit $\widehat{\theta}_l$ is approximated by Monte Carlo simulation, the probability in (11.10) being estimated by the drawing of an actual series of R bootstrap samples. Notice that prepivoting with this confidence set root then involves a nested, or iterated, bootstrap calculation. For instance, $\widehat{G}$, as given by (11.7), is approximated by the drawing of, say, S bootstrap samples Y^*: for each of these R (second-level) bootstrap samples must be drawn to approximate $u(Y^*, \widehat{\theta})$. A similar procedure is required in approximating $\widetilde{G}$, (11.9).

The basic assumption made by Lee and Young (2003) is that the root $u(Y, \theta)$ admits an asymptotic expansion of the form

$$u(Y, \theta) = \Phi(T) + \phi(T)\{n^{-1/2}r_1(\bar{Z}, T) + n^{-1}r_2(\bar{Z}, T) + \cdots\}, \tag{11.11}$$

where $T = (\widehat{\theta} - \theta)/\widehat{\sigma}$ is the studentised parameter estimate, asymptotically standard normal, as in Example 11.2 above, and where the precise specification of $\bar{Z} = n^{-1}\sum_{i=1}^{n} z_i(Y, F)$ and polynomials r_1, r_2 depend on the class of problem being considered. The basic result then holds under mild conditions on the choice of probability weights p_i. In particular, the conclusions hold for a whole class of distance measures, which generalise the Kullback–Leibler distance (Baggerly, 1998; Corcoran, 1998). The choice of distance measure is therefore largely irrelevant to the theoretical conclusion, allowing the use of well-developed algorithms (Owen, 2001) for construction of weighted bootstrap distributions $\widehat{F}_p$, as well as use of simple tilted forms of the empirical distribution function $\widehat{F}$, as described, for example, by DiCiccio and Romano (1990).

Lee and Young (2003) also consider the effects of successively iterating the prepivoting. They demonstrate that iterated weighted prepivoting accelerates the rate of convergence to zero of the bootstrap error, compared with the effect of iteration of the conventional bootstrap (Hall and Martin, 1988; Martin, 1990).

The same conclusions hold for testing. When testing a point null hypothesis $H_0 : \theta = \theta_0$, a one-sided test of nominal size α rejects H_0 if $u(Y, \theta_0) \leq \alpha$. If $u(Y, \theta_0)$ were exactly Uniform$(0, 1)$, the null rejection probability would be exactly α. To increase accuracy, weighted bootstrapping applied with $\theta = \theta_0$ reduces error by $O(n^{-1})$. Now, of course, constrained bootstrapping need only be carried out at the single parameter value θ_0, so computational complications over conventional bootstrapping are reduced.

Remark 4 DiCiccio *et al.* (2001) show that in the parametric context, and for the specific case $u(Y, \psi) = \Phi\{r_p(\psi)\}$, the coverage error of the confidence set is reduced by constrained prepivoting to $O(n^{-3/2})$. This same order of error as that obtained from the analytic adjustment r_a to the signed root statistic r_p is achieved without any need for analytic calculation, or specification of the ancillary required by r_a. A fuller theoretical analysis for confidence sets based on a general, asymptotically $N(0, 1)$, pivot $T(Y, \theta)$ shows this basic conclusion, of reduction of the coverage error from $O(n^{-\beta/2})$ for the initial confidence set root to $O(n^{-(\beta+1)/2})$ by conventional bootstrapping, and order $O(n^{-(\beta+2)/2})$ by constrained bootstrapping, to hold in considerable generality. See Lee and Young (2004).

Remark 5 To emphasise the conclusions reached from Illustrations 11.1 and 11.2 above, we note that, in the parametric context and in the absence of any nuisance parameter, confidence sets based on the constrained prepivoted root $\tilde{u}_1(Y, \psi)$ will always have *exactly* the desired coverage $1 - \alpha$, ignoring any error that results from a finite Monte Carlo simulation. The conventional bootstrap approach, based on $\widehat{u}_1(Y, \psi)$, will only yield exact inference if the initial root $u(Y, \psi)$ is exactly pivotal. There seems, therefore, strong arguments in favour of general adoption of constrained bootstrap schemes.

11.3 Data example: Bioequivalence

The following application of the non-parametric bootstrap is described by Efron and Tibshirani (1993), whose analysis we extend.

A drug company has applied each of three hormone supplement medicinal patches to eight patients suffering from a hormone deficiency. One of the three is 'Approved', having received approval from the US Food and Drug Administration (FDA). Another is a 'Placebo' containing no hormone. The third is 'New', being manufactured at a new facility, but otherwise intended to be identical to 'Approved'. The three wearings occur in random order for each patient and the blood level of the hormone is measured after each patch wearing: results are given in Table 11.2. The FDA requires proof of *bioequivalence* before approving sale of the product manufactured at the new facility.

Technically, let x be the difference between Approved and Placebo measurements on the same patient and let z be the difference between New and Approved:

$$x = \text{Approved} - \text{Placebo}, \quad z = \text{New} - \text{Approved}.$$

Let μ and ν be the expectations of x and z and let

$$\rho = \nu/\mu.$$

The FDA criterion for bioequivalence is that the new facility matches the old facility within 20% of the amount of hormone the old drug adds to the placebo blood levels, $|\rho| \leq 0.2$.

For the given data

$$(\bar{x}, \bar{z}) = (6342, -452) = (\widehat{\mu}, \widehat{\nu}),$$

giving a non-parametric estimate of the ratio ρ as

$$\widehat{\rho} = \widehat{\nu}/\widehat{\mu} = -0.071.$$

Table 11.2 *Bioequivalence data*

Subject	Placebo	Approved	New	x	z
1	9243	17649	16449	8406	−1200
2	9671	12013	14614	2342	2601
3	11792	19979	17274	8187	−2705
4	13357	21816	23798	8459	1982
5	9055	13850	12560	4795	−1290
6	6290	9806	10157	3516	351
7	12412	17208	16570	4796	−638
8	18806	29044	26325	10238	−2719

What about possible bias in this estimate? A simple bootstrap estimate of bias may be constructed as

$$\mathbb{E}^*\{\widehat{\rho}(Y^*)\} - \widehat{\rho}(Y),$$

where Y denotes the observed data, consisting of the eight pairs (x, z), and $\mathbb{E}^*$ denotes expectation under the drawing of bootstrap samples Y^* from the empirical distribution function $\widehat{F}$ of the observed data. In practice, this expectation is approximated by $R^{-1} \sum_{i=1}^{R} \widehat{\rho}(Y_i^*)$, where $Y_1^*, \ldots, Y_R^*$ denote an actual series of R bootstrap samples obtained by Monte Carlo simulation from $\widehat{F}$. Now, for bias estimation the rule of thumb is that R of order of a few hundreds is adequate.

For the data in Table 11.2, a bootstrap bias estimator constructed from $R = 500$ bootstrap samples gave the value 0.008, suggesting that bias is not a problem in this situation. Given the small sample size, we might however expect the estimator $\widehat{\rho}$ to be highly variable, and again simple bootstrapping can provide an estimate of this variability. The bootstrap estimator of the variance of the ratio estimator $\widehat{\rho}$ is

$$\text{var}^*\{\widehat{\rho}(Y^*)\},$$

the variance under sampling of bootstrap samples from $\widehat{F}$. Again, a Monte Carlo approximation is $R^{-1} \sum_{i=1}^{R} (\widehat{\rho}(Y_i^*) - \bar{\rho}^*)^2$, where $\bar{\rho}^* = R^{-1} \sum_{i=1}^{R} \widehat{\rho}(Y_i^*)$, with sensible R of the order of a few hundreds. For our data and $R = 500$, a variance estimator of 0.010 was obtained, so we estimate the standard error of $\widehat{\rho}$ to be 0.100.

The large estimate of standard error casts doubt on the bioequivalence requirement being satisfied. A crude confidence interval of nominal coverage 90% for ρ, the so-called 'standard interval' based on normal approximation to the distribution of $\widehat{\rho}$, is

$$\widehat{\rho} \pm 1.96 \times \text{estimated standard error},$$

here $(-0.267, 0.125)$, based on our standard error estimate.

In fact, in formal terms the explicit FDA bioequivalence requirement is that a 90% equitailed confidence interval of the form (11.2) for ρ lie within the range $(-0.2, 0.2)$. We use the percentile method to construct a nominal 90% confidence interval for ρ, using confidence set root $u(Y, \psi) = \text{Pr}^*\{\widehat{\rho}(Y^*) > \psi\}$.

Figure 11.1 illustrates $u(Y, \psi)$ as a function of ψ, based on $R = 9999$ bootstrap samples, for the data under study. The nominal 90% confidence interval (11.2) is $(-0.211, 0.115)$, suggesting that the bioequivalence criterion is violated.

But is this interval to be trusted? The accuracy, under repeated sampling from the underlying population, of the chosen confidence interval procedure is an important part of the way the FDA decision making operates, and we must therefore be concerned at the accuracy of our percentile method interval. Figure 11.1 shows also the confidence sets obtained from the prepivoted confidence set roots $\widehat{u}_1(Y, \psi)$ and $\tilde{u}_1(Y, \psi)$: in the construction of these sets the distribution functions $\widehat{G}$, as given by (11.7), and $\tilde{G}$, as given by (11.9), were also each constructed from $S = 9999$ bootstrap samples of the relevant conventional or weighted type. Conventional prepivoting yields a much wider confidence interval, $(-0.219, 0.306)$, suggesting clear violation of the criterion, but weighted prepivoting yields the confidence set $(-0.199, 0.124)$, not much different from the raw percentile interval. The conclusion is

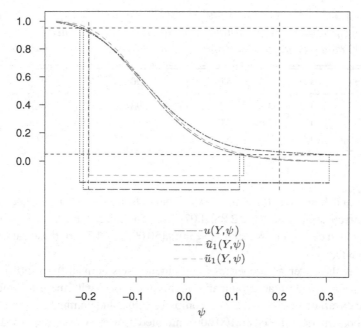

Figure 11.1 Two-sided confidence sets obtained from roots $u(Y, \psi)$, $\widehat{u}_1(Y, \psi)$ and $\tilde{u}_1(Y, \psi)$ for bioequivalence data

clear: we are uncertain whether the requirement is met, but the confidence set of greatest theoretical asymptotic accuracy suggests it just might be. Further investigation is appropriate.

11.4 Further numerical illustrations

Illustration 11.3 Exponential regression

Consider an exponential regression model in which $T_1, \ldots, T_n$ are independent, exponentially distributed lifetimes, with means of the form $E(T_i) = \exp(\beta + \xi z_i)$, with known covariates $z_1, \ldots, z_n$. Suppose that inference is required for the mean lifetime for covariate value z_0. Let the parameter of interest therefore be $\theta = \beta + \xi z_0$, with nuisance parameter ξ. The signed root likelihood ratio statistic is

$$r_p(\theta) = \mathrm{sgn}(\widehat{\theta} - \theta)[2n\{(\theta - \widehat{\theta}) + (\widehat{\xi}_\theta - \widehat{\xi})\bar{c} + n^{-1}\exp(-\theta)\sum_{i=1}^{n} T_i \exp(-\widehat{\xi}_\theta c_i) - 1\}]^{1/2},$$

where $c_i = z_i - z_0, i = 1, \ldots, n$ and $\bar{c} = n^{-1}\sum c_i$.

In this case, the calculations leading to the adjusted version r_a of r_p are readily performed. However, it is easily verified that r_p is exactly pivotal. To see this, substitute $T_i = \exp(\theta + \xi c_i)Y_i$, where the Y_i are independently, exponentially distributed with mean 1, and observe that the signed root statistic may be expressed as a (complicated) function of $Y_1, \ldots, Y_n$, and so has a distribution which does not depend on (θ, ξ). Therefore, even conventional bootstrapping yields the true sampling distribution, modulo simulation error. There is no need for constrained bootstrapping in this problem.

For numerical illustration, we consider data extracted from Example 6.3.2 of Lawless (1982), as analysed by DiCiccio *et al.* (2001). The $n = 5$ responses T_i are 156,

Table 11.3 *Coverages (%) of confidence sets for mean* $\theta = \exp(\beta + \xi z_0)$ *at* $z_0 = \log_{10}(50\,000)$ *in exponential regression example, estimated from* $100\,000$ *data sets of size* $n = 5$ *and using* $R = 4999$ *bootstrap replicates*

Nominal	1.0	2.5	5.0	10.0	90.0	95.0	97.5	99.0
$\Phi(r_p)$	1.7	4.1	7.8	14.6	93.7	97.1	98.7	99.5
$\Phi(r_a)$	1.0	2.6	5.0	10.1	90.0	95.0	97.5	99.0
Bootstrap	1.0	2.6	5.0	10.0	90.0	95.1	97.5	99.0

108, 143, 65 and 1, survival times in weeks of patients suffering from leukaemia, and the corresponding covariate values are 2.88, 4.02, 3.85, 5.0 and 5.0, the base-10 logarithms of initial white blood cell count. We take $z_0 = \log_{10}(50\,000) \simeq 4.7$. For these data, $\widehat{\theta} = 3.913$ and $\widehat{\xi} = -0.838$.

We compare the coverage properties of confidence sets derived from $\Phi(r_p)$, $\Phi(r_a)$ and bootstrapping for $n = 5$, in an exponential regression model with these parameter values and the fixed covariate values. Table 11.3 compares actual and nominal coverages provided by the three constructions, based on $100\,000$ simulated datasets. Coverages based on normal approximation to r_p are quite inaccurate, but normal approximation to r_a provides much more accurate inference, while bootstrap confidence sets (each based on $R = 4999$ bootstrap samples) also display coverages very close to nominal levels.

Other cases where it is easily verified that r_p is exactly pivotal, and therefore conventional bootstrapping of r_p will provide exact inference, include inference for the error variance in a normal-theory linear regression model, and the related Neyman–Scott problem, as described by Barndorff–Nielsen and Cox (1994: Example 4.2).

Illustration 11.4 Normal distributions with common mean

We consider now the problem of parametric inference for the mean, based on a series of independent normal samples with the same mean but different variances. Initially we consider a version of the Behrens–Fisher problem in which we observe $Y_{ij}, i = 1, 2, j = 1, \ldots, n_i$, independent $N(\theta, \sigma_i^2)$. The common mean θ is the parameter of interest, with orthogonal nuisance parameter $\xi = (\sigma_1, \sigma_2)$. Formally, this model is a (4,3) exponential family model. In such a model, the adjusted signed root statistic r_a is intractable, though readily computed approximations are available, such as the approximation $\tilde{r}_a$ detailed at (9.17): see also Skovgaard (1996), Severini (2000: Chapter 7); Barndorff-Nielsen and Chamberlin (1994).

We compare coverages of confidence sets derived from $\Phi(r_p)$, $\Phi(\tilde{r}_a)$, the conventional bootstrap, which bootstraps at the overall maximum likelihood estimator $(\widehat{\theta}, \widehat{\xi})$, and the constrained bootstrap, which uses bootstrapping at the constrained maximum likelihood estimator $(\theta, \widehat{\xi}_\theta)$, for $50\,000$ datasets from this model, with parameter values $\theta = 0$, $\sigma_1^2 = 1$, $\sigma_2^2 = 20$ and sample sizes $n_1 = n_2 = 5$. All bootstrap confidence sets are based on $R = 1999$ bootstrap samples. Also considered are the corresponding coverages obtained from $\Phi(W)$ and $\Phi(S)$ and their conventional and constrained bootstrap versions, where W and S are Wald and score statistics respectively, defined as the signed square roots of the statistics (8.23) and (8.22) respectively.

Table 11.4 *Coverages (%) of confidence intervals for Behrens–Fisher example,*
estimated from 50 000 data sets with bootstrap size R = 1999

Nominal	1.0	2.5	5.0	10.0	90.0	95.0	97.5	99.0
$\Phi(r_p)$	1.6	3.7	6.5	11.9	88.0	93.5	96.4	98.4
MLE bootstrap	0.9	2.3	4.8	10.1	90.1	95.3	97.8	99.2
Constrained MLE bootstrap	0.8	2.3	4.8	10.0	90.1	95.3	97.9	99.2
$\Phi(\tilde{r}_a)$	0.9	2.5	5.2	10.6	89.5	94.9	97.6	99.1
$\Phi(W)$	5.2	8.0	11.3	16.5	83.5	88.9	92.2	94.8
MLE bootstrap	0.8	2.3	4.7	10.1	90.1	95.2	97.8	99.2
Constrained MLE bootstrap	0.8	2.2	4.6	10.0	90.2	95.4	97.9	99.3
$\Phi(S)$	0.1	1.4	4.7	11.6	88.6	95.5	98.7	99.9
MLE bootstrap	1.1	2.4	5.0	10.0	90.1	95.2	97.7	99.0
Constrained MLE bootstrap	0.9	2.4	5.0	10.1	90.0	95.2	97.7	99.1

The coverage figures shown in Table 11.4 confirm that the simple bootstrap approach improves over asymptotic inference based on any of the statistics r_p, S, or W. Conventional bootstrapping yields very accurate inference for all three statistics: there are no discernible gains from using the constrained bootstrap. Overall, bootstrapping is very competitive in terms of accuracy when compared with $\tilde{r}_a$.

In the above situation, the nuisance parameter is two dimensional. As a more challenging case, we consider extending the above analysis to inference on the common mean, set equal to 0, of six normal distributions, with unequal variances $(\sigma_1^2, \ldots, \sigma_6^2)$, set equal to $(1.32, 1.93, 2.22, 2.19, 1.95, 0.11)$. These figures are the variances of data relating to measurements of strengths of six samples of cotton yarn, given in Example Q of Cox and Snell (1981). The inference is based on an independent sample of size 5 from each population. Table 11.5 provides figures corresponding to those in Table 11.4 for this regime. Now the bootstrap approach is clearly more accurate than the approach based on $\tilde{r}_a$, and it might be argued that now it is possible to discern advantages to the constrained bootstrap approach compared with the conventional bootstrap. Again, the constrained bootstrap works well when applied to the Wald and score statistics, casting some doubt on the practical significance of arguments of DiCiccio *et al.* (2001).

Illustration 11.5 Variance component model
As a further parametric illustration of a practically important inference problem, we consider the one-way random effects model considered by Skovgaard (1996) and DiCiccio *et al.* (2001). Here we have

$$Y_{ij} = \theta + \alpha_i + e_{ij}, \quad i = 1, \ldots, m; \ j = 1, \ldots, n_i,$$

where the α_is and the e_{ij}s are all independent normal random variables of mean 0 and variances σ_α^2 and σ_e^2 respectively. Inference is required for θ, both other parameters being treated as nuisance. If the group sizes $n_1, \ldots, n_m$ are not all equal the maximum likelihood estimators do not have closed-form expressions, and must be found iteratively. More importantly, ancillary statistics are not available to determine the analytic adjustment $v_p(\psi)$

Table 11.5 *Coverages (%) of confidence intervals for normal mean example,*
estimated from 50 000 data sets with bootstrap size $R = 1999$

Nominal	1.0	2.5	5.0	10.0	90.0	95.0	97.5	99.0
$\Phi(r_p)$	3.0	5.7	9.3	15.1	85.3	91.2	94.6	97.2
MLE bootstrap	1.1	2.7	5.2	10.2	90.3	95.0	97.5	98.9
Constrained MLE bootstrap	0.9	2.5	5.1	10.1	90.4	95.2	97.6	99.0
$\Phi(\tilde{r}_a)$	1.5	3.4	6.4	11.9	88.7	93.9	96.7	98.5
$\Phi(W)$	6.7	9.6	13.3	18.7	82.0	87.3	90.8	93.6
MLE bootstrap	1.1	2.7	5.3	10.2	90.2	95.0	97.4	98.9
Constrained MLE bootstrap	0.9	2.4	5.0	9.9	90.5	95.3	97.6	99.1
$\Phi(S)$	0.6	2.1	5.1	10.9	89.6	95.2	98.0	99.5
MLE bootstrap	1.2	2.7	5.2	10.1	90.4	95.1	97.4	98.8
Constrained MLE bootstrap	1.1	2.5	5.2	10.2	90.3	95.1	97.5	99.0

Table 11.6 *Coverages (%) of confidence intervals for random effects example,*
estimated from 50 000 data sets with bootstrap size $R = 1999$

Nominal	1.0	2.5	5.0	10.0	90.0	95.0	97.5	99.0
$\Phi(r_p)$	1.5	3.4	6.5	11.9	88.2	93.6	96.6	98.5
MLE bootstrap	1.0	2.5	5.1	10.2	90.0	95.0	97.6	99.1
Constrained MLE bootstrap	1.0	2.5	5.1	10.1	90.0	95.0	97.6	99.1
$\Phi(\tilde{r}_a)$	0.6	1.8	4.1	8.9	91.1	95.9	98.1	99.3
MLE bootstrap	1.0	2.5	5.1	10.2	90.0	95.0	97.6	99.1
Constrained MLE bootstrap	1.0	2.5	5.1	10.1	90.0	95.0	97.6	99.1
$\Phi(W)$	0.5	2.0	5.1	10.9	89.2	94.9	98.0	99.5
MLE bootstrap	1.0	2.5	5.1	10.2	90.0	95.0	97.6	99.1
Constrained MLE bootstrap	1.0	2.5	5.1	10.2	90.0	95.0	97.6	99.1
$\Phi(S)$	0.4	2.0	5.1	10.9	89.2	95.0	98.0	99.5
MLE bootstrap	1.0	2.5	5.1	10.2	90.0	95.0	97.6	99.1
Constrained MLE bootstrap	1.0	2.5	5.0	10.1	90.0	95.0	97.6	99.1

to the signed root likelihood ratio statistic $r_p(\psi)$, so again approximate forms such as $\tilde{r}_a(\psi)$ must be used.

We performed a simulation analogous to that described in Illustration 11.4 for the case $m = 10$, $n_i = i$, $\sigma_\alpha = 1$, $\sigma_e = 0.04$. Coverage figures obtained for the various confidence set constructions are given in Table 11.6, again as derived from a series of 50 000 simulations, with bootstrap confidence sets being based on $R = 1999$ bootstrap samples. Included in the study in this case are coverage figures obtained by applying both conventional and constrained prepivoting to the initial confidence set root $u(Y, \psi) = \Phi\{\tilde{r}_a(\psi)\}$: normal approximation to the distribution of $\tilde{r}_a$ itself yields poor coverage accuracy, and it is worthwhile considering prepivoting the confidence set root constructed from $\tilde{r}_a$. The effectiveness of bootstrapping is again apparent.

In summary, it is our general experience that analytic approaches based on r_a are typically highly accurate when the dimensionality of the nuisance parameter is small and r_a itself is readily constructed, as in, say, a full exponential family model, where no ancillary statistic is required. In such circumstances, the argument for bootstrapping then rests primarily on maintaining accuracy, while avoiding cumbersome analytic derivations. In more complicated settings, in particular when the nuisance parameter is high dimensional or analytic adjustments r_a must be approximated, the bootstrap approach is typically preferable in terms of both ease of implementation and accuracy. In all the parametric examples we have studied, it is striking that conventional bootstrapping already produces very accurate inference. Though constrained bootstrapping is advantageous from a theoretical perspective, in practice the gains over conventional bootstrapping may be slight or non-existent. Nevertheless, there is little risk of impaired repeated sampling performance relative to conventional bootstrapping, and implementation is straightforward.

In the non-parametric context, the situation is less clear-cut. In particular, it is unclear whether the theoretical benefits of weighted bootstrapping over conventional bootstrapping are realisable in any particular situation, or whether weighted bootstrapping might actually reduce finite sample accuracy, as our next illustration demonstrates. The primary difficulty that arises in the non-parametric context is that of choosing the re-weighting scheme most appropriately, a problem that does not arise in the parametric setting.

Illustration 11.6 Non-parametric inference for variance

We consider non-parametric inference for the variance $\theta = 0.363$ of a folded standard normal distribution $|N(0, 1)|$, for sample size $n = 50$.

From 20 000 datasets, we compared the coverage properties of confidence sets based on $u(Y, \psi) = \Phi\{(\widehat{\theta} - \psi)/\widehat{\sigma}\}$, with $\widehat{\theta}$ the sample variance and $\widehat{\sigma}^2$ an estimate of its asymptotic variance, and its conventional and weighted prepivoted forms $\widehat{u}_1(Y, \psi)$ and $\bar{u}_1(Y, \psi)$. Table 11.7 displays the coverages of the three intervals. Weighted bootstrapping here utilised the exponentially tilted distribution involving empirical influence values described by Davison and Hinkley (1997): see also DiCiccio and Romano (1990). Results for this, computationally simple, weighting procedure are very similar to those obtained from other, computationally less attractive, choices of construction of weighted bootstrap distribution.

Confidence sets based on $u(Y, \psi)$ are quite inaccurate, and substantial improvements are given by both conventional and weighted bootstrapping. Which of these is best

Table 11.7 *Coverages (%) of bootstrap confidence sets for the variance θ when F is the folded standard normal distribution, estimated from 20 000 data sets of size $n = 50$ and using $R = 4999$ bootstrap replicates; the root taken is $u(Y, \psi) = \Phi\{(\widehat{\theta} - \psi)/\widehat{\sigma}\}$ with $\widehat{\theta}$ sample variance*

Nominal	1.0	2.5	5.0	10.0	90.0	95.0	97.5	99.0
$u(Y, \psi)$	10.0	13.4	17.2	23.0	95.1	98.3	99.4	99.9
$\widehat{u}_1(Y, \psi)$	3.1	5.4	8.5	14.1	91.6	96.5	98.6	99.6
$\bar{u}_1(Y, \psi)$	6.0	8.7	12.1	17.2	90.7	95.9	98.1	99.4

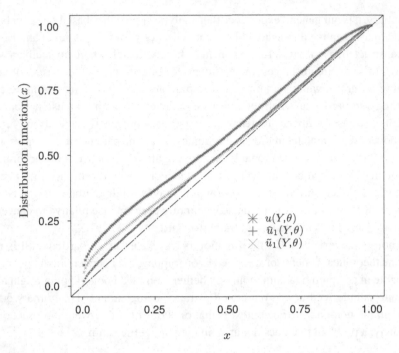

Figure 11.2 Prepivoting operation of the bootstrap, non-parametric inference for the variance of $|N(0, 1)|$

depends, however, on the required coverage level. Similar conclusions are seen in other non-parametric examples: see Example 3 of Davison *et al.* (2003) and the examples given by Lee and Young (2003).

Graphical illustration of the prepivoting operation of the bootstrap is provided in Figure 11.2, which shows the distribution functions, as estimated from the 20 000 datasets, of $u(Y, \theta)$, $\widehat{u}_1(Y, \theta)$ and $\tilde{u}_1(Y, \theta)$, with θ the true parameter value. The distribution of $u(Y, \theta)$ is distinctly *not* Un(0, 1), while both bootstrap schemes yield prepivoted roots which *are* close to uniform, except in the lower tail. There, the distribution function of the conventional prepivoted root is closer to uniform than that of the weighted prepivoted root. The coverage figures shown in Table 11.7 may, in principle, be read directly off the graph of the distribution functions of the three confidence set roots.

11.5 Conditional inference and the bootstrap

Our analysis so far is intended to demonstrate that, in both parametric and non-parametric settings, bootstrapping can yield highly accurate inference for a parameter of interest, when judged in terms of repeated sampling properties under the assumed model $F(y; \eta)$. There is, however, another consideration which should be taken into account in our analysis, relating to the Fisherian proposition that inference about a parameter θ should, in a parametric context, be based not on the original specified model $F(y; \eta)$, but instead on the derived model obtained by conditioning on an ancillary statistic. (Notions of ancillarity and conditioning in a non-parametric problem are much more difficult). Recall that the motivation is that the hypothetical repetitions, which are the basis of the inference in a frequentist analysis, should be as relevant as possible to the (unique) data being analysed, so that we should condition on aspects of the data that are uninformative about θ. Likelihood-based analytic approaches, such as those based on r_a, are explicitly designed to have conditional validity, given an ancillary statistic, as well as yielding improved distributional approximation. Bootstrap procedures, on the other hand, involve unconditional simulation from some empirical model, $\tilde{F}$. Yet, as we shall demonstrate, this unconditional approach realises, to a considerable extent, the goal of appropriate conditioning. There is, however, some evidence that bootstrap approaches are less accurate than analytic approaches when judged from this conditional, Fisherian, perspective, rather than from a purely frequentist viewpoint.

A central notion here is the idea of stability. In general, a statistical procedure is stable if it respects the principle of conditioning relative to any reasonable ancillary statistic, without requiring specification of the ancillary: see Barndorff-Nielsen and Cox (1994: Chapter 8). A statistic T is stable if its conditional distribution, given an arbitrary ancillary statistic a, and its unconditional distribution are approximately the same. Formally, the distribution of a statistic T is stable to first order if an asymptotic approximation to the conditional distribution of T given an arbitrary ancillary statistic a does not depend on a, to first order, that is ignoring terms of order $O(n^{-1/2})$. T is stable to second order if an asymptotic expansion of the conditional distribution of T given a does not depend on a, neglecting terms of the higher order $O(n^{-1})$.

For example, in a one-parameter problem, it may be shown that the normalised maximum likelihood estimator $T = \sqrt{i(\theta)}(\widehat{\theta} - \theta)$ is stable only to first order, as, in general, is the case for the score statistic. By contrast, the signed root likelihood ratio statistic $r(\theta)$ is stable to second order, as is the maximum likelihood estimator normalised by observed information $T_0 = \sqrt{j(\widehat{\theta})}(\widehat{\theta} - \theta)$. Similar results hold in the case of a vector parameter.

Illustration 11.1 (continued) Figure 11.3 shows the conditional density functions of three statistics, for samples of size $n = 10$ from the normal distribution $N(\theta, \theta^2)$, for two values of the conditioning ancillary, $a = 1, 1.5$. Case (a) refers to the maximum likelihood estimator normalised by the Fisher information $i(\theta)$, while cases (b) and (c) refer to the maximum likelihood estimator normalised by observed information and the signed root likelihood ratio statistic respectively. The stability of the latter two statistics is clear from the figure.

Inference based on a statistic with a distribution that is stable is particularly appealing, since the goals of conditioning on an ancillary statistic can be achieved using an unconditional procedure, such as bootstrapping! The key point here is that conclusions do not depend on the ancillary statistic used, and indeed precise specification of the ancillary is not required. So, bootstrapping a stable statistic is attractive from a Fisherian, as well as repeated sampling, perspective, in providing good approximation to a conditional analysis that may be awkward to carry out. One approach to conditional parametric bootstrapping in certain situations is through Metropolis–Hastings algorithms (Brazzale, 2000), but unconditional bootstrapping generally has to be used, for instance in circumstances where an appropriate ancillary is unavailable.

It is beyond the scope of this book to provide a rigorous analysis of the stability properties of bootstrapping. Instead, we make some general comments and consider some examples.

(a): normalised MLE, Fisher information

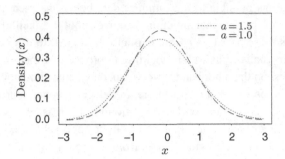

(b): normalised MLE, observed information

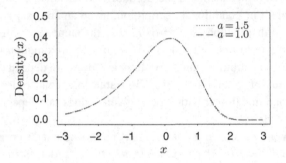

(c): signed root statistic

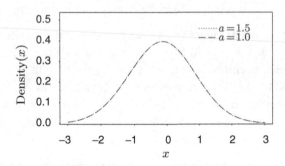

Figure 11.3 Stability of three statistics, normal distribution $N(\theta, \theta^2)$

If we consider a statistic T stable to second order, such as the signed root likelihood ratio, the unconditional and conditional distributions of T differ by $O(n^{-1})$. Appropriate bootstrapping, in particular use of the constrained bootstrap approach which has been the focus of this chapter, is specifically designed to approximate the unconditional distribution of T to error of order $O(n^{-1})$ at worst, and therefore approximates the conditional distribution which is appropriate for inference from a Fisherian perspective to this same level of error, $O(n^{-1})$. As we can see from our examples, this is often quite satisfactory, though in order of magnitude terms the conditional accuracy is less than the $O(n^{-3/2})$ accuracy of analytic procedures, such as r_a, which are specifically constructed to have such levels of conditional validity.

Illustration 11.1 (continued) Suppose that in our example of a normal distribution with known coefficient of variation we have $n = 10$, and $\widehat{\theta} = 3/2$ is observed, with observed ancillary $a = 1$. Consider testing the hypothesis $H_0 : \theta = \theta_0 = 1$, against the alternative that θ is larger, so that evidence against H_0 is provided by *large* values of $\widehat{\theta}$. To do so, we might calculate the signed root of the likelihood ratio statistic, $r(\theta_0) = 2.904$, by (11.8). Then a test of H_0 is carried out, in a Fisherian approach to significance testing, by computing the significance probability $\Pr_{\theta_0}\{r(\theta_0) \geq 2.904\} \equiv \Pr_{\theta_0}(\widehat{\theta} \geq 3/2)$: this turns out (parametric bootstrap!) to be 0.001222. By contrast, normal approximation to the distribution of $r(\theta_0)$ gives a significance probability of 0.001840 and normal approximation to the distribution of the adjusted signed root statistic $r_a(\theta_0)$ a value of 0.001252. But, the conditionality principle implies that the appropriate inference should actually be based on the conditional significance probability $\Pr_{\theta_0}\{r(\theta_0) \geq 2.904 | a = 1\}$. We know, from (9.18), the exact conditional density, given a, of $\widehat{\theta}$ and numerical integration of this gives an exact conditional significance probability of 0.001247. We see that the use of r_a provides a better approximation to the exact conditional significance probability than the unconditional bootstrap inference (which of course gives the answer 0.001222), though for any practical purpose the latter gives a quite satisfactory approximation.

A more complete picture is provided by Figure 11.4, where we compare the true conditional significance probabilities for testing the null hypothesis $\theta = \theta_0$, with approximate

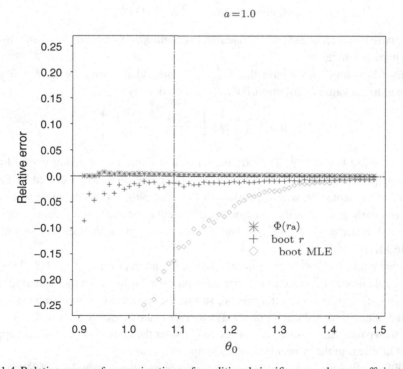

Figure 11.4 Relative errors of approximations of conditional significance values, coefficient of variation example

levels obtained from $\Phi(r_a)$ and by unconditional bootstrap estimation of the distribution of r, for a range of values of θ_0. We consider also use of the bootstrap to estimate the (unconditional) distribution of the maximum likelihood estimator normalised by observed information. In this context, significance values obtained from $\Phi(r)$ are quite inaccurate, as are those obtained by normal approximation to the distribution of the maximum likelihood estimator. The figure plots the relative error of each approximation, defined as the approximation minus the true conditional significance probability, expressed as a proportion of the true probability. Each bootstrap figure is based on $R = 10\,000\,000$ bootstrap samples. The conditional accuracy of the approximation obtained by unconditional bootstrapping of the distribution of r is impressive, though less than that obtained from normal approximation to the distribution of the analytic adjustment r_a, which gives excellent levels of conditional accuracy. This and similar examples favour applying the bootstrap to the signed root statistic r, rather than other statistics stable to second order. For reference, the vertical line shown in the figure corresponds to an exact conditional significance level of 1%. A similar picture is seen for other conditioning values of the ancillary a.

Illustration 11.7 Weibull distribution

We consider analysis of a dataset of size $n = 20$, discussed by Keating, Glaser and Ketchum (1990), concerning the time to failure, in hours, of pressure vessels subjected to a constant fixed pressure. The failure times are 274, 1661, 1787, 28.5, 236, 970, 1.7, 828, 0.75, 20.8, 458, 1278, 871, 290, 776, 363, 54.9, 126, 1311, 175. We suppose that $T_1, \ldots, T_n$ are independent, identically distributed observations from the Weibull density

$$f(t; \nu, \lambda) = \lambda \nu (\lambda t)^{\nu-1} \exp\{-(\lambda t)^{\nu}\} :$$

since the case $\nu = 1$ reduces to an exponential distribution, it is natural to consider inference for the shape parameter ν.

If we set $Y_i = \log(T_i)$, we have that $Y_1, \ldots, Y_n$ are independent, identically distributed from the extreme value distribution $EV(\mu, \theta)$, with density

$$f(y; \mu, \theta) = \frac{1}{\theta} \exp\left\{ \frac{y - \mu}{\theta} - e^{(y-\mu)/\theta} \right\},$$

where $\mu = -\log \lambda, \theta = \nu^{-1}$. The extreme value distribution is an example of a location-scale model, and we compare exact conditional inference for θ, as described by Pace and Salvan (1997: Chapter 7), with the results of bootstrapping. In this example, the adjusted signed root statistic r_a is easily constructed: the ancillary statistic is the configuration $a = (a_1, \ldots, a_n)$, where $a_i = (Y_i - \widehat{\mu})/\widehat{\theta}$. The form of $r_a(\theta)$ is given by Pace and Salvan (1997: Example 11.7).

An exact equi-tailed 90% conditional confidence interval for θ is (1.088, 2.081). The location-scale model structure of the inference problem ensures that the signed root likelihood ratio statistic $r_p(\theta)$ is exactly pivotal, so that the conventional bootstrap estimates the true (unconditional) distribution *exactly*. Though we make inference for θ in the presence of the nuisance parameter μ, there is no need to consider the constrained bootstrap approach, which is identical to the conventional bootstrap in this case.

The 90% confidence intervals obtained from the pivots $\Phi\{r_p(\theta)\}$ and $\Phi\{r_a(\theta)\}$ are respectively (1.041, 1.950) and (1.086, 2.077), confirming the conditional accuracy of the adjusted

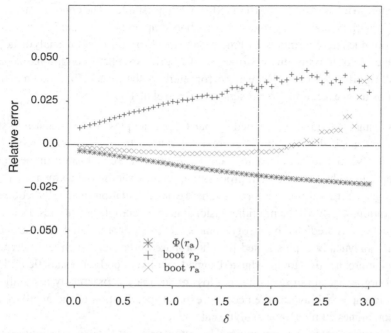

Figure 11.5 Relative errors of approximations of conditional significance values, exponential regression example

signed root statistic $r_a(\theta)$. Using $R = 4999$ bootstrap samples the confidence interval obtained from the bootstrap confidence set root $\widehat{u}_1\{\Phi(r_p(\theta))\}$ was calculated as $(1.085, 2.077)$.

The impressive agreement of the bootstrap interval with the exact confidence set is even more striking in an analysis of the first $n = 10$ observations in the data sample. The exact 90% conditional interval is then $(1.440, 3.702)$, with normal approximation to the unadjusted and adjusted signed root statistics yielding the intervals $(1.326, 3.217)$ and $(1.437, 3.685)$ respectively, the latter now *less* accurate than the interval obtained by bootstrapping r_p, which was obtained as $(1.440, 3.698)$.

Illustration 11.3 (continued) For the exponential regression example, exact conditional inference is described in Section 6.3.2 of Lawless (1982). This example displays the same essential features as the Weibull example, Illustration 11.7. The adjusted signed root statistic $r_a(\theta)$ is readily constructed, the unadjusted statistic $r_p(\theta)$ is exactly pivotal, so that the conventional bootstrap again estimates the true (unconditional) distribution exactly, and the exact inference is awkward, requiring numerical integration.

Based on the same five observations as considered previously, Figure 11.5 provides graphical comparison between exact conditional significance levels, given (exact) ancillary statistics $\log T_i - \widehat{\theta} - \widehat{\xi}c_i, i = 1, \ldots, n$, for testing the null hypothesis $H_0 : \theta = \theta_0$, with approximate levels obtained from $\Phi(r_a)$ and by bootstrapping the distribution of r_p. Now we consider testing H_0 against the alternative that θ is *smaller*, so that evidence against H_0 is provided by small values of $\widehat{\theta}$. Results are shown for a range of θ_0 of the form $\theta_0 = \widehat{\theta} + \delta$. Also shown are results obtained by bootstrapping the distribution of r_a. Again, the figure plots the relative error of each approximation, for a range of values of δ, and each bootstrap figure is based on $R = 10\,000\,000$ bootstrap samples. The conditional accuracy of the approximation obtained by unconditional bootstrapping of the distribution of r_p is less than that obtained from normal approximation to the distribution of the analytic adjustment r_a. However, bootstrapping the distribution of r_a gives excellent conditional accuracy, to very small significance levels. Once more, for reference the vertical line shown in the figure corresponds to an exact conditional significance level of 1%.

The circumstantial evidence provided by our three examples suggests that bootstrapping applied to a statistic that is stable to second order, in particular the signed root likelihood ratio statistic, yields a quite satisfactory approximation to the conclusions of the conditional inference required by the Fisherian proposition, while avoiding the awkward calculations demanded by exact conditional inference. These exact calculations generally require numerical integration, which will be multidimensional in more complex examples. Bootstrapping typically incurs, as predicted by theory, some slight loss of conditional accuracy compared with more analytically sophisticated likelihood-based proposals, such as inference based on normal approximation to the adjusted signed root likelihood ratio statistic r_a. However, we might argue that in our examples the loss of accuracy is practically insignificant, and the bootstrap approach avoids the need for explicit specification of the ancillary statistic, and the sometimes cumbersome analytic calculations.

In all of our examples, the conventional bootstrap provides the exact unconditional distribution of the signed root statistic r_p: this is true quite generally in location-scale, regression-scale models etc. In other cases, such as the curved exponential family context, the question of whether the constrained bootstrap approach is advantageous then arises. We have seen that the constrained bootstrap *is* more accurate for estimating the unconditional distribution, but the situation for estimation of the relevant conditional distribution is less clear. An important point, however, is that the choice between the two bootstrap approaches (conventional and constrained) need not be viewed as any more cumbersome than the choice between different approximations $\tilde{r}_a$ to the adjusted signed root statistic r_a that is typically required in such cases. Also, in this situation *either* bootstrap method, applied to a statistic stable to second order, approximates the conditional distribution relevant to the Fisherian proposition to the *same* level of error, $O(n^{-1})$, as does any $\tilde{r}_a$.

11.6 Problems

11.1 Let $x_1, x_2, \ldots, x_n$, with ordered values $x_{(1)} < x_{(2)} < \ldots < x_{(n)}$, be an independent, identically distributed sample from an unknown distribution F. Let $\widehat{\theta}(X_1, \ldots, X_n)$ be a statistic. A non-parametric bootstrap estimate of the standard deviation of $\widehat{\theta}$ is the standard deviation of $\widehat{\theta}(X^*)$ under the drawing of bootstrap samples X^* from the empirical distribution of the observed data sample. Recall that such a bootstrap sample is obtained by independently sampling, with replacement, from $\{x_1, \ldots, x_n\}$.

Suppose that $n = 2m - 1$ and that $\widehat{\theta}$ is $X_{(m)}$, the middle-order statistic. By showing that the bootstrap distribution is concentrated on the observed data points and obtaining an expression for the probability under the drawing of bootstrap samples of $\widehat{\theta}^* \equiv \widehat{\theta}(X^*)$ being equal to $x_{(k)}$, show that in this case the bootstrap estimate of standard deviation can be calculated theoretically, without simulation.

11.2 Given a dataset of n distinct values, show that the number of distinct bootstrap samples obtained by the non-parametric bootstrap scheme, as described in Problem 11.1, is

$$\binom{2n-1}{n}.$$

11.3 Let $X_1, X_2, \ldots, X_n$ be independent, identically distributed from a uniform distribution on $(0, \theta)$. What is the maximum likelihood estimator $\widehat{\theta}$ of θ? What is the asymptotic distribution of $n(\theta - \widehat{\theta})/\theta$? Let $\widehat{\theta}^*$ denote the value of $\widehat{\theta}$ computed from a bootstrap sample obtained by sampling with replacement from the data. By considering

$$\mathrm{Pr}^* \left\{ \frac{n(\widehat{\theta} - \widehat{\theta}^*)}{\widehat{\theta}} = 0 \right\},$$

show that the non-parametric bootstrap fails to estimate the distribution of $n(\theta - \widehat{\theta})/\theta$ consistently.

11.4 Let $X_1, \ldots, X_n$ be an independent, identically distributed sample from a distribution F. Suppose the parameter of interest is the square of the mean of F, $\theta = \mu^2$, and let θ be estimated by $\widehat{\theta} = \bar{X}^2$, where $\bar{X} = n^{-1} \sum_{i=1}^{n} X_i$ is the sample mean.

Show that $\widehat{\theta}$ is biased as an estimator of θ, and derive an explicit expression for the non-parametric bootstrap estimator, $b_{BOOT}(\widehat{\theta}) \equiv \mathbb{E}^*\{\widehat{\theta}(X^*)\} - \widehat{\theta}$, of the bias of $\widehat{\theta}$, where X^* denotes a generic bootstrap sample obtained by independently sampling with replacement from $\{X_1, \ldots, X_n\}$, and the expectation is taken with respect to the drawing of such samples.

A *bias-corrected* estimator $\widehat{\theta}_1$ is defined by

$$\widehat{\theta}_1 = \widehat{\theta} - b_{BOOT}(\widehat{\theta}). \tag{11.12}$$

Find the form of $\widehat{\theta}_1$ and compare its bias with that of $\widehat{\theta}$.

The bias-correction is iterated to produce a sequence of estimators $\widehat{\theta}_1, \widehat{\theta}_2, \ldots$, with

$$\widehat{\theta}_j = \widehat{\theta}_{j-1} - b_{BOOT}(\widehat{\theta}_{j-1}),$$

where $b_{BOOT}(\widehat{\theta}_{j-1}) \equiv \mathbb{E}^*\{\widehat{\theta}_{j-1}(X^*)\} - \widehat{\theta}$ is a bootstrap estimator of the bias of $\widehat{\theta}_{j-1}$. Show that $\widehat{\theta}_j = \bar{X}^2 - (n-1)^{-1}(1 - n^{-j})\widehat{\sigma}^2$, where $\widehat{\sigma}^2 = n^{-1} \sum_{i=1}^{n} (X_i - \bar{X})^2$, and that as $j \to \infty$, we have $\widehat{\theta}_j \to \widehat{\theta}_\infty$, where $\widehat{\theta}_\infty$ is unbiased for θ.

11.5 Suppose that in Problem 11.4 bootstrapping is done instead parametrically. Compare the form of the bias-corrected estimator (11.12) with the minimum variance unbiased estimator of θ in the cases of: (i) the normal distribution $N(\mu, 1)$; (ii) the exponential distribution of mean μ.

11.6 Show that in Problem 11.4, the bootstrap estimator of bias, $b_{BOOT}(\widehat{\theta})$, is biased as an estimator of the true bias of $\widehat{\theta}$.

Let $\widehat{\theta}_i$ be the value of the estimator $\widehat{\theta}(X_1, \ldots, X_{i-1}, X_{i+1}, \ldots, X_n)$ computed from the sample of size $n - 1$ obtained by deleting X_i from the full sample, and let $\widehat{\theta}. = n^{-1} \sum_{i=1}^{n} \widehat{\theta}_i$. The *jackknife* estimator of the bias of $\widehat{\theta}$ is

$$b_J = (n-1)(\widehat{\theta}. - \widehat{\theta}).$$

Derive the jackknife estimator in the case $\widehat{\theta} = \bar{X}^2$, and show that, as an estimator of the true bias of $\widehat{\theta}$, it is unbiased.

11.7 Let $Y = \{Y_1, \ldots, Y_n\}$ be an independent, identically distributed sample from $N(\mu, \sigma^2)$, and let $\widehat{\mu}$ and $\widehat{\sigma}^2$ denote the maximum likelihood estimators of μ and σ^2.

What is the asymptotic distribution of $s_n(\sigma^2) = \sqrt{n}(\widehat{\sigma}^2 - \sigma^2)$? Show that this asymptotic distribution leads to a confidence set root (11.3) of the form

$$u(Y, \psi) = \Phi\left\{\sqrt{n/2}\left(1 - \frac{\psi}{\widehat{\sigma}^2}\right)\right\}.$$

What is the exact (finite sample) distribution of $s_n(\sigma^2)$? Show that the confidence set root (11.4) is of the form

$$u(Y, \psi) = G\left\{n(2 - \psi/\widehat{\sigma}^2)\right\},$$

where G is the distribution function of the chi-squared distribution with $n - 1$ degrees of freedom.

Show that, for both initial confidence set roots, the (conventional) bootstrap confidence set root is of the form

$$\widehat{u}_1(Y, \psi) = G(n\widehat{\sigma}^2/\psi),$$

and that the corresponding bootstrap confidence set has exactly the desired coverage. Confirm that this is true also for the constrained bootstrap.

11.8 Let $Y_1, \ldots, Y_n$ be independent inverse Gaussian observations each with probability density function

$$f(y; \phi, \gamma) = \frac{\sqrt{\phi}}{\sqrt{2\pi}} e^{\sqrt{\phi\gamma}} y^{-3/2} \exp\{-\frac{1}{2}(\phi y^{-1} + \gamma y)\}.$$

Take the parameter of interest as ϕ, with γ as nuisance parameter.

Show that the profile log-likelihood function for ϕ is of the form

$$l_p(\phi) = \frac{n}{2}(\log \phi - \phi\widehat{\phi}^{-1}),$$

and find the forms of the signed root likelihood ratio statistic, $r_p(\phi)$, and of the adjusted signed root likelihood ratio statistic $r_a(\phi)$.

Using the distributional result that $n\phi\widehat{\phi}^{-1}$ is distributed as chi-squared with $n - 1$ degrees of freedom, examine numerically the accuracy of the standard normal approximation to the distributions of $r_p(\phi)$ and $r_a(\phi)$, for a range of sample sizes n.

Perform a numerical study to examine the coverage accuracy of confidence sets obtained by using the conventional and constrained bootstrap confidence set roots $\widehat{u}_1(Y, \psi)$ and $\tilde{u}_1(Y, \psi)$ obtained from the initial confidence set root $u(Y, \psi) = \Phi\{r_p(\psi)\}$.

11.9 The Weibull distribution considered in Illustration 11.7 can be extended to include regressor variables. The most commonly used model is one in which the density function of lifetime T, given regressor variables z, is of the form

$$f(t; z, \nu) = \frac{\nu}{\alpha(z)}\left(\frac{t}{\alpha(z)}\right)^{\nu-1} \exp\left[-\left(\frac{t}{\alpha(z)}\right)^{\nu}\right]. \tag{11.13}$$

Then the density function of $Y = \log T$, given z, is of extreme value form

$$f(y; z, \sigma) = \frac{1}{\sigma} \exp\left[\frac{y - \mu(z)}{\sigma} - \exp\left\{\frac{y - \mu(z)}{\sigma}\right\}\right],$$

where $\mu(z) = \log \alpha(z)$.

Consider the case $\mu(z) = \alpha + \beta z$. Let $\widehat{\alpha}$, $\widehat{\beta}$ and $\widehat{\sigma}$ denote the maximum likelihood estimators from a random sample $y_1, \ldots, y_n$ corresponding to fixed covariate values $z_1, \ldots, z_n$.

Obtain the joint probability density function of $W_1 = (\widehat{\alpha} - \alpha)/\widehat{\sigma}$, $W_2 = (\widehat{\beta} - \beta)/\widehat{\sigma}$ and $W_3 = \widehat{\sigma}/\sigma$, given the ancillary statistics $a_i = (y_i - \widehat{\alpha} - \widehat{\beta} z_i)/\widehat{\sigma}$. Verify that w_1 can be integrated out of this density analytically, to give the joint probability density function of W_2 and W_3, given a, as

$$k(a, z) w_3^{n-2} \exp\left(w_3 \sum_{i=1}^{n} a_i\right) \Big/ \left(\sum_{i=1}^{n} e^{a_i w_3 + z_i w_2 w_3}\right)^n.$$

Discuss how a (conditional) confidence interval for σ may be obtained from the marginal distribution of W_3. [Note that evaluation of this confidence interval will require a double numerical integration.]

Consider again the leukaemia survival data considered in Illustration 11.3. Modelling the survival times as in (11.13), construct an exact 90% conditional confidence interval for v. Compare the exact confidence set with that obtained by unconditional bootstrapping of an appropriate statistic. Was the assumption of an exponential distribution used in Illustration 11.3 justified?

Bibliography

Aitchison, J. (1975) Goodness of prediction fit. *Biometrika* **62**, 547–54.

Aitchison, J. and Dunsmore, I. R. (1975) *Statistical Prediction Analysis.* Cambridge: Cambridge University Press.

Anderson, T. W. (1984) *An Introduction to Multivariate Statistical Analysis.* New York: Wiley. (This is a standard reference on multivariate analysis.)

Baggerly, K. A. (1998) Empirical likelihood as a goodness of fit measure. *Biometrika* **85**, 535–47.

Barndorff-Nielsen, O. E. (1983) On a formula for the distribution of the maximum likelihood estimator. *Biometrika* **70**, 343–65.

  (1986) Inference on full or partial parameters based on the standardized signed log likelihood ratio. *Biometrika* **73**, 307–22.

  (1990) Approximate interval probabilities. *J.R. Statist. Soc. B* **52**, 485–96.

Barndorff-Nielsen, O. E. and Chamberlin, S. R. (1994) Stable and invariant adjusted directed likelihoods. *Biometrika* **81**, 485–500.

Barndorff-Nielsen, O. E. and Cox, D. R. (1979) Edgeworth and saddlepoint approximations with statistical applications (with discussion). *J.R. Statist. Soc. B* **41**, 279–312.

  (1989) *Asymptotic Techniques for Use in Statistics.* London: Chapman & Hall.

  (1994) *Inference and Asymptotics.* London: Chapman & Hall.

  (1996) Prediction and asymptotics. *Bernoulli* **2**, 319–40.

Bayes, T. (1763) An essay towards solving a problem in the doctrine of chances. *Phil. Trans. Roy. Soc.* **53**, 370–418.

Beran, R. J. (1987) Prepivoting to reduce level error of confidence sets. *Biometrika* **74**, 457–68.

  (1988) Prepivoting test statistics: a bootstrap view of asymptotic refinements. *J. Amer. Statist. Assoc.* **83**, 687–97.

Berger, J. O. (1985) *Statistical Decision Theory and Bayesian Analysis.* New York: Springer Verlag (Second Edition).

Berger, J. O. and Pericchi, L. (1996) The intrinsic Bayes factor for model selection and prediction. *J. Amer. Statist. Assoc.* **91**, 109–22.

Berger, J. O. and Sellke, T. (1987) Testing a point null hypothesis: the irreconcilability of P values and evidence. *J. Amer. Statist. Assoc.* **82**, 112–22.

Besag, J. E. and Green, P. J. (1993) Spatial statistics and Bayesian computation. *J.R. Statist. Soc. B* **55**, 25–37. (An effective review of Gibbs sampling etc.)

Bjørnstad, J. F. (1990) Predictive likelihood: a review (with discussion). *Stat. Sci.* **5**, 242–65.

Blyth, C. R. (1951) On minimax statistical decision procedures and their admissibility. *Ann. Math. Stat.* **22**, 22–42.

Brazzale, A. R. (2000) Practical small-sample parametric inference. Ph.D. thesis, Department of Mathematics, Swiss Federal Institute of Technology, Lausanne, Switzerland.

Butler, R. W. (1986) Predictive likelihood inference with applications (with discussion). *J.R. Statist. Soc. B* **48**, 1–38.

  (1989) Approximate predictive pivots and densities. *Biometrika* **76**, 489–501.

  (1990) Discussion of Bjørnstad (1990). *Stat. Sci.* **5**, 255–9.

Carlin, B. P., Gelfand, A. E. and Smith, A. F. M. (1992) Hierarchical Bayesian analysis of changepoint problems. *Appl. Stat.* **41**, 389–405.

Carpenter, J. (1999) Test inversion bootstrap confidence intervals. *J.R. Statist. Soc. B* **61**, 159–72.

Casella, G. and Berger, R. L. (1990) *Statistical Inference*. Pacific Grove, California: Wadsworth & Brooks/Cole. (Discusses much of the material of the current text and provides much useful introductory material on probability background.)

Corcoran, S. A. (1998) Bartlett adjustment of empirical discrepancy statistics. *Biometrika* **85**, 967–72.

Cox, D. R. (1975) Prediction intervals and empirical Bayes confidence intervals. In *Perspectives in Probability and Statistics*, ed. J. Gani. London: Academic Press, pp. 47–55.

  (1986) Discusson of Butler (1986). *J.R. Statist. Soc. B* **48**, 27.

Cox, D. R. and Hinkley, D. V. (1974) *Theoretical Statistics*. London: Chapman & Hall. (Classic text.)

Cox, D. R. and Reid, N. (1987) Parameter orthogonality and approximate conditional inference (with discussion). *J.R. Statist. Soc. B* **49**, 1–39.

Cox, D. R. and Snell, E. J. (1981) *Applied Statistics: Principles and Examples*. London: Chapman & Hall.

Cramér, H. (1946) *Mathematical Methods of Statistics*. Princeton, NJ: Princeton University Press. (Classical reference, in particular on such things as asymptotics of maximum likelihood, regularity conditions.)

Daniels, H. E. (1954) Saddlepoint approximations in statistics. *Ann. Math. Stat.* **25**, 631–50.

Datta, G. S., Mukerjee, R., Ghosh, M. and Sweeting, T. J. (2000) Bayesian prediction with approximate frequentist validity. *Ann. Stat.* **28**, 1414–26.

Davison, A. C. (1986) Approximate predictive likelihood. *Biometrika* **73**, 323–32.

  (2003) *Statistical Models*. Cambridge: Cambridge University Press.

Davison, A. C. and Hinkley, D. V. (1997) *Bootstrap Methods and Their Application*. Cambridge: Cambridge University Press.

Davison, A. C., Hinkley, D. V. and Young, G. A. (2003) Recent developments in bootstrap methodology. *Stat. Sci.* **18**, 141–57.

Dawid, A. P. (1982) The well-calibrated Bayesian (with discussion). *J. Amer. Statist. Assoc.* **77**, 605–13. (This and the next two references are concerned with how one assesses whether subjective probabilities are consistent with observed data.)

  (1984) Statistical theory: the prequential approach (with discussion). *J.R. Statist. Soc. A* **147**, 278–92.

  (1986) Probability forecasting. In *Encyclopedia of Statistical Sciences*, Vol. 7, eds. S. Kotz, N. L. Johnson and C. B. Read. New York: Wiley-Interscience, pp. 210–18.

de Finetti, B. (1974 and 1975) *Theory of Probability*, Vols. 1 and 2. New York: Wiley.

De Groot, M. H. (1970) *Optimal Statistical Decisions*. New York: McGraw-Hill. (Another standard reference on decision theory, concentrating on the Bayesian viewpoint.)

DiCiccio, T. J. and Martin, M. A. (1993) Simple modifications for signed roots of likelihood ratio statistics. *J.R. Statist. Soc. B* **55**, 305–316.

DiCiccio, T. J., Martin, M. A. and Stern, S. E. (2001) Simple and accurate one-sided inference from signed roots of likelihood ratios. *Can. J. Stat.* **29**, 67–76.

DiCiccio, T. J. and Romano, J. P. (1990) Nonparametric confidence limits by resampling methods and least favorable families. *Int. Stat. Rev.* **58**, 59–76.

(1995) On bootstrap procedures for second-order accurate confidence limits in parametric models. *Statistica Sinica* **5**, 141–60.

Donoho, D. L., Johnstone, I. M., Kerkyacharian, K. and Picard, D. (1995) Wavelet shrinkage: asymptotia? (with discussion). *J.R. Statist. Soc. B* **57**, 301–69. (Mentioned here as a reference on 'modern' minimax theory.)

Efron, B. (1979) Bootstrap methods: another look at the jackknife. *Ann. Stat.* **7**, 1–26.

(1992) Introduction to James and Stein (1961) 'Estimation with quadratic loss'. In *Breakthroughs in Statistics*, Vol. 1, eds. S. Kotz and N. L. Johnson. New York: Springer Verlag, pp. 437–42.

(1998) R.A. Fisher in the 21st Century (with discussion). *Stat. Sci.* **13**, 95–122.

(2003a) Robbins, empirical Bayes and microarrays. *Ann. Stat.* **31**, 366–78.

(2003b) Second thoughts on the bootstrap. *Stat. Sci.* **18**, 135–40. (Second thoughts, but not second doubts.)

(2004) Large-scale simultaneous hypothesis testing: the choice of a null hypothesis. *J. Amer. Statist. Assoc.* **99**, 96–104.

Efron, B. and Hinkley, D. V. (1978) Assessing the accuracy of the maximum likelihood estimator: observed versus expected Fisher information (with discussion). *Biometrika* **65**, 457–87.

Efron, B. and Morris, C. M. (1975) Data analysis using Stein's estimator and its generalizations. *J. Amer. Statist. Assoc.* **70**, 311–19.

(1977) Stein's paradox in statistics. *Sci. Am.* **236**, 119–27.

Efron, B. and Tibshirani, R. J. (1993) *An Introduction to the Bootstrap*. London: Chapman & Hall.

Efron, B., Tibshirani, R. J., Storey, J. D. and Tusher, V. (2001) Empirical Bayes analysis of a microarray experiment. *J. Amer. Statist. Assoc.* **96**, 1151–60.

Feller, W. (1971) *An Introduction to Probability Theory*, Vol. 2. New York: Wiley (Second Edition).

Ferguson, T. S. (1967) *Mathematical Statistics: A Decision-Theoretic Approach*. New York; Academic Press. (A classical reference work.)

Fisher, R. A. (1922) On the mathematical foundations of theoretical statistics. *Phil. Trans. Roy. Soc. A* **222**, 309–68. (This and the next reference are often debated as the most important single paper in statistical theory.)

(1925) Theory of statistical estimation. *Proc. Camb. Phil. Soc.* **22**, 700–25.

(1934) Two new properties of mathematical likelihood. *Proc. R. Soc. Lond. A* **144**, 285–307. (The origin of the conditional inference procedure for location-scale families, and therefore of much of the Fisherian viewpoint.)

(1990) *Statistical Methods, Experimental Design and Scientific Inference*. Oxford: Clarendon Press. (A relatively recent reprint of three of Fisher's best known works on statistics.)

Gamerman, D. (1997) *Markov Chain Monte Carlo: Stochastic Simulation for Bayesian Inference*. London: CRC Press.

Garthwaite, P. H. and Buckland, S. T. (1992) Generating Monte Carlo confidence intervals by the Robbins–Monro process. *Appl. Stat.* **41**, 159–71.

Geisser, S. (1993) *Predictive Inference: An Introduction*. New York: Chapman & Hall.

Gelman, A., Carlin, J. B., Stern, H. S. and Rubin, D. B. (2003) *Bayesian Data Analysis*. London: CRC Press (Second Edition).

Geman, S. and Geman, D. (1984) Stochastic relaxation, Gibbs distributions and the Bayesian restoration of images. *IEEE Trans. Pattern Anal. Mach. Intell.* **6**, 721–41. (The paper which introduced Gibbs sampling.)

Geyer, C. J. (1992) Practical Markov chain Monte Carlo. *Stat. Sci.* **7**, 473–82. (Useful review of Gibbs sampling and related techniques.)

Gilks, W. R., Richardson, S. and Spiegelhalter, D. J. (eds.) (1996) *Markov Chain Monte Carlo in Practice*. London: CRC Press.

Goldstein, H. and Spiegelhalter, D. J. (1996) League tables and their limitations: statistical issues in comparisons of institutional performance (with discussion). *J.R. Statist. Soc. A* **159**, 385–443.

Guttman, I. (1970) *Statistical Tolerance Regions: Classical and Bayesian*. London: Griffin.

Hall, P. (1992) *The Bootstrap and Edgeworth Expansion*. New York: Springer Verlag.

Hall, P. and Martin, M. A. (1988) On bootstrap resampling and iteration. *Biometrika* **75**, 661–71.

Hall, P., Peng, L. and Tajvidi, N. (1999) On prediction intervals based on predictive likelihood or bootstrap methods. *Biometrika* **86**, 871–80.

Hall, P. and Presnell, B. (1999a) Intentionally biased bootstrap methods. *J.R. Statist. Soc. B* **61**, 143–58.

(1999b) Biased bootstrap methods for reducing the effects of contamination. *J.R. Statist. Soc. B* **61**, 661–80.

(1999c) Density estimation under constraints. *J. Comput. Graph. Stat.* **8**, 259–77.

Harris, I. R. (1989) Predictive fit for natural exponential families. *Biometrika* **76**, 675–84.

Hastings, W. K. (1970) Monte Carlo sampling methods using Markov chains and their applications. *Biometrika* **57**, 97–109.

Hinkley, D. V. (1979) Predictive likelihood. *Ann. Stat.* **7**, 718–28.

James, W. and Stein, C. (1961) Estimation with quadratic loss. *Proc. Fourth Berk. Symp. Math. Statist. Probab.* **1**, 361–79. Berkeley, CA: University of California Press.

Jeffreys, H. (1939) *Theory of Probability*. Oxford: Oxford University Press. (Third Edition, 1961.)

Kass, R. E. and Raftery, A. E. (1995) Bayes factors. *J. Amer. Statist. Assoc.* **90**, 773–95. (Accessible review paper.)

Keating, J. P., Glaser, R. E. and Ketchum, N. S. (1990) Testing hypotheses about the shape parameter of a gamma distribution. *Technometrics* **32**, 67–82.

Komaki, F. (1996) On asymptotic properties of predictive distributions. *Biometrika* **83**, 299–313.

Laplace, P. S. (1812), *Théorie Analytique des Probabilités*. Paris: Courcier.

Lauritzen, S. L. (1974) Sufficiency, prediction and extreme models. *Scand. J. Stat.* **1**, 128–34.

Lawless, J. (1982) *Statistical Models and Methods for Lifetime Data*. New York: Wiley.

Lee, S. M. S. and Young, G. A. (2003) Prepivoting by weighted bootstrap iterations. *Biometrika* **90**, 393–410.

(2004) Bootstrapping in the presence of nuisance parameters. Preprint, University of Cambridge.

Lehmann, E. L. (1986) *Testing Statistical Hypotheses*. New York: Wiley (Second Edition). (The original first edition, long a classic in the field, was published in 1959.)

Lehmann, E. L. and Casella, G. (1998) *Theory of Point Estimation*. New York: Springer Verlag. (This is a very much updated version of Lehmann's 1983 book of the same title.)

Lehmann, E. L. and Scheffé, H. (1950) Completeness, similar regions and unbiased estimation. *Sankhyā A* **10**, 305–40.

Leonard, T. (1982) Comment on 'A simple predictive density function' by M. Lejeune and G. D. Faulkenberry. *J. Amer. Statist. Assoc.* **77**, 657–8.

Lindgren, B. W. (1960) *Statistical Theory*. New York: Macmillan. (Classical text at a general introductory level.)

Lindley, D. V. (1971a) *Bayesian Statistics, A Review*. Philadelphia: SIAM. (Well-written exposition of the Bayesian viewpoint.)

(1971b), *Making Decisions*. London: Wiley.

Lugannani, R. and Rice, S. (1980) Saddlepoint approximations for the distribution of the sum of independent random variables. *Adv. Appl. Probab.* **12**, 475–90.

Mardia, P. V., Kent, J. T. and Bibby, J. N. (1979) *Multivariate Analysis*. London: Academic Press. (Another standard reference on multivariate analysis.)

Martin, M. A. (1990) On bootstrap iteration for coverage correction in confidence intervals. *J. Amer. Statist. Assoc.* **85**, 1105–18.

Mathiasen, P. E. (1979) Prediction functions. *Scand. J. Stat.* **6**, 1–21.

Metropolis, N., Rosenbluth, A. W., Rosenbluth, M. N., Teller, A. H. and Teller, E. (1953) Equations of state calculations by fast computing machines. *J. of Chem. Phys.* **21**, 1087–92.

Neyman, J. and Pearson, E. S. (1933) On the problem of the most efficient tests of statistical hypotheses. *Phil. Trans. Roy. Soc. A* **231**, 289–337.

Owen, A. B. (1988) Empirical likelihood ratio confidence intervals for a single functional. *Biometrika* **75**, 237–49.

(2001) *Empirical Likelihood*. Boca Raton: Chapman & Hall/CRC.

Pace, L. and Salvan, A. (1997) *Principles of Statistical Inference from a Neo-Fisherian Perspective.* Singapore: World Scientific.

Patefield, W. M. (1977) On the maximized likelihood function. *Sankhyā B* **39**, 92–6.

Rao, C. R. (1973) *Linear Statistical Inference and Its Applications*. New York: Wiley (Second Edition). (A classic reference work.)

Robert, C. P. and Casella, G. (1999) *Monte Carlo Statistical Methods*. New York: Springer Verlag.

Roberts, G. O., Gelman, A. and Gilks, W. R. (1997) Weak convergence and optimal scaling of random walk Metropolis algorithms. *Ann. Appl. Probab.* **7**, 110–20.

Savage, L. J. (1954) *The Foundations of Statistics*. New York: Wiley. (Of historical importance.)

Severini, T. A. (2000) *Likelihood Methods in Statistics*. Oxford: Clarendon Press.

Shao, J. and Tu, D. (1995) *The Jackknife and Bootstrap*. New York: Springer Verlag.

Skovgaard, I. M. (1996) An explicit large-deviation approximation to one-parameter tests. *Bernoulli* **2**, 145–65.

Smith, A. F. M. and Roberts, G. O. (1993) Bayesian computation via the Gibbs sampler and related Markov chain Monte Carlo methods (with discussion). *J.R. Statist. Soc. B* **55**, 3–23. (Good general review of simulation-based methodology for Bayesian statistics.)

Smith, A. F. M. and Spiegelhalter, D. J. (1980) Bayes factors and choice criteria for linear models. *J.R. Statist. Soc. B* **42**, 213–20.

Smith, J. Q. (1988) *Decision Analysis: A Bayesian Approach*. London: Chapman & Hall.

Smith, R. L. (1997) Predictive inference, rare events and hierarchical models. Preprint, University of North Carolina.

(1999) Bayesian and frequentist approaches to parametric predictive inference (with discussion). In *Bayesian Statistics* 6, eds. J. M. Bernardo, J. O. Berger, A. P. Dawid and A. F. M. Smith. Oxford: Oxford University Press, pp. 589–612.

Stein, C. (1956) Inadmissibility of the usual estimator for the mean of a multivariate normal distribution. *Proc. Third Berk. Symp. Math. Statist. Probab.* **1**, 197–206. Berkeley, CA: University of California Press.

Stein, C. (1981) Estimation of the mean of a multivariate normal distribution. *Ann. Stat.* **9**, 1135–51.

Tierney, L. (1994) Markov chains for exploring posterior distributions. *Ann. Stat.* **22**, 1701–28.

Tierney, L. and Kadane, J. B. (1986) Accurate approximations for posterior moments and marginal densities. *J. Amer. Statist. Assoc.* **81**, 82–6.

Tierney, L., Kass, R. E. and Kadane, J. B. (1989) Approximate marginal densities of nonlinear functions. *Biometrika* **76**, 425–33.

Wald, A. (1950) *Statistical Decision Functions*. New York: Wiley.

Index

Printed in the United States
By Bookmasters